Offshore Sedimentary Basins of India

Geology, Tectonics and Hydrocarbon Prospects

Offshore Sedimentary Basins of India

Geology, Tectonics and Hydrocarbon Prospects

B.T.V. Seshavataram

Director G&G HERAMEC Limited,
Hyderabad.

BSP **BS Publications**
A unit of **BSP Books Pvt. Ltd.**

4-4-309/316, Giriraj Lane, Sultan Bazar,
Hyderabad - 500 095
Phone : 040 - 23445605, 23445688

Offshore Sedimentary Basins of India ***Geology, Tectonics and Hydrocarbon Prospects*** **by B.T.V. Seshavataram**

Published by

BSP **BS Publications**

A unit of **BSP Books Pvt. Ltd.**

4-4-309/316, Giriraj Lane, Sultan Bazar,
Hyderabad - 500 095
Phone : 040 - 23445605, 23445688
e-mail : info@bspbooks.net

ISBN: 978-93-52300-83-9 (HB)

This compendium is Dedicated to

My Late Parents,

Late Professor, C. Mahadevan,
Professor, Principal and Head of Department of Geology,
Andhra University, Waltair (A.P. India)

&

To my colleague Late Sri. P.V. Ananthanarayanan, Eminent Geologist, ONGC who are my mentors and inspired me throughout my career.

Contents

Foreword .... (ix)
Preface .... (xvii)
Acknowledgements .... (xix)
About the Author .... (xxi)

Chapter - 1

Introduction 1

Chapter - 2

Sedimentary Basins of India 2

Chapter - 3

Classification of West and East Coast Indian Offshore Basins 9

Chapter - 4

Plate Tectonics with Specific Reference to Indian Plate Movements 11

4.1 Indian Plate Movement in General 11
4.2 Tectonic Evolution 12
4.3 Western Continental Margin 13
4.4 Eastern Continental Margin 14

Chapter - 5

Western Offshore Basins 16

5.1. Kutch – Saurashtra Basin 17
5.2 Kutch Basin 17
5.3 Saurashtra Basin 28
5.4 Bombay Offshore 32
5.5 Kerala-Konkan-Lakshadweep Basin 78

Chapter - 6

The East-Coast Offshore Sedimentary Basins **91**

6.1 Cauvery Basin 91

6.2 Palar Basin 106

6.3 Krishna-Godavari Basin 110

6.4 Mahanadi Offshore Basin 136

6.5 Bengal Basin 145

6.6 Andaman and Nicobar Islands 165

Chapter - 7

Deep Water Exploration **174**

7.1 East Coast Basins 178

7.2 Western Offshore 185

Chapter - 8

Gas Hydrates **187**

Appendix **195**

References **197**

Index **203**

D.J. Pandian, IAS
Principal Secretary

Energy & Petrochemicals Department
Government of Gujarat
Block No. 5, Sachivalaya,
Gandhinagar – 382 010 INDIA
Phone: + 91-79-23250771 Fax: + 97-79-23250797
E-mail: secepd@gujarat.gov.in

Foreword

16th February 2013

The author is having an excellent varied Experience of over 40 years in the upstream sector of Oil Industry, having exposure to various facets of Exploration, Development and Production for Oil and Gas. I am glad that he could place on record his experience and lucid ideas together, to bridge the gap between the demand and supply of Energy sector.

It is quite fitting to compare, the set up of various sedimentary Basins with oil/gas producing structures of Eastern and Western Offshore India, to a gold necklace studded with Diamonds, Gems and precious stones, adored on a very pretty queen's Neck.

It is very important to adopt an Exploration Strategy for unconventional gas resource like Shale Gas, Gas Hydrates, Coal Bed Methane and Coal Gasification, while intensifying the Exploration Thrust on Conventional Energy Sources Oil, Gas and Coal. The chapter with discussion on "Gas Hydrates and National Gas Hydrate Program" is quite interesting and it definitely throws some light on unconventional gas energy, which is the future vital Energy source, capable of bridging the gap between "Demand and Supply" for the future energy security of the nation.

Globally many exploration companies are venturing more and more into hostile environments and logistically difficult areas in search of hydrocarbons. The author had given an excellent account on Deep Water and Ultra Deep Water Hydrocarbon Exploration and Production Status.

The role of plate Tectonics, the Northward movement of Indian plate for the evolution of passive continental margins and offshore sediments is clearly explained and it is commandable.

The book would not only serve as a textbook for the young Geoscientists Pursuing Petroleum Geology/Petroleum Technology Courses, but also serve as a reference book for all E & P personals enagaged in the Hydrocarbon upstream sector.

I congratulate the author for his efforts and endeavors in bringing this volume. I wish him all the best.

(D.J. Pandian)

शाम व्यास राव
निदेशक (अन्वेषण)
S.V. Rao
Director (Exploration)

ऑयल एण्ड नेचुरल गैस कॉरपोरेशन लिमिटेड
रजिस्टर्ड ऑफिस: जीवन भारती, टॉवर-II, 7वाँ तल,
124, इंदिरा चौक, नई दिल्ली-110001 (भारत)
Oil and Natural Gas Corporation Limited
Regd. Office : Jeevan Bharti, Tower-II, 7th Floor,
124, Indira Chowk, New Delhi -110001 (India)
Tel. : +91-11-23314549, Fax : +91-11-23312012

Foreword

The book is a lucid and well-structured compilation of general technical information on Geology, Geophysics, Reservoir and Production data of Offshore Sedimentary Basins of India.

The sedimentary basins of India have received attention of Geoscientists for hydrocarbon exploration since the middle of last century. The major hydrocarbon resource potential belongs to Offshore Sedimentary Basins. (Hence, it is quite pertinent to bring out a book on Offshore Sedimentary Basins of India).

The author has begin with and explained very lucidly the role of plate tectonics and northward flight/movement of the Indian plate in the tectonic evolution of continental margins vis-à-vis the formation of Mesozoic-Tertiary Pericratonic rift basins along the Western and Eastern Offshore of India.

Each offshore basin has been discussed with respect to its Geological set up, Tectonics, Sedimentation History, Depositional Environment, Geochemistry and Petroleum Habitat in a manner which gives an overall idea of the evolution of petroleum system of the basin.

The brief discussion on exploration history along with the reservoir geology and production testing data has thrown some light and direction on further exploration thrust required in the different Basins.

As the days of easy oil are approaching its limit and demand of energy is ever increasing, future hydrocarbon exploration is mainly focused in the new frontier area i.e., Deep waters worldwide. The Govt. of India has offered a number of deep water blocks through its revamped NELP rounds which yielded results. The author has given the right thrust on the recent discoveries made in deep water blocks by National, Private and Joint Venture consortiums and their ongoing exploratory efforts.

The discoveries of Offshore Bangladesh, Myanmar and Sumatra and comparison of their geological similarity with the offshore basins of India will surely generate special interest of the readers.

India has already established physical presence of Methane Gas Hydrates in deepwater of Krishna-Godavari, Mahanadi and Andaman basins. It is ranked third only after USA and Japan as regards progress made in gas hydrates R & D. The chapter on Gas Hydrates and the National Gas Hydrate Programme (NGHP) will evoke new interest in this unconventional future hydrocarbon resource.

The book provides a comprehensive overview of the Offshore Sedimentary Basins of India and their E & P activities which will be quite useful and act as a ready reckoner to the students and persons associated with upstream sector of the Hydrocarbon Industry.

(S.V. Rao)
Director (Exploration)

सत्यमेव जयते

Giridhar Aramane
Joint Secretary

भारत सरकार
पेट्रोलियम एवं प्राकृतिक गैस मंत्रालय
शास्त्री भवन, नई दिल्ली - 110001
उपभोक्ता पिन कोड - 110115
Government of India
Ministry of Petroleum & Natural Gas
Shastri Bhawan, New Delhi - 110 001
Customer Pin Code - 110115

Foreword

The subject "Geology, Tectonics and Hydrocarbon Prospects, Offshore Sedimentary Basins of India" is vast and encompasses whole gamut of E & P upstream sector. Up to date and authentic information on various geological processes can result in successful E & P endeavors, enabling to take appropriate timely decisions.

I am glad that the author having had, rich experience of over 50 years in the upstream sector of Hydrocarbon Industry could place on record his experience and his lucid ideas together and attempted to bridge the gaps between many geological uncertainties. Author's insightful impetus on various categorized sedimentary basins for future exploration and suggestions on deep water and ultra deep water endeavors are commendable. The chapter with the discussion of "Gas Hydrates and National Gas Hydrate Programme (NGHP)" is quite interesting and throws some light on unconventional gas energy. Gas hydrates may be the future vital energy source which may help in bridging the gas between demand and supply. Gas Hydrates may play dynamic important role for the Future Energy Security of the Nation.

The book would serve as a reference book not only for the Young Geoscientists but also to E & P Professionals engaged in the upstream sector of Hydrocarbon Industry.

I congratulate the author and wish him all the best.

(Giridhar Aramane)

Foreword

While Geology is considered as one of the important subjects of earth sciences, unlike other sciences Imagination and Intuition are essential to a Professional Geologist. A common man thinks of geology in terms of landscapes too vast to fully comprehend-volcanoes, mountain ranges, and canyons created by forces beyond our control. Most of the geological processes are dead slow and in one's lifetime some may experience volcanic eruptions and earth quakes without ever knowing the reasons there of. Transportation of sediments by river systems, burial of marine life, and multitude of organic reactions set by pressures and temperatures, generation of hydrocarbons in the source rocks and their migration to reservoir rocks- all these things do not happen overnight. They take millions of years. However by perceiving those changes, the geologist is able to recognize the shallow mineral wealth and fossil fuels quite easily. But to locate deep reservoir rocks containing hydrocarbons, one has to trace the geological history into the very distant past, correlate it with the present day evidences gathered by various scientific methods, and project an intuitive but logical scenario for exploration. Altogether a different platform is necessary to view and understand petroleum geology in conjunction with other scientific data acquisition. The said attributes of intuition and imagination make one a successful Petroleum Geologist. **Mr. *B.T.V. Seshavataram,*** the author of this book is one such professional.

The author who retired in a senior management position from Oil and Natural Gas Corporation Limited of India, the premier Exploration and Production Company of India had an excellent and varied experience of over 40 years in the upstream sector of Oil Industry. He had an exposure to various facets of exploration, development and production for Oil and Gas. He worked with zeal throughout his career. He had been deployed in almost all the Offshore Basins of India and his understanding is thorough and experience invaluable. After retiring instead of keeping his rich experience on a back burner, he had a burning desire to share and dispel his knowledge for the benefit of budding geo-scientists and Post Graduate students pursuing degrees in Petroleum Geology/Technology. The result is this compendium which is expertly written. The book gives a detailed overview of the exploration activities in Indian Offshore Basins and serves as a referral means for those who venture into Offshore Exploration.

(P.V. Ramana Rao)

Preface

Being a Geoscientist with rich experience in upstream sector of Petroleum Industry in various facets of exploration, development and production for Oil and Gas, I had ventured to bring out this book on "Geology, Tectonics and Hydrocarbon Prospects of Offshore Sedimentary Basins of India". The information and data are carefully collected, synthesized, interpreted and presented in such a way that it will be like a handbook for all practicing Geoscientists in Hydrocarbon upstream sector and will be of immense help to the Post Graduate students, pursuing the Petroleum Geology/Petroleum Technology courses. The author co-opted M/S Krishna Mohan S., Suman T., Mallikarjuna Rao S.A., Naidu M.A., Subramanyam P. and Sanyasirao K., Young Geo-scientists of M/S Heramac Ltd., in finalization of this volume.

Some of the views and interpretations given in the course of this compilation are exclusive of the author but not necessarily to the organizations in which he served or imparting his consultations.

In spite of my full devotion and careful handling of data in presenting this book, it is quite possible, that some minor discrepancies might have crept in inadvertently. Readers are honestly requested to point out unhesitatingly the same if any, and to offer suggestions, for further inclusion and improvement of subsequent editions.

B. T. V. Seshavataram

Hyderabad
1-10-2011

Acknowledgements

The Author is extremely grateful and indebted to his Parent Organization, *Oil & Natural Gas Corporation Limited* of India.

The Author is extremely thankful to all the distinguished authors/publishers and other agencies for giving their consent to reproduce some figures and relevant data from their published books/articles. This may please be considered as a personal and individual acknowledgement. I am indebted to and humbly acknowledge all the organizations like Oil & Natural Gas Corporation, Director General of Hydrocarbons, Ministry of Petroleum and Natural Gas, Oil India, Cairn Energy, Reliance Industries, GSPCL for their valuable published data, useful while strengthening the evolution and evaluation of various offshore basins and their hydrocarbon prospectivity. Deeply appreciate and acknowledge M/s. Infraline Energy and the authors Dr. V. K. Rao and Yogesh Garg for their compilation and publishing a wonderful reference book on "Oil and Gas Exploration and Production in India" (2009).

The Author is doubly grateful to the Director General of DGH, Noida, India, for his keen interest in this type of work, his valuable suggestions, which are having immense importance in improving the subject of the book and also readily permitting the use of the literature, plates and figures wherever necessary from the published works of DGH, taking into consideration that such information will be useful to the wide spectrum of hydrocarbon industry. Special thanks to Sri S. K. Sakhtawat for critically going through the draft, for his valuable suggestions and constructive comments on various subjects/chapters of the book. Thanks are also due to S/Sri. S. Chawla and Amitav Roy, Malcolm Lall and other officers of DGH for their kind cooperation, valuable suggestions. Special thanks are to Sri R. K. Sinha for his immense help and interest and forwarding the permission of the competent authority in the publication of this book.

The Author is extremely thankful to Shri. S.V. Rao, Director-Exploration, ONGC for his valuable guidance, discussions and suggestions in finalizing the various chapters of the book and also for permitting the use of various published literatures of ONGC, wherever necessary, without which, it would have been very difficult to compile and synthesize such a huge volume of data for proper presentation to the readers.

The Author is extremely grateful to Shri. N.K. Verma, MD, OVL, ONGC for his interest in the subject and for his valuable suggestions and comments for improving the contents of book and its updation.

The Author expresses his gratitude to the Management, Heramec Ltd., Hyderabad, for all the help and facilities extended for publishing this book and is extremely grateful to S/Sri Pranav Kapadia, Malav Mehta, Ilesh Shah and P V Trivedi, Directors GNRL; Sri Anand Mudigonda, Director Sigma Oil and Gas Pvt Ltd.; for their keen interest in this project and cooperation extended from time to time. Thanks are also due to Sri. P. Bharat Sastry, GM (Technical), Heramec Ltd., for his interest in the publication and valuable suggestions from time to time.

The Author co-opted S/Sri. S. Krishna Mohan, P. Subrahmanyam, T. Suman, M.A.Naidu, S.A. Mallikarjuna Rao and K. Sanyasirao, Geoscientists of Heramec Ltd., who had shown tremendous interest in this publication, actively participated in data collection, synthesizing and organizing the data. All are personally acknowledged and separate inlet is included in the publication.

Special thanks and appreciation are due to Smt. A. Sirisha Rani of Heramec Ltd., for her keen interest, attitude and professional skills in bringing out this book.

-Author

The following are co-opted

S/Sri. S. Krishna Mohan, Geophysicist

P. Subrahmanyam, Geologist

T. Suman, Geologist

M.A. Naidu, Geophysicist

S.A. Mallikarjuna Rao, Geologist

K. Sanyasirao, Geophysicist

About the Author

Mr. B.T.V. Seshavataram, born on April 07, 1943 in West Godavari District, Andhra Pradesh. Pursued his B.Sc (Hons) Special Course in Geology in 1962 and M.Sc (Geology) in 1963 from Andhra University, Visakhapatnam, India. Successfully completed Post Graduate Diploma in Business Administration from Annamalai University, Tamilnadu during 1987. Joined in the Oil and Natural Gas Commission, as an Executive in October 1963. Underwent training in Exploration & Exploitation of Hydrocarbons from USSR during 1972 and from Paris, France during 1983.

Mr. Seshavataram had served Oil and Natural Gas Corporation for over 37 years in various capacities dealing with various facets of Hydrocarbon Exploration and Exploitation. Worked as field Geologist in deep interiors / reserved forests of Assam, Mizoram, Tripura, Nagaland, Andaman and Nicobar islands for mapping various Sedimentary Basins / Structures. Worked in almost all the Sedimentary Basins of India (both Onshore & Offshore) in key positions and had been instrumental in co-coordinating Geological mapping, Basin Evaluation, Strategic Planning, Resource Identification etc. Associated with Directorate General of Hydrocarbons (DGH / MoPNG) as an Exploration Head and contributed substantial technical back up for carving out various Exploration Blocks (both Onshore and Offshore) for NELP rounds. Actively associated with M/s. Western Geophysical, London during the year 1997 and generated a report on the interpretation of speculative seismic surveys in Eastern Offshore of India which ultimately formed the vital input in carving out various offshore blocks, some of which are already proved as major Oil/Gas producers. He had number of technical reports/publications to his credit, also presented number of technical papers in various seminars and conferences. He had an excellent aptitude for teaching and training the young Geoscientists for the upstream Hydrocarbon sector. In fact, trained many young Geoscientists working under him and the same enabled them to acquire domain knowledge in their respective fields of E & P industry.

Mr. Seshavataram, having extensive industry experience and with his incessant quest for sharing his experience, currently offering services to PSC operator M/s. Heramec Limited, Hyderabad in the capacity as Director (G&G). Toured extensively around the Globe, in evaluating the various diversified Sedimentary Basins in search of Hydrocarbons in Myanmar, Indonesia, Timor, N.W. Australia, Newzealand, Yemen and Syria and so on, helping the company to bid, participate and obtain various foreign ventures.

CHAPTER - 1

INTRODUCTION

India is having a very special distinction, wherein the setup of various sedimentary basins, very rich in hydrocarbons both on Eastern and Western Offshore India can be compared to a invaluable gold necklace studded with Gems and Precious Stones adored on a Pretty Queen's neck.

A careful attempt has been made to integrate the valuable data on various Indian offshore basins of both East and West Coasts. To understand the basin evolution of this type, the author feels that the reader should have a basic knowledge on the Plate Tectonics, especially with reference to Indian Plate Movements, with varied divergent passive continental margins and also convergent margins in geological times in the formations of sedimentary basins. Accordingly, a very comprehensive chapter is included on the Plate Tectonics and also a map on Tectonics of India depicting various tectonic elements.

In the process of integrating valuable data in bringing out this volume, an attempt has been made to give an alternative interpretation wherever it is required, for better understanding the tectonics, basin evolution and inturn to focus on the future exploration needs of the country.

One special chapter is included on the demanding alternate source of energy which is drawing the attention of the whole world that is the 'Gas Hydrates'. Favourably Indian Offshore Basins are very rich in gas hydrates. Incidentally, the estimated amount of natural gas in the gas hydrate accumulations of the world greatly exceeds the volume of all the known conventional gas resources. Gas hydrate reserves are estimated to be more than double the known reserves of fossil fuel, as per C.N.R.Rao, founder and Honorary President of Jawaharlal Nehru Centre of advance research and Kuznetsov, Director of Organic Chemistry, Russia. Rightly, Government of India is focusing and concentrating its efforts on 'Gas Hydrate – Exploration & Production' and a road map is drawn to achieve the goal in the very near future.

Since the scope and subject of this book is quite large and open, only a short brief with few details are given on the producing fields and also on oil/gas discoveries.

CHAPTER - 2

SEDIMENTARY BASINS OF INDIA

The sedimentary basins of India onland and offshore up to 200 m isobaths have an aerial extent of about 1.79 million Sq. Kms. An additional area of about 1.35 million sq. kms. has been estimated for the deep waters beyond 200 m isobaths. The total area thus works out to be 3.14 million Sq. Kms. All together, 26 major sedimentary basins are grouped into four categories based on the geological history, tectonic evolution, accumulated sediments and on their present status of exploration and prospectivity (Fig. 2.1). The categorization was earlier made by ONGC, essentially to prioritise their exploratory efforts. These sedimentary basins have been in age from Proterozoic to Cenozoic. The classification of the four categories / groups is as under.

Category I Basins

Category I basins are proved petroliferous basins with commercial production: Basins included are Bombay Offshore, Cambay Basin, Cauvery Basin, Krishna-Godavari Basin, Assam Shelf, Tripura-Cachar Basin and Rajasthan Basin.

Category II Basins

Category II basins are sedimentary basins with known occurrence of hydrocarbons but not on commercial production.

Such as Kutch, Mahanadi, Andaman & Nicobar, Himalayan Foot Hills and Bengal Basins.

Category III Basins

Category III basins are those, in which significant shows of hydrocarbons have not yet been found, but which on general geological considerations are assumed to be prospective.

Himalayan Foreland, Ganga, Vindhyan, Saurashtra and Kerala-Konkan fall under this category.

Category IV Basins

Category IV basins are those, which on analogy with similar hydrocarbon producing basins in the world are deemed to be prospective. The hydrocarbon potential of such basins in India is still uncertain and yet to be established.

Karewa, Spiti-Zanskar, Satpura, Rewa- Damodar, Narmada, Deccan Synclise, Bhima-Kaladgi, Pranhita Godavari, Bastar and Chhattisgarh basins are under this category.

Besides the above four categories, Indian deep water basins located beyond 400 m bathymetry are exclusively categorised.

The geology and prospectivity of these basins are getting unfolded as exploratory efforts, are intensified in these areas in the last decade and half. In the process, some of the basins which are in the lower category have been upgraded to the next category depending on the exploration efforts and success. For e.g. Cauvery, Krishna-Godavari, Tripura-Cachar, Rajasthan basins which were originally placed in Category-III were upgraded to Category-I. The detailed break-up of sedimentary basins as per the categorisation is shown in the Table 2.1.

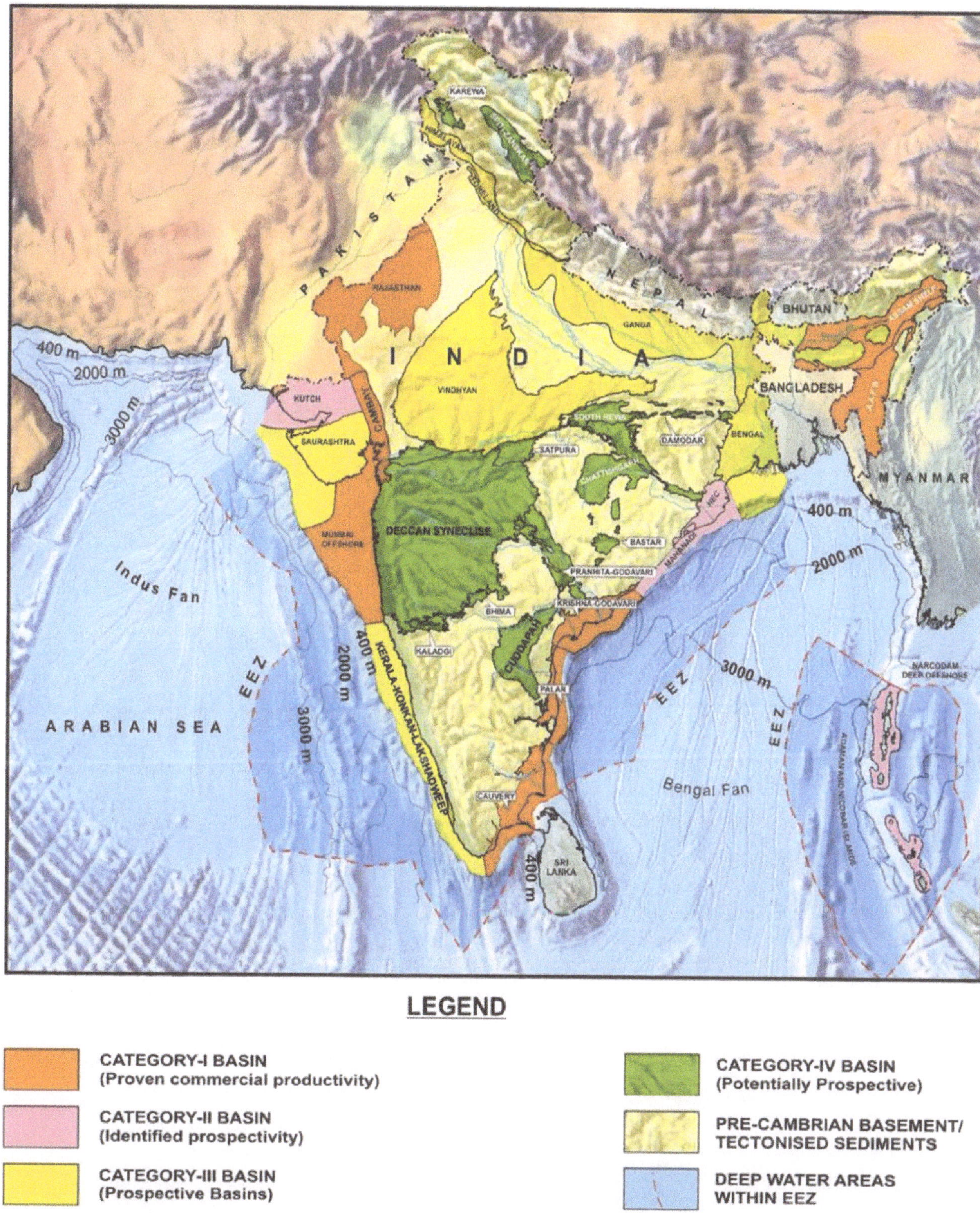

(Courtesy DGH)

Fig. 2.1

Table 2.1: Categorisation of Sedimentary Basins

Category	Basin	Basinal Area (Sq. Km)		
		Onland	Offshore	Total
UPTO 200 M ISOBATH				
I	Cambay	51,000	2,500	53,500
	Assam Shelf	56,000	-	56,000
	Mumbai Offshore	-	116,000	116,000
	Krishna Godavari	28,000	24,000	52,000
	Cauvery	25,000	30,000	55,000
	Assam-Arakan Fold Belt	60,000	-	60,000
	Rajasthan	126,000	-	126,000
	SUB TOTAL	**346,000**	**172,500**	**518,500**
II	Kutch	35,000	13,000	48,000
	Mahanadi-NEC	55,000	14,000	69,000
	Andaman-Nicobar	6,000	41,000	47,000
	SUB TOTAL	**96,000**	**68,000**	**164,000**
III	Himalayan Foreland	30,000	-	30,000
	Ganga	186,000	-	186,000
	Vindhyan	162,000	-	162,000
	Saurashtra	52,000	28,000	80,000
	Kerala-Konkan-Lakshadweep	-	94,000	94,000
	Bengal	57,000	32,000	89,000
	SUB TOTAL	**487,000**	**154,000**	**641,000**
IV	Karewa	3,700	-	3,700
	Spiti-Zanskar	22,000	-	22,000
	Satpura-South Rewa-Damodar	46,000	-	46,000
	Narmada	17,000	-	17,000
	Deccan Syneclise	273,000	-	273,000
	Bhima-Kaladgi	8,500	-	8,500
	Cuddapah	39,000	-	39,000
	Pranhita-Godavari	15,000	-	15,000
	Bastar	5,000	-	5,000
	Chhattisgarh	32,000	-	32,000
	SUB TOTAL	**461,000**	**-**	**461,200**
	TOTAL	**1,390,200**	**394,500**	**1,784,700**
Deep waters	Kori-Comorin 85°E Narcodam	-	-	1,350,000
GRAND TOTAL		**-**	**-**	**3,134,700**

The total hydrocarbon resources inclusive of deep waters are estimated at 28 billion tons of oil and oil equivalent of gas. As on April 2008, an initial inplace of 9.13 billion tons, of which recoverable reserves of 3.4 billion tons have been established indicating an upgradation of 32% of resources to reserve category. With more and more exploration efforts further upgradation can be achieved. However, the resources as estimated by DGH stands at 32 billion tons (O + OEG) – Table 2.2.

Table 2.2: Basin-wise Hydrocarbon Resources

(MMT)			
Basin	**Offshore**	**Onland**	**Total**
Mumbai	9,190	-	9,910
Assam-Arakan Fold Belt	-	3,180	3,180
Cambay	-	2,050	2,050
Upper Assam	-	1,860	1,860
Krishna-Godavari	555	575	1,130
Cauvery	270	430	700
Rajasthan	-	380	380
Kutch	550	210	760
Andaman-Nicobar	180	-	180
Kerala-Konkan	660	-	660
Saurashtra Offshore	280	-	280
Ganga Valley	-	230	230
Bengal	30	160	190
Himalayan Foreland	-	150	150
Mahanadi	100	45	145
Deepwater *	7000	-	7,000
GRAND TOTAL	**18,815**	**9,270**	**28,085**

(After DGH)

*Hydrocarbon resources in Deepwater off East Coast are likely to increase further by about 400 MMT as per interpretation results of reconnaissance surveys carried out by DGH.

These resource potentials do not include the gas hydrate resources. The total prognosticated gas resources from the gas hydrates in the country is placed additionally at 1894 TCM.

Indian Stratigraphy

The stratigraphy of Indian basins in general as established by different authors is indicated in Table 2.3 & 2.4.

Table 2.3: General Geological Succession in different parts of India and BURMA Prof. M.S. Krishnan, 1943

Standard Scale	Northern Himalaya	Baluchistan Arc	Salt Range & Potwar	Kashmir-Hazara	Simla-Garhwal	Assam	Burma	Peninsula	Coastal Areas
Recent	Alluvium	Sands, Loess	Alluvium, Loess	Alluvium & C	Alluvium & C	Alluvium, River gravels	Alluvium, River gravels	Alluvium Khadar	Alluvium sands
Pleistocene	Older Alluvium, gravels, moraines	Older alluvium, sands	Older alluvium, sands	Older alluvium, moraines Upper Karewa	Older alluvium, moraines	Older alluvium, gravels	Older alluvium, River terraces	Older alluvium, Bhangar, Cave deposits, river terraces	Raised beaches, Porbander stone etc.
Pliocene Upper Miocene	......	Siwalik Sy. Manchhar Sy.	Siwalik Sy.	Lower Karewa Siwalik Sy.	Siwalik Sy.	Dihing S.	Irrawaddy Sy.	....	Siwalik Sy. Karaikal B. Quilon B. Cuddalore Sst. Etc.
Middle Miocene Lower Miocene	.....	Mekran Sy. Flysch, Bugti B. Gaj S.	Murree S.	Murree S. Fatehjang zone	Kasauli B. Dagshai B.	Tipam S. Surma S.	Upper Pegu S.	....	Gaj s.
Oligocene	......	Nari S., Khojak Shales	....	....	....	Barali S. (in part).	Lower Pegu S.	....	Nari S. Dwaraka B.
Eocene	Eocene	Kirthar S. Laki S. Ranikot S.	Chharat S. Kirthar S. Laki S. Ranikot S.	Chharat S. Hill Lst. Ranikot S.	Subathu B.	Barail S. Jaintia S. Sylhet Lst Disang S. Cherra Sst.	Yaw and Pondaung Stages. Tabyin, Tilin, and Laungshe stages	..... Deccan Trap and Intertrappeans, Lametas	Kirthar S. Laki S. Ranikot S.
Upper Cretaceous	Upper Cretaceous Chikkim S.	Cardita Beaumonti B., Pab Sst.	Upper Cretaceous	Upper Cretaceous	......	Disang S. (in part) Upper Cretaceous	Axial S. (in part) Negrais S.	Deccan Trap and Intertrappeans, Lametas	Bagh B. Trichinopoly Cretaceous
Lower Cretaceous	Giumal Sst.	Parh Lst.	Lower Cretaceous	Giumal S. Orbitolina Lst.	.....	....	Lower Cretaceous	Umia B.	Umia B.

Table 2.3: *Contd...*

Standard Scale	Northern Himalaya	Baluchistan Arc	Salt Range & Potwar	Kashmir-Hazara	Simla-Garhwal	Assam	Burma	Peninsula	Coastal Areas
Jurassic	Spiti Shales Koto Lst.	Massive Lst. Crinoidal Lst.	Jurassic	Spiti shale Jurassic	? Tal. S.	.....	Namyau B. ? Loi-an B.	Jubbulpore Kota Rajmahal	Katrol Chari Patcham
Triassic	Triassic	U. Triassic	Triassic	Triassic	? Shali Lst. ? Krol S.	.....	Napeng B.	Maleri Pachmarhi Panchet	......
Permian ...	Productus B. Syringothyris Lst.	Permo-Carboniferous	Productus Lst. Speckled Sst.	Agglomertic Slates, Zewan B.	? Infra-Krol.	.....	Upper Plateau Lst. Moulmein Lst.	Raniganj S. Barren Measures, Barakar S.	......
Up. Carboniferous	......	......	Conularia and Eurydesma B. Boulder Bed	Agglomeratic sh. Gangamopteris B. Tanakki Boulder B.	...... ?Blaini boulder Bed.	...... Subansiri B.	?	Umaria B. Talchir boulder B.	......
Middle & L. Carboniferous	Po Series Fenestella Sh.	......	......	Fenestella sh.	......	......	L. Plateau Lst.	......	......
Devonian	Muth Qtzt.	......	....	Muth Qtzt Devonian (Chitral)	? Jaunsar S.	......	Padaukpin Lst. Wetwin Sh.	......	......
Silurian ...	Silurian	......	....	Silurian	......	......	Zebingy S. Namshim S.	......	......
Ordovician	Ordovician	......	....	Ordovician	......	......	Naungkangyi S. Mawson S.	......	......
Cambrian	Cambrian	......	Cambrian	Cambrian	? Deoban Lst.	......	Bawdwin volcanic ?Mergui S.	Vindhyan Sy.	......
Algonkian	Haimanta Sy.	......	Attock Slates	Dogra Slates	Simla Slates	Buxa S. Daling S. Darjeeling S.	Chaung. Magyi S.	Cuddapah Sy. Delhi Sy.	......
Archaean	Vaikrita Sy.	......	?	Salkhala S. Gneisses	Chail S. Jotogh S.	Shillong s. Gneisses	Mogok S. Gneisses	Dharwar Sy. Aravalli Sy. Etc. Gneisses	......

International statigraphic nomenclature, as established by International Commission on Statigraphy is shown in Table 2.5.

Table 2.5: International Statigraphic Chart

International Commission on Stratigraphy

Eonothem Eon	Erathem Era	System Period	Series Epoch	Stage Age	Age Ma
Phanerozoic	Cenozoic	Quaternary *	Holocene		0.0117
			Pleistocene	Upper	0.126
				"Ionian"	0.781
				Calabrian	1.806
				Gelasian	2.588
		Neogene	Pliocene	Piacenzian	3.600
				Zanclean	5.332
			Miocene	Messinian	7.246
				Tortonian	11.608
				Serravallian	13.82
				Langhian	15.97
				Burdigalian	20.43
				Aquitanian	23.03
		Paleogene	Oligocene	Chattian	28.4 ±0.1
				Rupelian	33.9 ±0.1
			Eocene	Priabonian	37.2 ±0.1
				Bartonian	40.4 ±0.2
				Lutetian	48.6 ±0.2
				Ypresian	55.8 ±0.2
			Paleocene	Thanetian	58.7 ±0.2
				Selandian	~ 61.1
				Danian	65.5 ±0.3
	Mesozoic	Cretaceous	Upper	Maastrichtian	70.6 ±0.6
				Campanian	83.5 ±0.7
				Santonian	85.8 ±0.7
				Coniacian	~ 88.6
				Turonian	93.6 ±0.8
				Cenomanian	99.6 ±0.9
			Lower	Albian	112.0 ±1.0
				Aptian	125.0 ±1.0
				Barremian	130.0 ±1.5
				Hauterivian	~ 133.9
				Valanginian	140.2 ±3.0
				Berriasian	145.5 ±4.0

Eonothem Eon	Erathem Era	System Period	Series Epoch	Stage Age	Age Ma
					145.5 ±4.0
Phanerozoic	Mesozoic	Jurassic	Upper	Tithonian	150.8 ±4.0
				Kimmeridgian	~ 155.6
				Oxfordian	161.2 ±4.0
			Middle	Callovian	164.7 ±4.0
				Bathonian	167.7 ±3.5
				Bajocian	171.6 ±3.0
				Aalenian	175.6 ±2.0
			Lower	Toarcian	183.0 ±1.5
				Pliensbachian	189.6 ±1.5
				Sinemurian	196.5 ±1.0
				Hettangian	199.6 ±0.6
		Triassic	Upper	Rhaetian	203.6 ±1.5
				Norian	216.5 ±2.0
				Carnian	~ 228.7
			Middle	Ladinian	237.0 ±2.0
				Anisian	~ 245.9
			Lower	Olenekian	~ 249.5
				Induan	251.0 ±0.4
	Paleozoic	Permian	Lopingian	Changhsingian	253.8 ±0.7
				Wuchiapingian	260.4 ±0.7
			Guadalupian	Capitanian	265.8 ±0.7
				Wordian	268.0 ±0.7
				Roadian	270.6 ±0.7
			Cisuralian	Kungurian	275.6 ±0.7
				Artinskian	284.4 ±0.7
				Sakmarian	294.6 ±0.8
				Asselian	299.0 ±0.8
		Carboniferous	Pennsylvanian Upper	Gzhelian	303.4 ±0.9
				Kasimovian	307.2 ±1.0
			Pennsylvanian Middle	Moscovian	311.7 ±1.1
			Pennsylvanian Lower	Bashkirian	318.1 ±1.3
			Mississippian Upper	Serpukhovian	328.3 ±1.6
			Mississippian Middle	Visean	345.3 ±2.1
			Mississippian Lower	Tournaisian	359.2 ±2.5

Eonothem Eon	Erathem Era	System Period	Series Epoch	Stage Age	Age Ma
					359.2 ±2.5
Phanerozoic	Paleozoic	Devonian	Upper	Famennian	374.5 ±2.6
				Frasnian	385.3 ±2.6
			Middle	Givetian	391.8 ±2.7
				Eifelian	397.5 ±2.7
			Lower	Emsian	407.0 ±2.8
				Pragian	411.2 ±2.8
				Lochkovian	416.0 ±2.8
		Silurian	Pridoli		418.7 ±2.7
			Ludlow	Ludfordian	421.3 ±2.6
				Gorstian	422.9 ±2.5
			Wenlock	Homerian	426.2 ±2.4
				Sheinwoodian	428.2 ±2.3
			Llandovery	Telychian	436.0 ±1.9
				Aeronian	439.0 ±1.8
				Rhuddanian	443.7 ±1.5
		Ordovician	Upper	Hirnantian	445.6 ±1.5
				Katian	455.8 ±1.6
				Sandbian	460.9 ±1.6
			Middle	Darriwilian	468.1 ±1.6
				Dapingian	471.8 ±1.6
			Lower	Floian	478.6 ±1.7
				Tremadocian	488.3 ±1.7
		Cambrian	Furongian	*Stage 10*	~ 492 *
				Stage 9	~ 496 *
				Paibian	~ 499
			Series 3	Guzhangian	~ 503
				Drumian	~ 506.5
				Stage 5	~ 510 *
			Series 2	*Stage 4*	~ 515 *
				Stage 3	~ 521 *
			Terreneuvian	*Stage 2*	~ 528 *
				Fortunian	542.0 ±1.0

Jagdish Pandey, 1998

Table 2.4

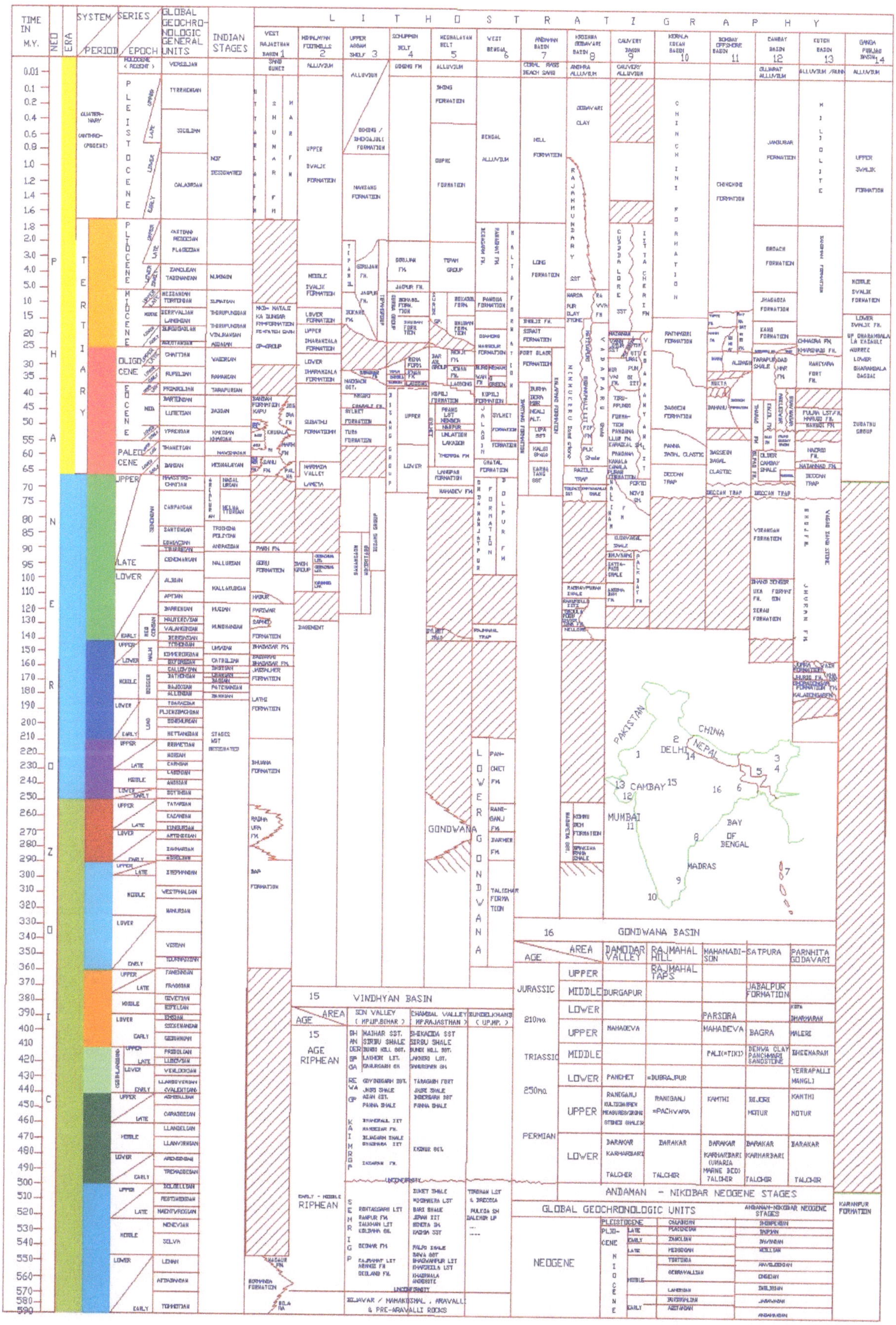

CHAPTER - 3

CLASSIFICATION OF WEST AND EAST COAST INDIAN OFFSHORE BASINS

A: West Coast Offshore Basins

The sedimentary basins in the western continental shelf of India cover an area of about 2,55000 Sq.Kms. up to 200 m isobaths, extending into the deeper parts beyond the continental shelf and up to the continental margin (200-3000 m) isobath covering an additional area of 6,44,840 Sq.Kms.

The sedimentary basins comprising the western offshore (Fig. 3.1) are:

1. Kutch-Saurashtra
2. Bombay Offshore
    - (i) Tapti-Daman,
    - (ii) Bombay High Platform (including DCS area)
    - (iii) Heera-Panna-Bassein-Neelam Blocks
    - (iv) Ratna
3. Kerala-Konkan
    - (i) Konkan Block
    - (ii) Kasargod Block
    - (iii) Kerala offshore Block
    - (iv) Lakshadweep Block

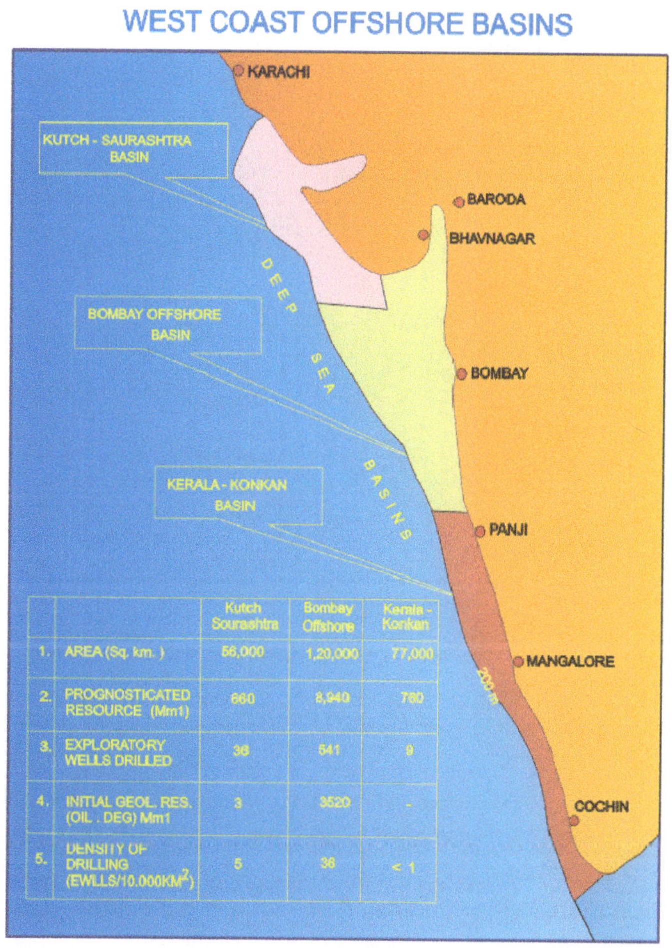

		Kutch Sourashtra	Bombay Offshore	Kerala - Konkan
1.	AREA (Sq. km.)	56,000	1,20,000	77,000
2.	PROGNOSTICATED RESOURCE (Mm1)	660	8,940	780
3.	EXPLORATORY WELLS DRILLED	36	541	9
4.	INITIAL GEOL. RES. (OIL . DEG) Mm1	3	3520	-
5.	DENSITY OF DRILLING (EWLLS/10.000KM^2)	5	36	< 1

(Courtesy: **ONGC)**

Fig. 3.1

Kutch-Saurashtra basins are having corresponding contiguous onshore basins like Kutch onshore and Saurashtra onshore. Bombay Offshore and Kerala-Konkan Ooffshore sedimentary basins do not have the corresponding onshore sedimentary basins. However, the Cambay Gulf (Tapti-Daman block) situated Northeast of Bombay High platforms, extends into onshore Cambay basin.

B: East Coast Basins

The sedimentary Basins in the eastern continues shelf of India cover an area of approx 1,50,000 sq. kms upto 200 m isobath, extending into deeper parts beyond continental shelf (200 to 3000 m water depth) isobath covering an additional area of approx. 7,00,000 sq. kms.

The East Coast offshore sedimentary basins are classified (Fig. 3.2) from South to North, as below:

1. Cauvery Basin
   - (i) Gulf of Mannar Block
   - (ii) Ramnad Palk Bay Block and
   - (iii) Cauvery Sub-basin
2. Palar Basin
3. Krishna-Godavari Basin
4. Mahanadi including NEC Block
5. Bengal Basin
6. Andaman Nicobar Basin

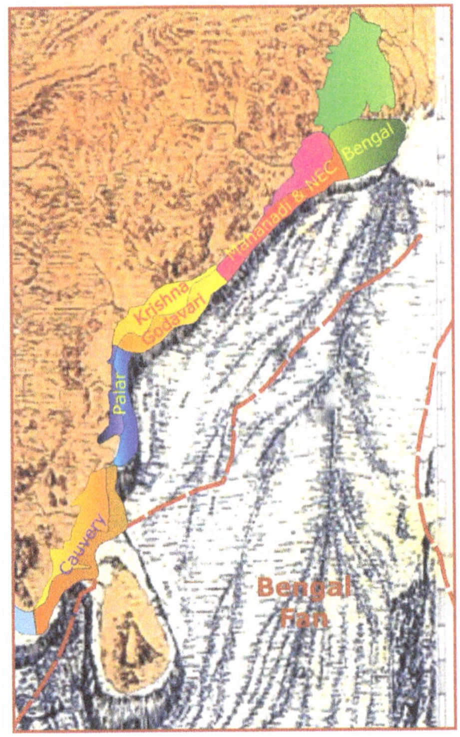

Fig. 3.2

CHAPTER - 4

PLATE TECTONICS WITH SPECIFIC REFERENCE TO INDIAN PLATE MOVEMENTS

4.1 Indian Plate Movement in General

To understand the offshore Indian sedimentary basins and their evolution, it is utmost necessary to have some knowledge about plate tectonics with special reference to Indian plate movement (Fig. 4.1) and it's bearing on the evolution of offshore sedimentary basins.

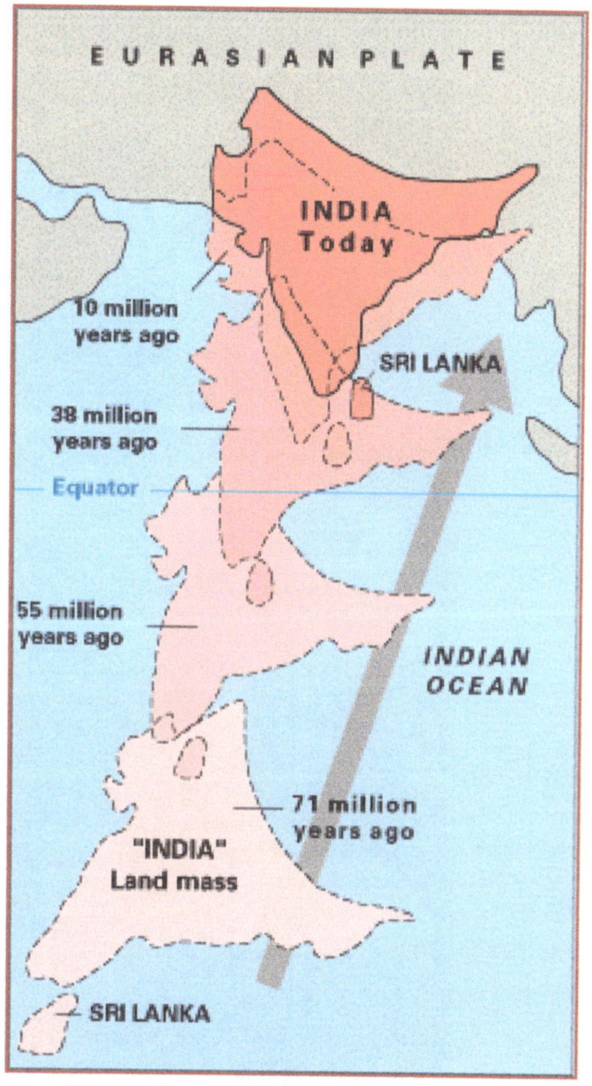

Fig. 4.1

The Indian craton was once a part of the super continent of "Pangaea", and at that time India was attached to Medagaskar and Southern Africa on the South west coast and Australia along the East coast (Fig. 4.2). During the Jurassic period (160 Ma}, rifting caused Pangaea to break apart into two super continents, namely Gondwana (to the south) and Laurasia (to the north). The Indian craton remained attached to Gondwana until the super continent began to rift apart, about in the early Cretaceous. (125 Ma.) The Indian plate then drifted northwards towards the Eurasian plate, at a pace, that is the fastest movement of any known plate. It is generally believed that the Indian plate separated from Medagaskar about 90 Ma. However, some Biogeographical and Geological evidences suggest that the connection between Medagaskar and Africa was retained at the time, when the Indian plate collided at, with the Eurasian plate about 50 Ma. This orogeny which is continuing today is related to closure of the Tethys Ocean. The closure of this ocean which created the Alps Mountains in Europe, and the Caucasus range in western Asia, created Himalayan Mountains and the Tibetan plateau in South Asia. The current orogenic event is causing parts of the Asian continent to deform westward and eastward on either side of the tethyian orogeny. Concurrently with this provision, the Indian plate sutured on the adjacent Australian plate, forming a new larger plate, the Indo- Australian Plate.

4.2 Tectonic Evolution

Due to continental drift, the Indian plate split from Medagaskar and collided with the Eurasian plate resulting in the formation of the Himalayas.

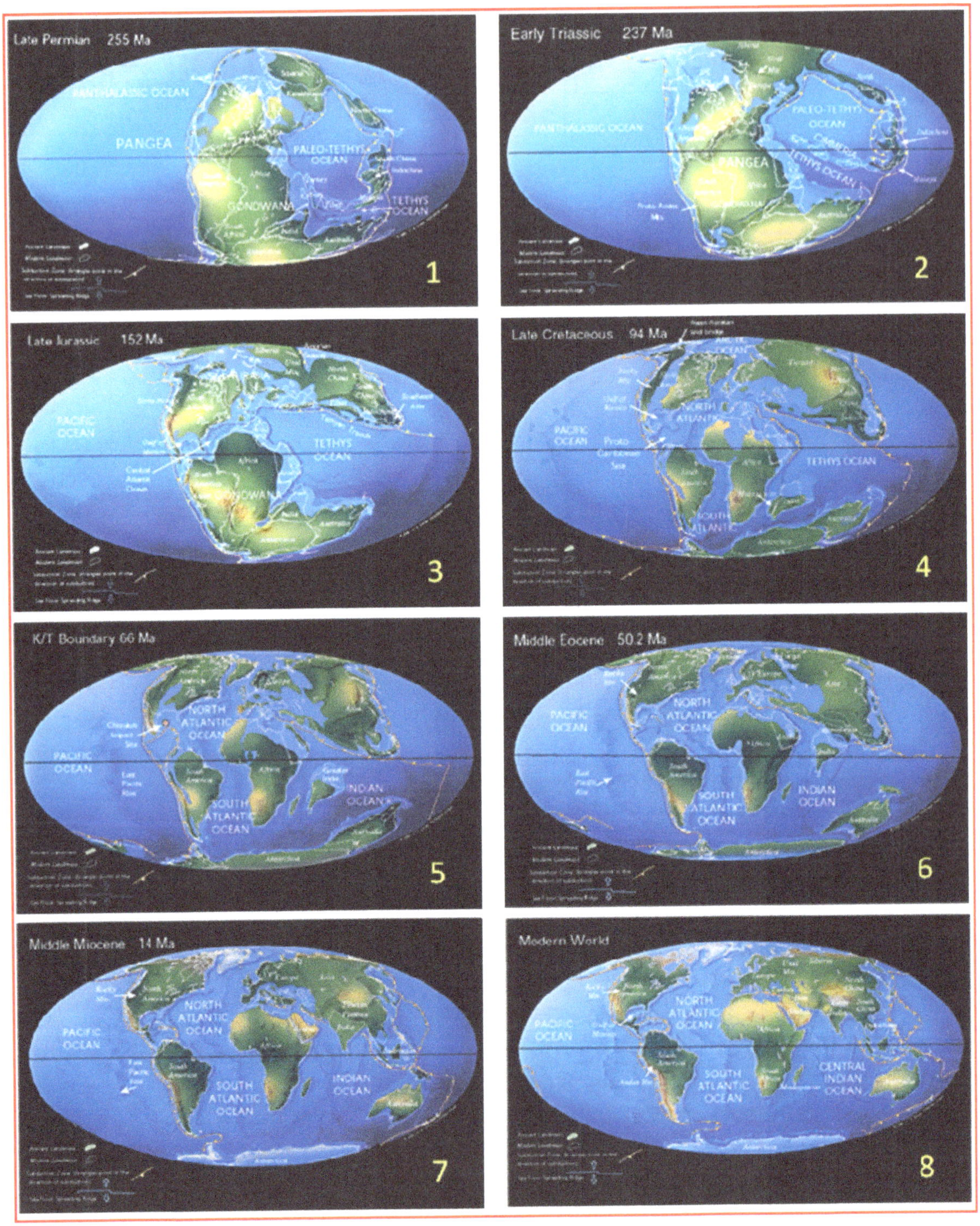

Source: Christopher R. Scotese (PALEOMAP Project) cscotese @gmail.com

Fig. 4.2

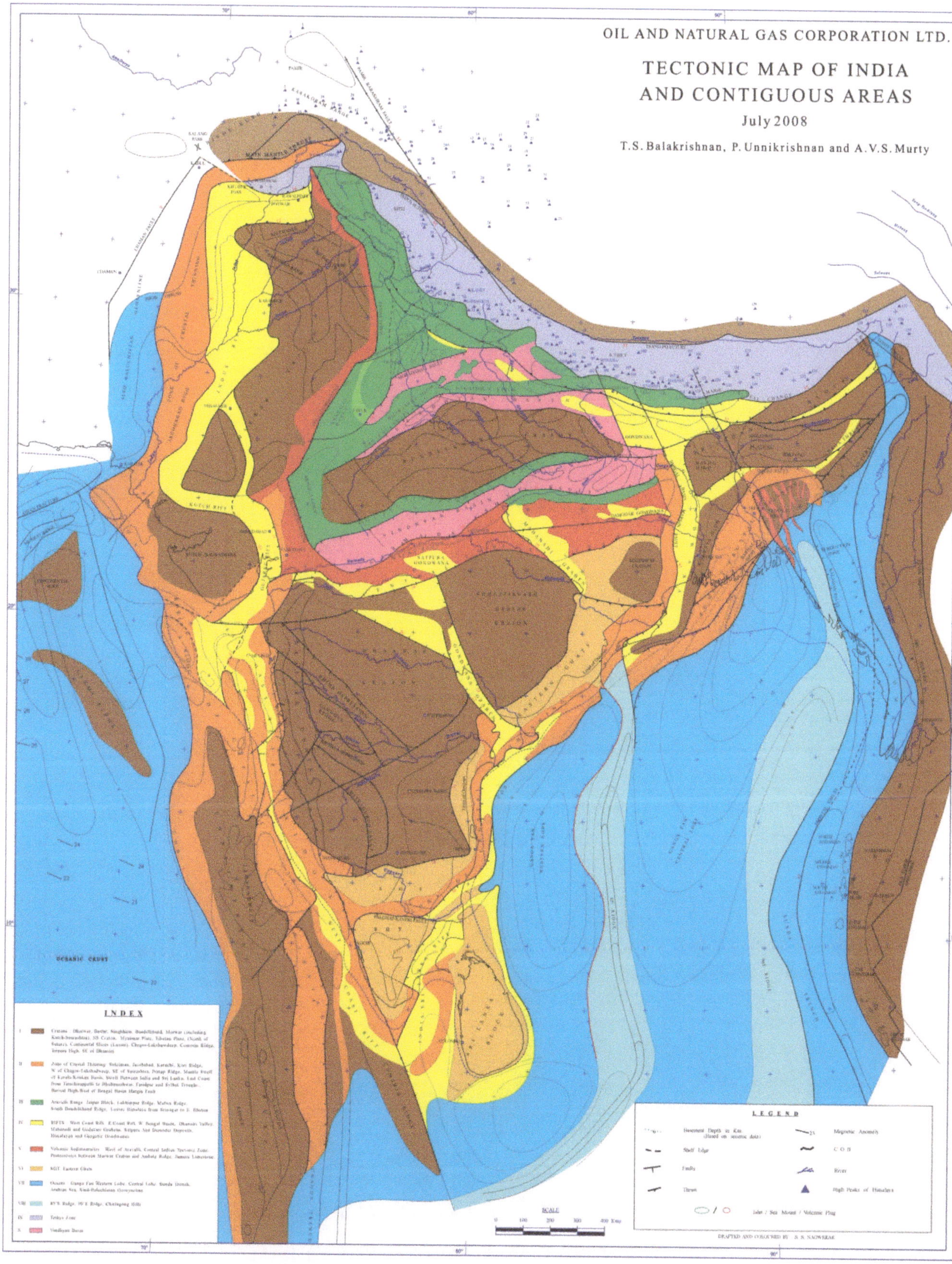
OIL AND NATURAL GAS CORPORATION LTD.
TECTONIC MAP OF INDIA
AND CONTIGUOUS AREAS
July 2008
T.S. Balakrishnan, P. Unnikrishnan and A.V.S. Murty
INDEX
LEGEND
OCEANIC CRUST

Plate Tectonics

The earliest phase of tectonic evolution was marked by the cooling and solidification of the upper curst of the earth's surface in the Archaean era (prior to 2-5 billion years) which is represented by the exposure of Gneisses and granites especially on the peninsula. These form the core of the Indian craton. The Aravalli range is the remnant of an early Proterozoic orogen called the Aravalli-Delhi orogen that joined the two older segments that make up the Indian craton. It extends approximately 500 Kms from its Northern end to isolated hills and rock ridges near Delhi.

Minor igneous intrusions, deformation (folding and faulting) and subsequent metamorphism of the Aravalli Mountains represent the main phase of orogenesis. The erosion of the mountains and further deformation of the sediments of the Dharwarian group (Bijawars) mark the second phase. The volcanic activities and intrusions, associated with the second phase are recorded in the composition of these sediments.

Early to Late Proterozoic Calcareous and Arenaceous deposits which correspond to humid and semi arid climatic regimes were deposited in the Cuddapah and Vindhyan basins. These basins, which border or lie within the existing crystalline basement, were uplifted during the Cambrian (500 Ma.). The sediments are generally un-deformed and in many places have preserved their original horizontal stratifications. The Vindhyans are believed to have been deposited between 1700 and 650 Ma.

Early Paleozoic rocks found in the Himalayas consist of southerly derived sediments eroded from the crystalline craton and deposited on the Indian platform.

In the Late Paleozoic, Permocarboniferous, glaciations left extensive Glacio-fluvial deposits across central India, in the new basins created by sag / normal faulting. These tillites and glacially derived sediments are designated the Gondwana series. The sediments are overlain by rocks resulting from a Permian marine transgression (270 Ma).

The Late Paleozoic coincided with the deformation and drift of the Gondwana super- continent. To this drift, the uplift of the Vindhyan sediments and the deposition of northern peripheral sediments in the Himalayan Sea can be attributed.

During the Jurassic, Pangea began to rift apart and large grabens formed in central India, filling with upper Jurassic and Lower Cretaceous sandstones and conglomerates.

By the Late Cretaceous, India had separated from Australia and Africa and was moving northwards towards Asia. At that time, prior to Deccan eruptions, uplift in southern India resulted in sedimentation in the adjacent Nascent Indian Ocean. Exposures of these rocks occur along the southern Indian coast at Pondicherry and in Tamil Nadu.

At the close of the Mesozoic, one of the greatest volcanic eruptions in earth's history occurred, the 'Deccan lava flows' covering more than 5, 00,000 Sq.Kms area, marks the final break from Gondwana. In the Early Tertiary, the first phase of the Himalayan orogeny, the Karakoram occurred. The Himalayan orogeny has continued to the present day and it is still in the progress.

4.3 Western Continental Margin

As commonly perceived, the Gondwana land had remained a single super continent, during late Cretaceous and Paleozoic times. Its breakup had began with successive rifting (Fig. 4.2) between Madagascar and Somalia around 150Ma i.e. beginning of Middle-Jurassic that had been predated by a long period of intra-cratonic rifting, during almost the whole of Paleozoic. The spreading between Somalia – Mozamique and

Madagascar and southward of Sri Lanka had commenced in Ordovician - Late Jurassic i.e about 160 Ma. This led to the separation of East Gondwana from Africa – Arabia. By Late Cretaceous (Cenomonion) i.e about **100Ma** ago the Greater India (India Continent, Madagascar and Sheychelles) got completely separated from Africa, Arabia and Antarctica and began its northward drift.

Following this, at around 87-88 Ma i.e. Concion (Late Cretaceous), Madagascar got separated from Indo-Sheychelles block with intense volcanic eruption along eastern margin. It continued with the spreading and opening at Mascarene basin between 83 and 64 Ma (Late Cretaceous-Early Paleocene) and almost simultaneously opening of Laxmi Basin got initiated that led to the separation of Sheychelles- Laxmi, and probably Laccadive continental fragments form western continental margin of India. Subsequently Deccan volcanism which is placed around 65-67Ma (Maastrichtian-Cretaceous) is being related by many earlier workers to the initiation of the Re Union Hot Spot. It had probably triggered the final breakup event around Durian (Early Paleocene) i.e., 63 Ma with the spreading axis. This event led to the breakup of Sheychelles micro continent from the Laxmi Ridge and its drifting to the south relative to the Indian continent.

Thus, it can be seen that rifting played the most significant role in shaping the present day configuration of the Western Continental Margin. This western margin was repeatedly subjected to external tectonics in cretaceous and early tertiary with separation of several variably sized continental fragments. Extensively acquired seismic data also reveals a number of rift related structural features. All these observations lead us to regard the Indian western continental margin as a rifted volcanic continental margin.

4.4 Eastern Continental Margin

The Gondwana paleo-geographic reconstruction shows juxtaposition of Antarctica to the east coast of Peninsular India with Australia lying further to the east. The eastern continental margin of India is a rifted passive margin evolved in response to continental rifting. Indian Craton is circumscribed by rifted grabens and marginal sag basins. The east-west trending Narmada-Son Tectonic Lineament, an important line of discontinuity across the Indian shield is a mid-continental rift system, which divides the shield into two halves- a northern 'Foreland block' that now forms the Himalayan Foreland Region and a southern' Peninsular Block'. The process was initiated along the Satpura weak zone during Late Carboniferous / Triassic time due to crustal stretching that also caused co-genetic rifting of Mahanadi and Godavari grabens along NW-SE tectonic trend. During the Early Rift Stage (Permo-Triassic), down warping of the Eastern Margin formed a series of northeast-southwest trending faults following an older set of Archaean fault lineaments 'The Eastern Ghat Trend'. Initial break-up of the Indian Craton was caused by rift initiation possibly by the formation of a series of triple junctions forming two sets of cross trending grabens.

Northwesterly paleo - current directions during Permo-Carboniferous and Triassic times, marine intercalations and deltaic facies of Godavari and Mahanadi grabens suggest the presence of a sea to the north of these grabens. This and the formation of depocenters at the meeting point of these grabens with the Narmada-Son rift (Satpura and Son) prompted to reasonably infer the presence of a seaway along this major rift. This rift episode seems to have ended by Late Triassic when the seaway was probably filled up and uplifted, as the Indian Plate had started drifting away. The failed arms of these rifts forming aulacogens have been exploited by the Mahanadi, Krishna, Godavari and Cauvery river systems.

During Late Jurassic to Early Cretaceous period, India – Australia plates rotated from Antarctica with continental extension between Australia and Antarctica and a right lateral transform movement along the southern part of the Coromandal coast (east coast) margin of India (to the south of Krishna-Godavari Basin) and Indian plate got separated from Antarctica. Thus the initial breakup of the Indian Plate from Australia-Antarctica possibly occurred prior to Early Neocomian, creating the proto Bay of Bengal Ocean with the

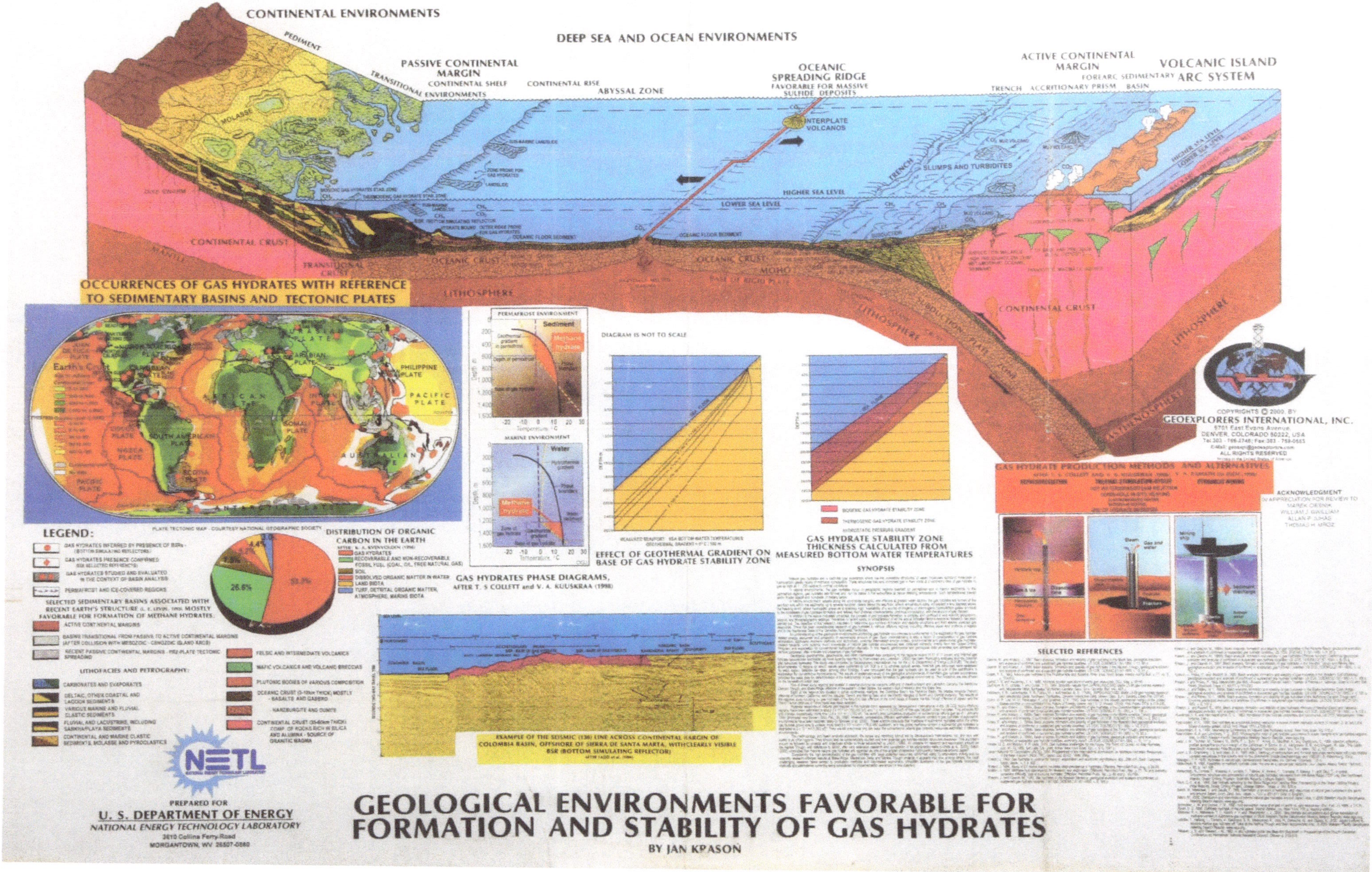
CONTINENTAL ENVIRONMENTS
DEEP SEA AND OCEAN ENVIRONMENTS
PEDIMENT
PASSIVE CONTINENTAL MARGIN
CONTINENTAL SHELF
TRANSITIONAL ENVIRONMENTS
CONTINENTAL RISE
ABYSSAL ZONE
OCEANIC SPREADING RIDGE
FAVORABLE FOR MASSIVE SULFIDE DEPOSITS
INTERPLATE VOLCANOS
ACTIVE CONTINENTAL MARGIN
TRENCH
ACCRITIONARY PRISM
FOREARC SEDIMENTARY BASIN
VOLCANIC ISLAND ARC SYSTEM
MOLASSE
SLUMPS AND TURBIDITES
HIGHER SEA LEVEL
LOWER SEA LEVEL
CONTINENTAL CRUST
TRANSITIONAL CRUST
OCEANIC CRUST
MOHO
LITHOSPHERE
SUBDUCTION PLATE ZONE
ASTHENOSPHERE
OCCURRENCES OF GAS HYDRATES WITH REFERENCE TO SEDIMENTARY BASINS AND TECTONIC PLATES
PHILIPPINE PLATE
PACIFIC PLATE
SOUTH AMERICAN PLATE
NAZCA PLATE
SCOTIA PLATE
SOMALI PLATE
PLATE TECTONIC MAP - COURTESY NATIONAL GEOGRAPHIC SOCIETY
LEGEND:
GAS HYDRATES INFERRED BY PRESENCE OF BSRs - (BOTTOM SIMULATING REFLECTORS)
GAS HYDRATES PRESENCE CONFIRMED
GAS HYDRATES STUDIED AND EVALUATED IN THE CONTEXT OF BASIN ANALYSIS
PERMAFROST AND ICE-COVERED REGIONS
SELECTED SEDIMENTARY BASINS ASSOCIATED WITH RECENT EARTH'S STRUCTURE MOSTLY FAVORABLE FOR FORMATION OF METHANE HYDRATES:
ACTIVE CONTINENTAL MARGINS
BASINS TRANSITIONAL FROM PASSIVE TO ACTIVE CONTINENTAL MARGINS (AFTER COLLISION WITH MESOZOIC - CENOZOIC ISLAND ARCS)
RECENT PASSIVE CONTINENTAL MARGINS - PRE-PLATE TECTONIC SPREADING
LITHOFACIES AND PETROGRAPHY:
CARBONATES AND EVAPORATES
DELTAIC, OTHER COASTAL AND LAGOON SEDIMENTS
VARIOUS MARINE AND FLUVIAL CLASTIC SEDIMENTS
FLUVIAL AND LACUSTRINE, INCLUDING SABKHA/PLAYA SEDIMENTS
CONTINENTAL AND MARINE CLASTIC SEDIMENTS, MOLASSE AND PYROCLASTICS
FELSIC AND INTERMEDIATE VOLCANICS
MAFIC VOLCANICS AND VOLCANIC BRECCIAS
PLUTONIC BODIES OF VARIOUS COMPOSITION
OCEANIC CRUST (3-10km THICK) MOSTLY - BASALTS AND GABBRO
HARZBURGITE AND DUNITE
CONTINENTAL CRUST (35-60km THICK) COMP. OF ROCKS RICH IN SILICA AND ALUMINA - SOURCE OF GRANITIC MAGMA
DISTRIBUTION OF ORGANIC CARBON IN THE EARTH
GAS HYDRATES
RECOVERABLE AND NON-RECOVERABLE FOSSIL FUEL (COAL, OIL, FREE NATURAL GAS)
SOIL
DISSOLVED ORGANIC MATTER IN WATER
LAND BIOTA
TURF, DETRITAL ORGANIC MATTER, ATMOSPHERE, MARINE BIOTA
53.3%
26.6%
PERMAFROST ENVIRONMENT
Sediment
Methane hydrate
MARINE ENVIRONMENT
Water
Temperature, °C
GAS HYDRATES PHASE DIAGRAMS, AFTER T. S COLLETT and V. A. KUUSKRAA (1998)
DIAGRAM IS NOT TO SCALE
EFFECT OF GEOTHERMAL GRADIENT ON BASE OF GAS HYDRATE STABILITY ZONE
GAS HYDRATE STABILITY ZONE THICKNESS CALCULATED FROM MEASURED BOTTOM WATER TEMPERATURES
SYNOPSIS
EXAMPLE OF THE SEISMIC (138) LINE ACROSS CONTINENTAL MARGIN OF COLOMBIA BASIN, OFFSHORE OF SIERRA DE SANTA MARTA, WITH CLEARLY VISIBLE BSR (BOTTOM SIMULATING REFLECTOR)
GAS HYDRATE PRODUCTION METHODS AND ALTERNATIVES
SELECTED REFERENCES
COPYRIGHTS © 2000, BY GEOEXPLORERS INTERNATIONAL, INC.
5701 East Evans Avenue
DENVER, COLORADO 80222, USA
ALL RIGHTS RESERVED
ACKNOWLEDGMENT
IN APPRECIATION FOR REVIEW TO
MAREK OBSNIK
WILLIAM J. GAILLIAM
ALLAN P. JUHAS
THOMAS H. MROZ
NETL
PREPARED FOR
U. S. DEPARTMENT OF ENERGY
NATIONAL ENERGY TECHNOLOGY LABORATORY
3610 Collins Ferry Road
MORGANTOWN, WV 26507-0880
GEOLOGICAL ENVIRONMENTS FAVORABLE FOR FORMATION AND STABILITY OF GAS HYDRATES
BY JAN KRASON

onset of sea-floor spreading. Since Cretaceous, the Indian Plate rotated northwestward. Development of a new spreading center (?) initiated spreading in north-south direction. It was during Late Albian that the paleo slope was reversed in these graben basins with the southeasterly tilt of the peninsular block. This event coincided with the first basin wide marine transgression during Cenomanian and continued up to Maastrichtian. The post-rift thermal subsidence continued throughout the remainder of Cretaceous and Tertiary. An extensive uplift of that major rift and erosion followed to the end of Cretaceous prior to the volcanic episode of Deccan trap. The northward drift continued till it had initial (soft) collision with Eurasia in Paleocene. Sea level fall in Upper Oligocene followed this. The southeasterly/ easterly tilt of the Peninsula had resulted in the increase of fluvial energies of various fluvial systems leading to pouring of substantial sediment loads into the adjoining sea. In Early Miocene, crustal shortening continued as the Indian Plate continued its northward drift with the resultant further increase in fluvial activity. Himalayan upliftment and the deposition of sediments to the south by river Ganges and Brahmaputra commenced in Miocene age. Thus Miocene and younger times witnessed very significant increase in the sediment outpouring into the offshore, extending the basins into the deeper waters.

The evolution of East Coast of India as a passive margin set up resulted in the formation of a number of peri-cratonic basins- Bengal, Mahanadi, Krishna-Godavari, Pennar and Palar, while Cauvery basin developed as an intra-cratonic basin.

CHAPTER - 5

WESTERN OFFSHORE BASINS

Six contiguous, NW-SE trending major tectonic elements are identified in the western offshore/ continental margin of India (Fig. 5.1). These are

1. The shelf with horst and graben complex.
2. Kori-Comorin depression
3. Kori-Comorin Ridge
4. Laxmi-Laccadive Depression
5. Laxmi-Laccadive Ridge
6. Arabian Abyssal Plains

Three Pre-Cambrian Orogenic trends i.e. the NNW-SSE Dharwar trend, NE-SW Aravalli trend and the ENE-WSW Satpura trend dominate the structural fabric of western offshore. Reactivation of these Ancient trends during and after rifting, determined the shape, extent and subsidence histories of the various horsts and grabens in the shelf and also the structural and sedimentary styles of various sedimentary basins. Further, the shelf is differentiated into a number of sub-basins by the transverse basement Arches and other related features.

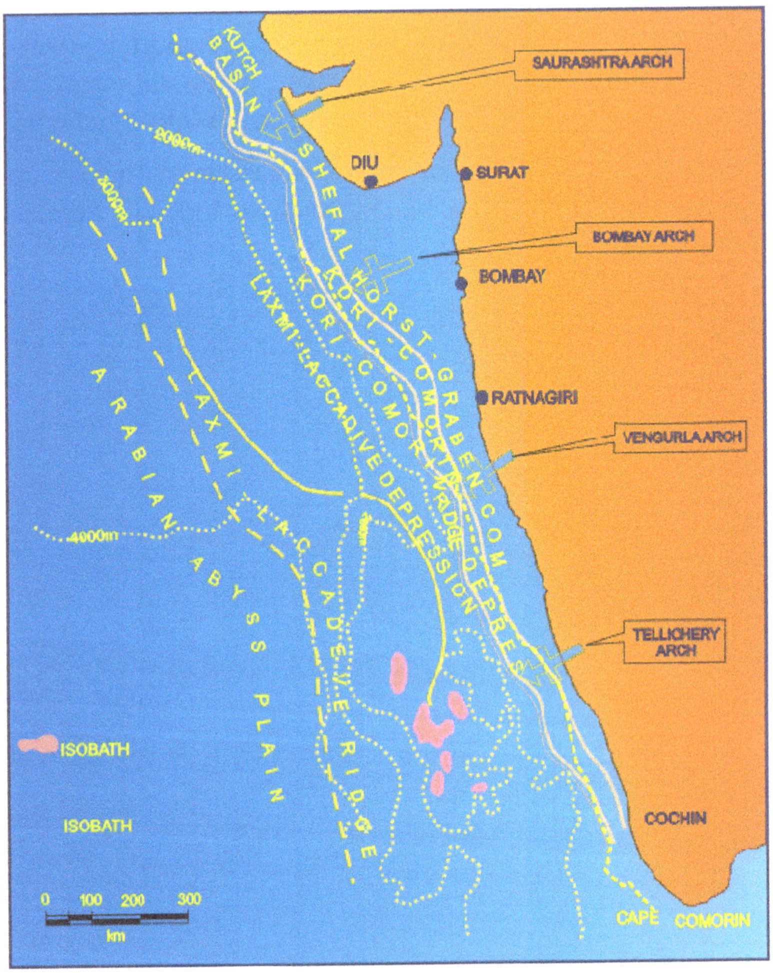

TECTONIC FRAMEWORK OF WESTERN MARGIN OF INDIA

Fig. 5.1

Southwesterly plunging Saurashtra Arch separates the Northern Kutch basin from the southern Bombay offshore basin. Westerly plunging Bombay Arch separates the Tapti-Daman basin to the North and Heera-Panna, Mukta and Ratnagiri blocks to the South. Ratnagiri basin is separated from the Konkan basin by

Southwesterly plunging Vengurla Arch. The Konkan basin is separated from Kerala basin by Southwesterly plunging Tellicherry Arch (Fig. 5.1).

The NW-SE trending shelf is separated from Cori-Comorin Ridge by Cori-Comorin depression, and the Kori-Comorin Ridge is separated from Laxmi Laccadive Ridge by a long linear depression named Laxmi - Laccadive Depression. The continental shelf with average shelf break occurrence at about 200 m water depth is quite broad i.e. about 350 Kms off Bombay, but it narrows down to 60 Kms off Cochin in the South and to 80 Kms off Porbandar in the North. From all the evidences so far collected, it appears that the continental margin of Bombay has a characteristic expression of convergent plate boundary.

The prognosticated resources of oil and oil equivalent gas estimated at 10,800 MMT for the west coast up to 200 m isobaths of which about 3550 MMT have been converted into geological reserves (about 34%). Still 2/3rd of the resources are yet to be upgraded. About 5.3 billion tons of prognosticated resources of oil and oil equivalent gas have been estimated exclusively for deep water basins beyond 200 m isobath.

5.1 Kutch – Saurashtra Basin

5.2 Kutch Basin

The Kutch basin, covering parts of Kutch mainland and Rann of Kutch encompasses an area of 42,000 sq.kms and extends westwards into the offshore where it covers an additional area of 26,000 sq.kms. up to 200 m isobaths (Fig. 5.1.a.1).

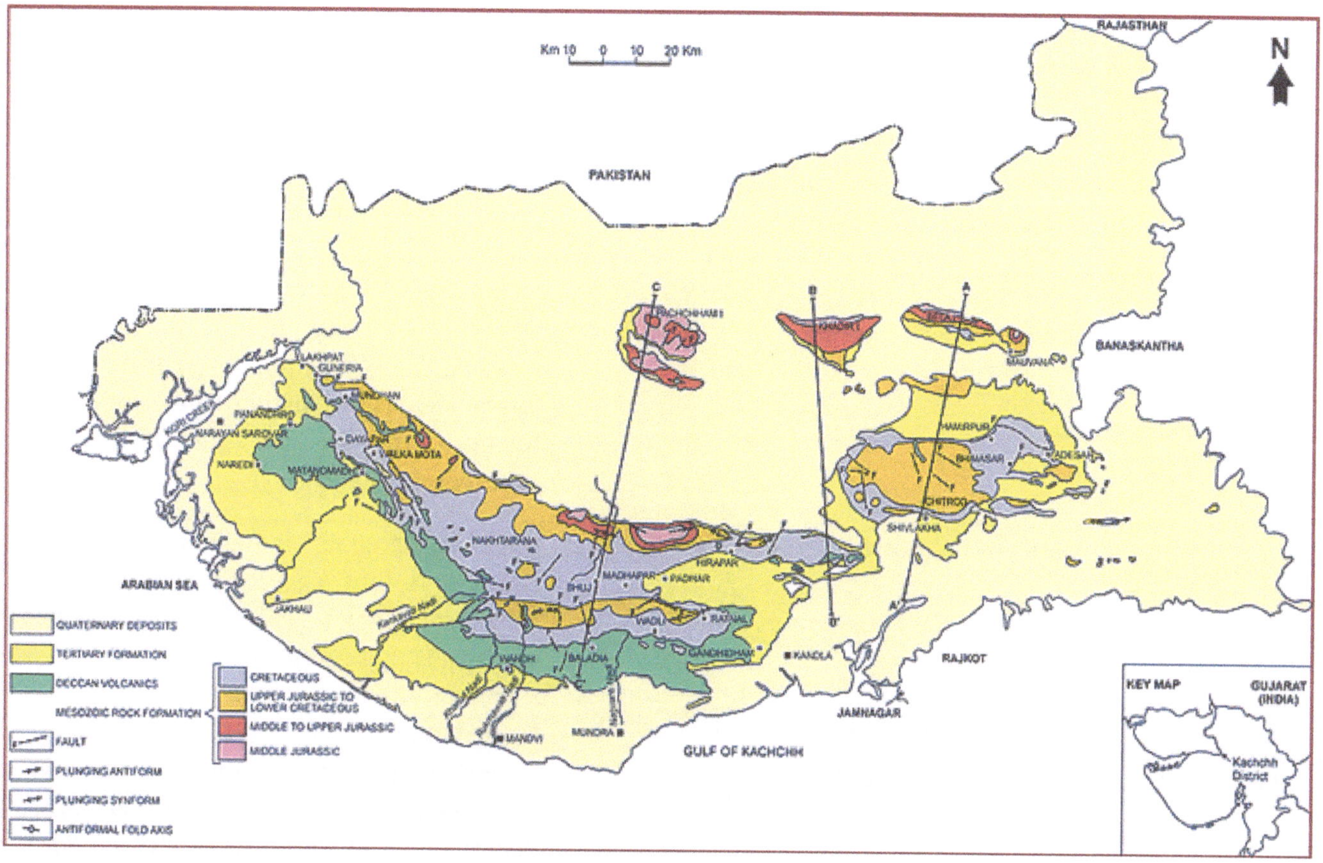

(Courtesy ONGC)

Fig. 5.1.a.1

Geologically Kutch basin is contiguous with the lower part of the lower Indus basin of Pakistan where commercial quantity of oil and gas are discovered. Exploration activities by ONGC started in the basin as far back as 1957.

Miocene and Paleogene carbonates form a platform over the shelf in the offshore basin, deepening towards Southwest. Mesozoic section was encountered below thick volcanic trap in some of the wells drilled on the platform area. The area in the platform margin show a belt of steep dips associated with faults and has been designated as hinge zone. The generalized stratigraphy encountered on the platform and to the west of hinge zone is indicated in the Table 5.1.a.1.

Table 5.1.a.1

Epoch	Age	Series	Stage	Lithology
				Rann Silt, Alluvium, Blown sand etc. Partially consolidated Conglomerates, Grits, Kankars etc., Oyester Beds, Raised Beaches, Terrace beds etc.
Holocene to Pleistocene		Moliolite Formation		Pellatoidal calcareous sandstones, Sandy pelmicrites, and sandy oomicrites
Pliocene to Pleistocene		Kankawati Formation		Grey sandstones, pink fossiliferous calc, grits and conglomerates with subordinate shales.
Miocene	Helvetian to Burdigalian		Vinjhan Stage	Grey and khaki clay with fossiliferous marls.
	Aquitanian	Khari Series	Aida Stage	Variegated siltstones
Oligocene	Chattian	Disconformity (Feeble)	Warior Stage	Banded cream coloured fossilifereous silty marls, silty shales, and glauconitic sandstones and green sand beds.
	Lattorfian to Rupelian	Ber Moti Series	Ramania Stage	Dirty white to yellowish banded marlites and impure limestones with glauconitic pellites, calcareous claystones. Presence of glauconite pellets very charecterstic.
Eocene	Lutetian	Paraconformity	Babia Stage	Upper 75 – 140 ft: Cream and buff massivesilty limeston, packed with foraminifera (silty biolithitesand biomicrites) Lower 45 ft: Greenish grey fossiliferous calcareous clay and shales.
	Yepresian	Berwali Series	Kakdi Stage	Upper 45 ft:grey shales and red laterites. Lower 87 ft: Brown gypseous Shales & green glauconitic shales with thin fossiliferous marls and mud balls. Locally, gypseous shales with red ochre and black shales with lignite beds occur.
Paleocene		Disconformity Madh Series		Lateritic Conglomerates, Laterites, Bauxites, Tuffaceous shales, Sandstones & Grits, Bentonitic and ferruginous clays with volcanic ash.

Table 5.1.a.1 *Contd...*

Epoch	Age	Series	Stage	Lithology
Early Cretaceous		Deccan Trap		Dark Green Basalts: Alternating flows of columnar and amugdolooidal basalt. Locally intertrappean beds are present.
Late Cretaceous		Unconformity Madhwan Formation Dhrangandhra Formation		

A schematic geological cross section across Kutch basin from West to East covering offshore shelf part to Banni graben onland is shown in the Figs. 5.1.a.2 and 5.1.a.3.

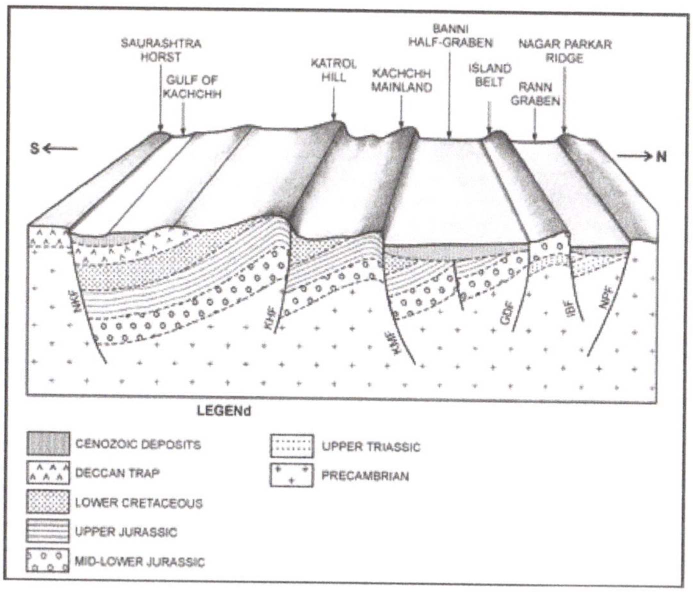

(Courtesy ONGC)

Fig. 5.1.a.2

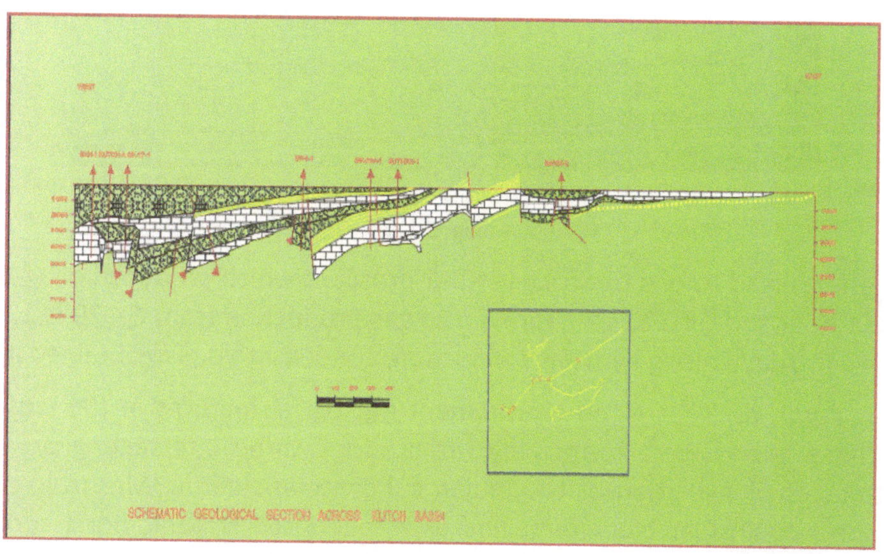

(Courtesy ONGC)

Fig. 5.1.a.3

The entire offshore area was initially covered by M/s. Reading and Bates under an exploration contract during 1974. The company carried out Gravity Magnetic Surveys and seismic surveys to the tune of 10,800 LKM and drilled one well GKH-1 to a target depth of 4575 m without success. Subsequent to the relinquishment of the contract, ONGC acquired additional 20,478 LKM of Seismic data. In the process, 46 prospects had been identified for exploratory drilling, of which about 15 prospects had been in the offshore part of the basin (Fig. 5.1.a.4).

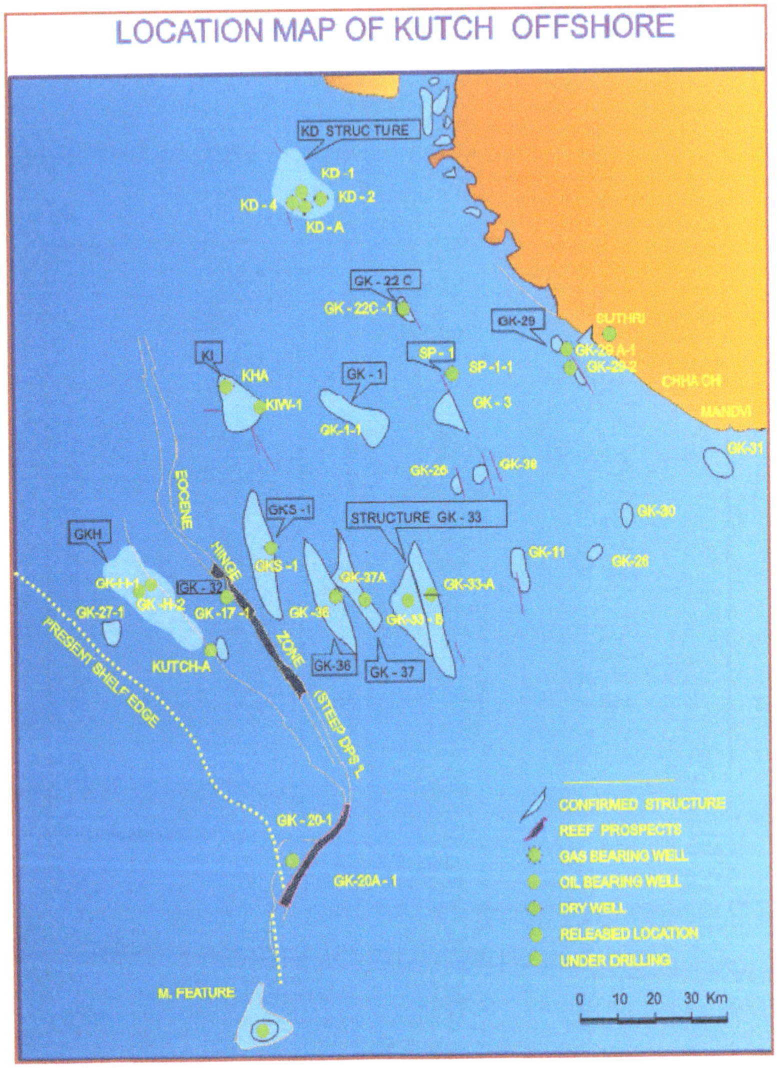

(Courtesy ONGC)

Fig. 5.1.a.4

The exploration efforts lead to one oil strike in KD-1 prospect which produced oil @ 172 barrels per day and two gas finds in GK-29 and GK-22C structures. The gas production from GK-29 is 1,24, 330 m^3/day and from GK-22C 2,39,980 m^3/day. Along with gas production, condensate is also obtained in these wells.

Narmada-Cambay-Kutch basins are three prominent geological features in the western peripheries of Indian peninsula. These basins were formed by rifting along three intersecting pre-Cambrian orogenic trends at different stages of evolution of the Indian subcontinent during Mesozoic times (Biswas, 1982 Fig. 5.1.a.5); Saurashtra peninsular block remained as a foundered cratonic horst surrounded by the rift grabens. Filling up of Kutch & Cambay graben and tilting of Saurastra block westward in the Early Cretaceous resulted in the development of extensive depositional platform for the early cretaceous fluvio-deltaic sediments.

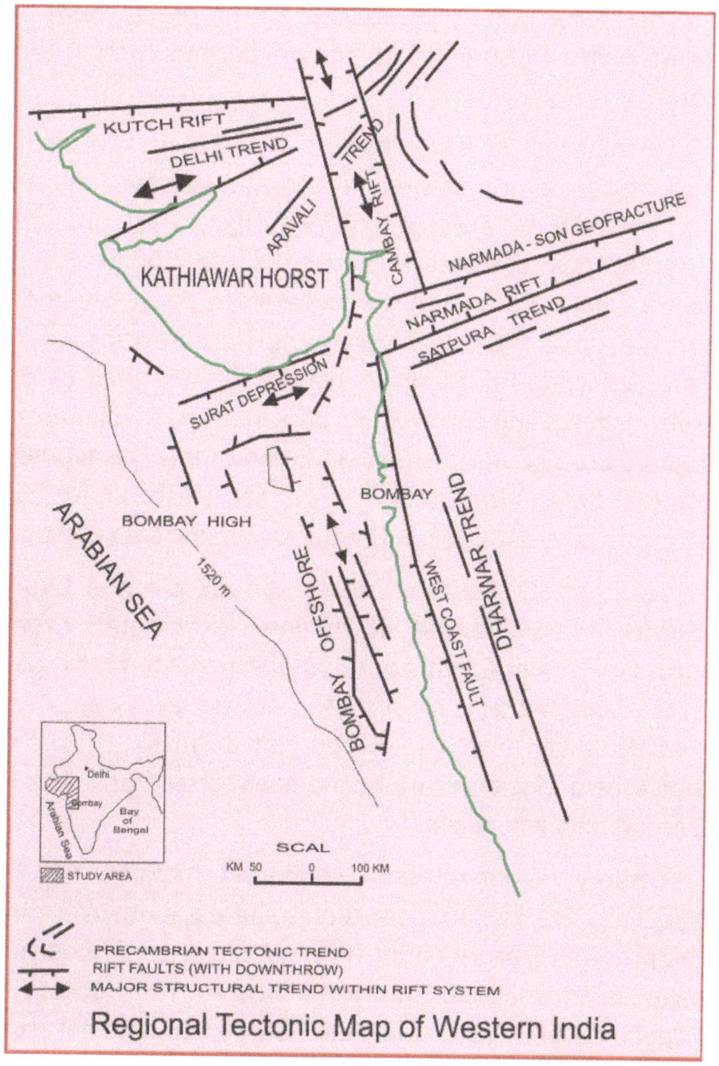

Regional Tectonic Map of Western India

(After S.K Biswas)

Fig. 5.1.a.5

West coast of India and associated peri cratonic rift basins are styled by a conjugate system of the principal pre-Cambrian orogenic trends and subsequent reactions in Mesozoic and Tertiary times.

1. The ENE – WSW, Narmada – Son lineament which parallels the Satpura orogenic trend is major tectonic boundary (West, 1962, Mathur .et .al 1966) - a geofracture dividing the Indian shield into a southern peninsular block & Northern foreland block. Narmada graben is formed in the western part of this lineament by faults parallel to it.
2. The west coast fault responsible for shaping the present coast line follows the NNW- SSE – Dharwar trend. Its northward extension in across Naramada graben forms the onshore Cambay graben. Thus the Gulf of Cambay is considered as Tri-Junction with Cambay graben as the aborted rift/the failed arm.
3. The third important orogenic trend in this region is the NE – SW Delhi – Aravali trend.

The western continent shelf is caracterised by a series of extension faults sub parallel to Dharwar rift, giving rise to a series of narrow linear horsts & Grabens typical of a passive margin (Partsch, 1978). These basin developments & its tectonic evolutions are interrelated to major events of plate tectonics.

It was further brought out that the rifting developed sequentially from north to south around saurastra horst. Kutch basin was formed in early Jurassic at the time of rifting of Africa and India, followed by cambay basin in early cretaceous and Narmada in late cretaceous. Saurastra remains as positive area between Kutch and offshore Narmada. Thus mid Jurassic to Early Cretaceous stratigraphic section is present in Kutch, where as Saurastra has Lower Cretaceous and Narmada has Upper Cretaceous section.

In Kutch-Saurastra-Cambay areas major tectonic episodes accompanied by tectonic volcanism (Deccan trap) in Late Cretaceous, Early Paleocene time terminated the Mesozoic sediments.

The sedimentary basin of Kutch includes the whole Kutch and the western part of Banskantha district of North Gujarat. The basin is an East- west embayment opening out and deepening into the Arabian Sea on the west.

The basin framework consists of the embayment between Nagar-Parker uplift in the North, Kathiawar uplift in the south and Radhanpur- Barmar Arch in the East with a median high occurring along the hinge zone.

To the west of hinge zone, the sediments thicken considerably with lot of facies variations. The E-W trending structural axis plunges to the west into the Arabian Sea. These East-West Mesozoic trends extend almost up to the present day shelf edge, which are completely masked by the wide spread volcanic activity (the thickness of Deccan traps varies from 100 m to over 1000 m), over which, the NW-SE trending Neogene tectonic play, parallel to the present day coast line and Dharwar trend play a dominant role. In other words, the Mesozoic trend underlying the Deccan traps will be oblique to the early Neogene trend. The Seismic data acquisition parameters have to be designed carefully to get the information below the thick volcanic trap. For getting the desired information, special array of source, capable of injecting high pressured and high volume of compressed air is required. This would also involve an extensive experimentation to design optimal field parameters. Hence, for Mesozoic exploration, seismic data acquisition is needed, to obtain interpretable data below the thick volcanic trap and also to map the reliable structural styles and events.

Sedimentation during Mesozoic took place in two main phases starting with a transgressive phase in Middle Jurassic during which fine clastics and carbonates were deposited. A regressive phase (Upper Jurassic – Lower Cretaceous), where increased rates of sedimentation with good detritous supply, formed a thick clastic wedge, mostly sand stones with minor shale. It is evident from the sediment characteristics that the basin was a sheltered gulf during Mesozoic times. The wedge - Arkose association in the North indicates deposition in a Yoked basin conditions. The Mesozoic sedimentation ended with late cretaceous Deccan trap volcanism.

Tertiary sedimentation continued rapidly till Middle Miocene. The Tertiary depocentre shifted South-West ward in the present offshore basin. Well marked unconfirmed surfaces have been mapped in early Eocene, Late Eocene, Late Oligocene and Late Miocene with denudational surfaces (Biswas, 1973). At the peak of transgression, shale and limestone alternations of Khari series (Vinjkan shale) were deposited. Post Miocene was a period of regression when Pliocene deltaic and shallow marine sands were deposited followed by a quaternary transgression. The generalized stratigraphy of Kutch is shown in Table-5.1.a.1 for onland basin and Fig. 5.1.a.6 for offshore basin.

The saurashtra basin is bounded by the Kutch and Cambay rift basins to the North & East and by Surat depression to the south. The entire area forms a horst block, characteristic of lower cretaceous sedime-ntation followed by thick 'Deccan Trap' volcanic activity overlain by thin Neogene. The structure and tectonic style of the basin is largely controlled by the NE-SW trending Precambrian Aravali trend.

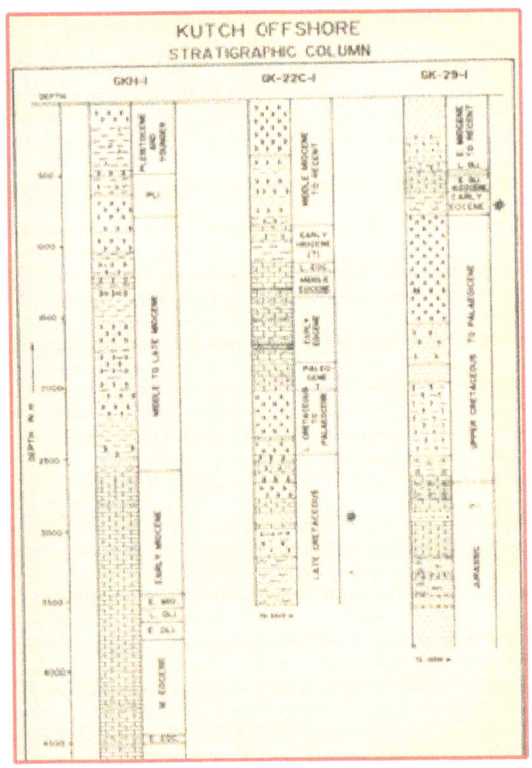

(Courtesy ONGC)

Fig. 5.1.a.6

The petroleum prospects in Kutch are dependent on two major Hydrocarbon plays i.e., the Mesozoic and Tertiary. The Mesozoic play is masked by thick sequence of volcanic rocks (Deccan Trap). The upper Jurassic and lower cretaceous formations have very good source rock potential for generation of hydrocarbons in the basin. Tertiary sedimentation is very much restricted on the onshore part of the basin, whereas it progrades and thickens over 2500 to 3000 mt in the offshore part of the Kutch-Saurastra basin.

The well drilled in Kutch offshore on 'H' prospect encountered a very thick carbonate sequence (> 2000 mt) from middle Miocene to upper Paleocene. 'H' structure is a large anticlinal closure, flanked by faults. (Fig 5.1.a.7)

In the upper part of Miocene section occasional oil and gas shows were recorded in carbonates. In Eocene section oil and gas shows are very sporadic. The 'H' feature was drilled on the shelf flanking the shelf margin zone.

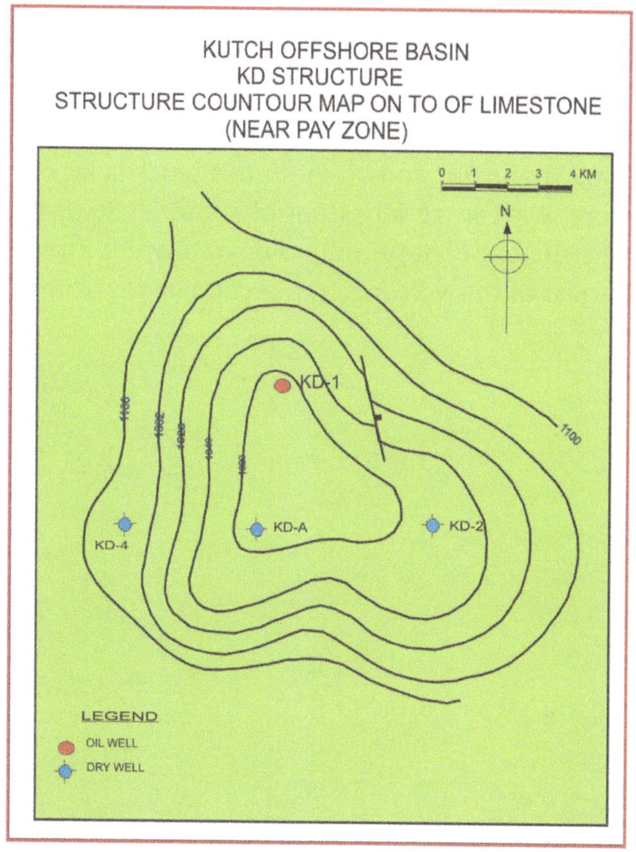

(Courtesy ONGC)

Fig. 5.1.a.7

KD

The KD structure is located about 45 km away from the coast. The areal extent is 125 sq.km and the vertical closure is about 90 m at the lower Eocene level. a) This is a domal structure. The discovery well was drilled in 1984. Oil has been struck in Eocene section. Subsequently three more exploratory wells were drilled to delineate the field and all proved to be dry. Depth contour map at a reflection close to Eocene pay zone is shown in Fig. 5.1.a.8. The first well drilled on KD structure proved to be oil well in the Eocene limestone. Further delineation / development plans were not drawn.

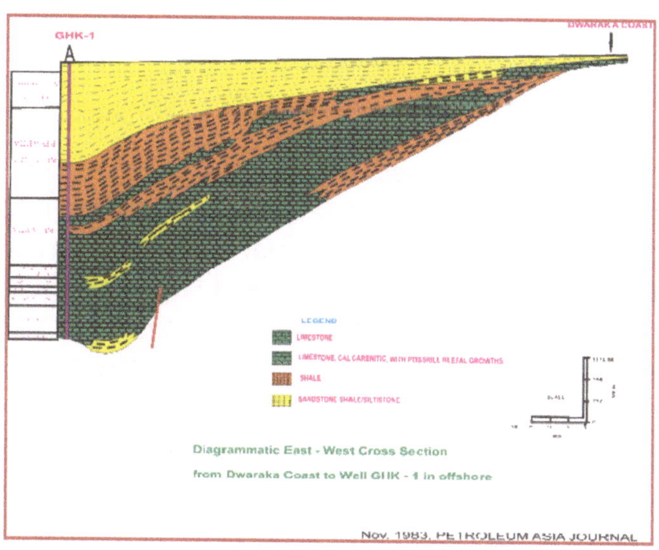

(Courtesy ONGC)

Fig. 5.1.a.8

GK-22C

GK-22 structure is located in a water depth of 15 m and is about 35 kms west of Jakhau. GK-22C proved to be gas bearing in Cretaceous sandstone reservoir below a thick volcanic trap section. This gas strike being the first in the Mesozoic, is very significant and opens up new vistas for Mesozoic exploration in Kutch basin. The major constraint in this block is lack of reliable reflections below volcanic trap. This is a fault closure, with an aerial extent of about 25 Sq.Kms. and a vertical closure of 60 msec. at the trap level. The time structure contour map prepared within Upper Cretaceous near pay zone is placed at Fig. 5.1.a.9. The thickness of this volcanic trap section varies from 150 m. in the North to 1800 m. in the South.

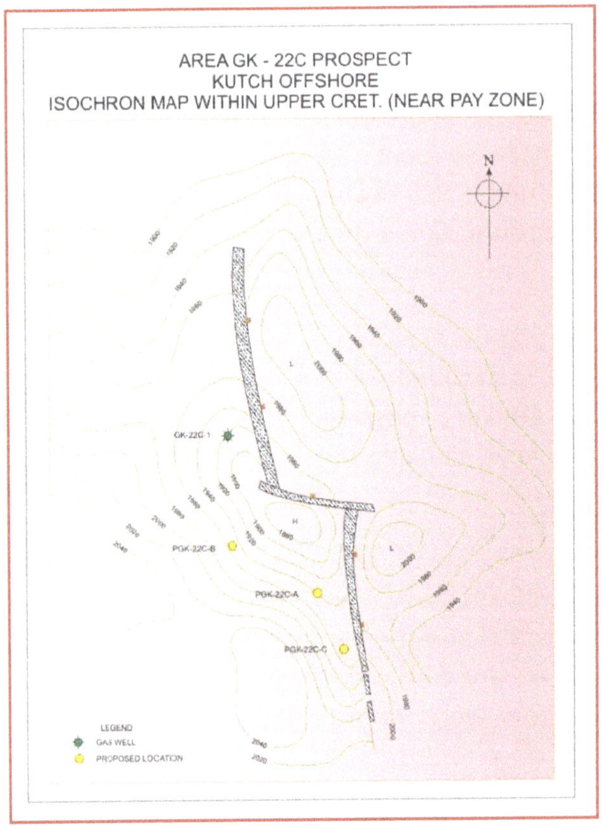

(CourtesyONGC)

Fig. 5.1.a.9

GK-29

GK-29A structure is located in Kutch offshore basin 8 Kms away from the coast at a water depth of 15 m. Commercial quantity of gas flowed from the lower Paleocene-Eocene sandstone reservoir. The GK-29 structure is a NW-SE trending anticline, bounded by a major fault on eastern side with down throw to the east. The area of the structure is 75 Sq.Kms, with amplitude of 60 msec. Structure contour map at the top of Trap level is placed at Fig. 5.1.a.10. Teleseis surveys are also required to connect the onshore and the shallow offshore areas for exploration. This will also facilitate to connect and interpret the structural and tectonic styles prevailed in the offshore and onshore parts especially below the volcanic trap section, which is expected to be different from the overlying Tertiary structural style and tectonics.

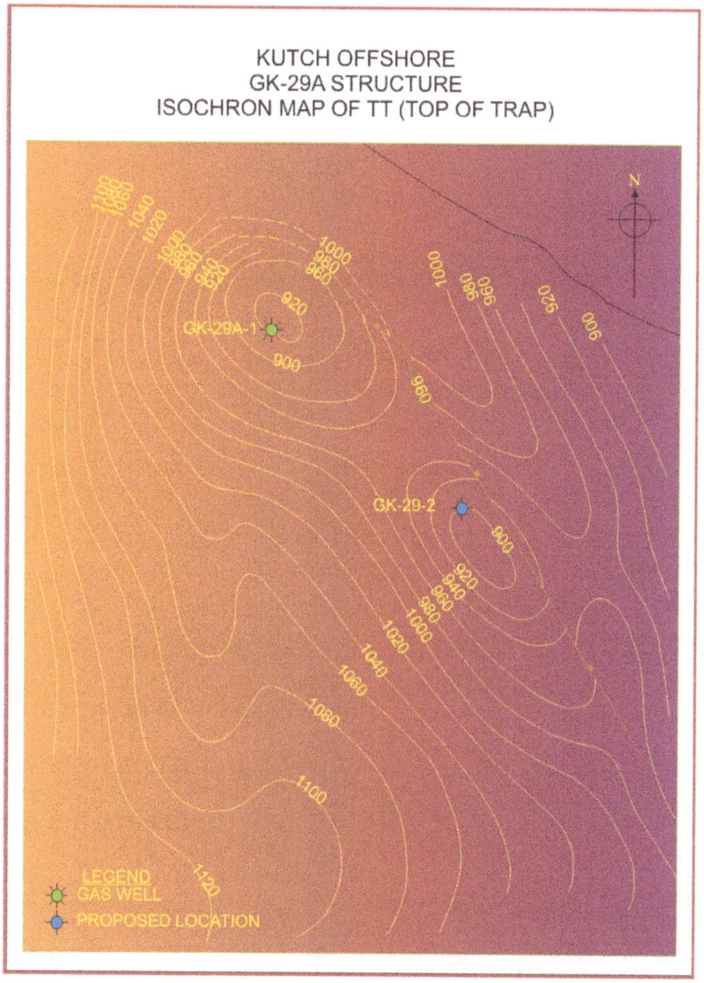

(Courtecy ONGC)

Fig. 5.1.a.10

Exploration and Production Status

Basin area : 28,000 sq. Km.

Seismic data : 2D -154,974 LKM

3D- 9,729 Sq. Km.

Data acquired by	2D (LKM)	3D (Sq. Km.)
DGH	10,218	-
ONGC	115012	5490
Pvt. /JV	29744	4239

Other Surveys : Gravity - 27,395 LKM

: Magnetic - 14,691 LKM

: Satellite Gravity Survey

Well Data Number of Exploratory Wells : 21

Hydrocarbon Shows : Gas shows in Miocene carbonates and sandstones in 3 wells, and in Cretaceous sandstones in 1 well

Basin Type : Western continental margin basin

Reserves & Production Status:

1. **Hydrocarbon Resource Potential** : Onland: 210 MMT
   Offshore: 550 MMT
   Total: 760 MMT

2. **Hydrocarbon Reserves:**

S.No	Structure	Geological Reserves (MMT)	Recoverable Reserves (MMT)
I	KD Structure	Oil: 2.39 (Z cat.) Gas: 50.2 (Z cat.)	-
II	GK-22C	Free Gas: 241	168.2
III	GK-29	Free Gas: 475.9	285.5

3. **Production: Nil**

Blocks on Offer (NELP I to VIII)

The Blocks offered and awarded under production sharing contract from NELP I to VIII are shown in the Fig. 5.1.a.11 and Table 5.1.a.2.

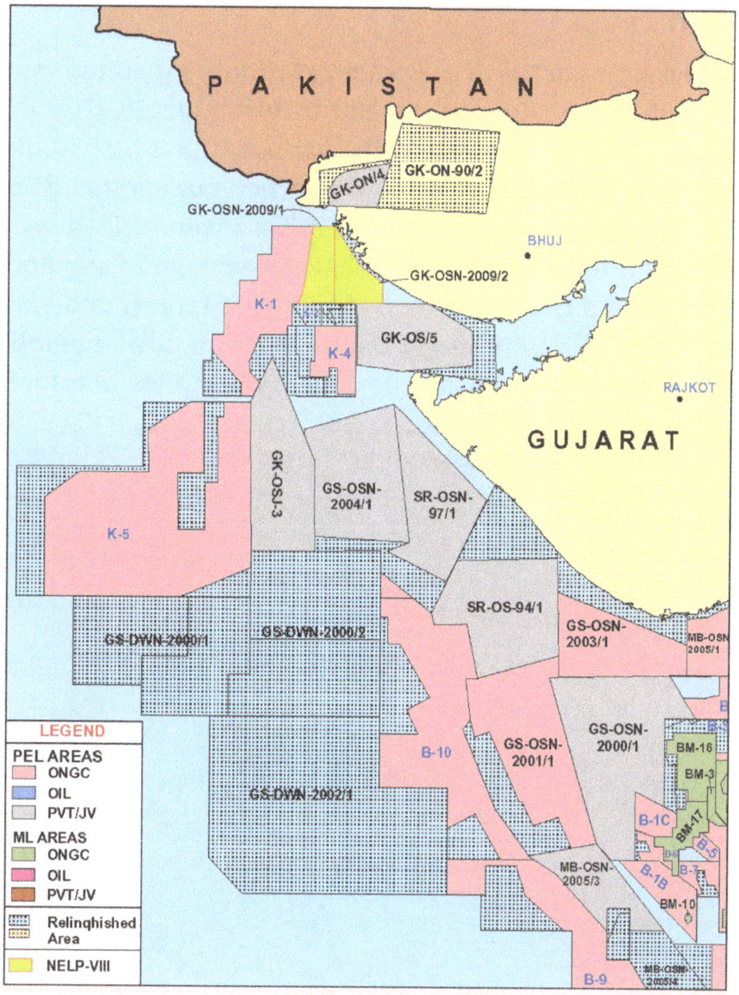

(Courtesy DGH)

Fig. 5.1.a.11

Table 5.1.a.2

NELP	Blocks offered	Blocks awarded	Awardees	Date of signing/ Effective date	Present area (sq.km)	Status
I		GK-OSN-97/1	RIL (100%-OP)	12-Apr-2000/	Relinquished	
II						
III						
IV						
V						
VI						
Vii						
Viii	GK-OSN-2009/1	GK-OSN-2009/1	ONGC (40%-OP), GSPC (20%), AWEL (20%), IOC (20%)	30-Jun-2010	1264	
	GK-OSN-2009/2	GK-OSN-2009/2	ONGC (40%- OP), Adani Welspun Exploration Ltd (30%), IOCL (30%)	30-Jun-2010	1242	

5.3 Saurashtra Basin

Saurashtra offshore block is lying to the North of Bombay platform, separated from Surat depression by Diu arch. It covers an area of 30,000 Sq.Kms up to 200 m water depth (Fig. 5.1.b.1). The National Oil Companies, ONGC and OIL are activity engaged in the various exploration activities in the basin. ONGC carried out seismic surveys of about 29,024 LKM and OIL carried out another 8730 LKM of seismic surveys. ONGC drilled 11 wells, OIL drilled 3 wells and another 3 wells were drilled by Chevoron Co. So far, no hydrocarbons are encountered in this basin. Mesozoic, Paleocene and Early Eocene section appear to be the targets for exploration. But to prove paleozoic and mesozoic targets underlying the thick volcanic trap section in the basin, super deep wells with target depth >5000 m. are required. To reach the Mesozoic prospect, huge thick section of volcanic trap has to be drilled, where over pressures are also expected.

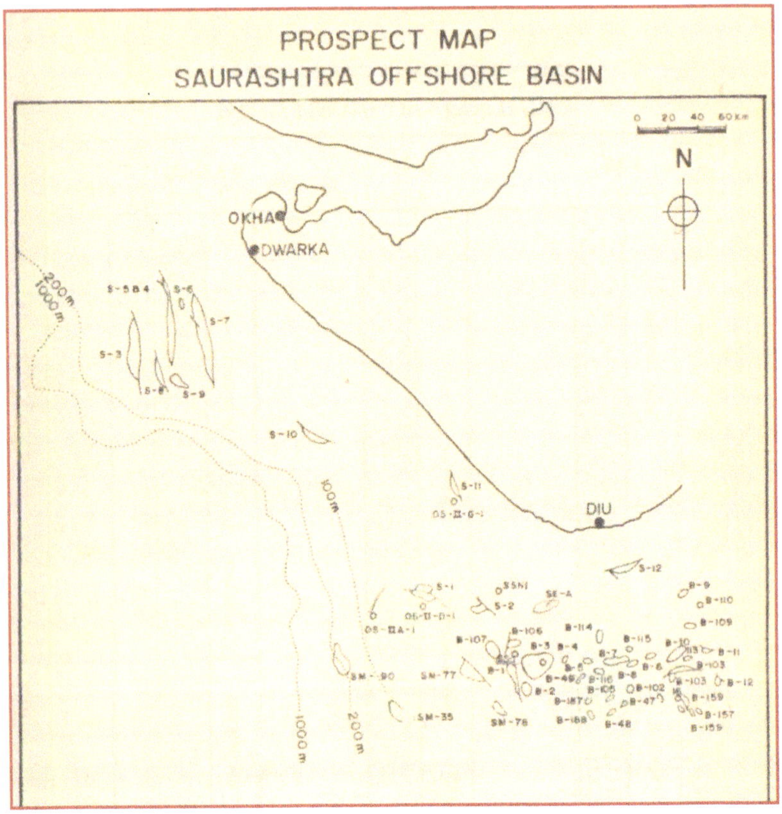

(Courtecy ONGC)

Fig. 5.1.b.1

The southern portion of saurashtra basin was awarded to M/S Chevron oil comp, India during 1973-81 and during 1973-79 Chevron carried out 2D seismic survey on a 10/15 km grid. A well SS-1 was drilled to 4500 m (Fig. 5.1.b.2). It encountered Paleocene carbonates in the lower most drilled section. Nearly 2300 m. of middle Miocene consists of thick carbonates and minor shale. In the lower Eocene 90 mts. of fine-grained sand stone and silt stone were penetrated. The SS-1 structure is aligned along the satpura trend (Fig. 5.1.b.1). Wedge out prospects on the southern flank of saurashtra are considered the potential targets for exploration.

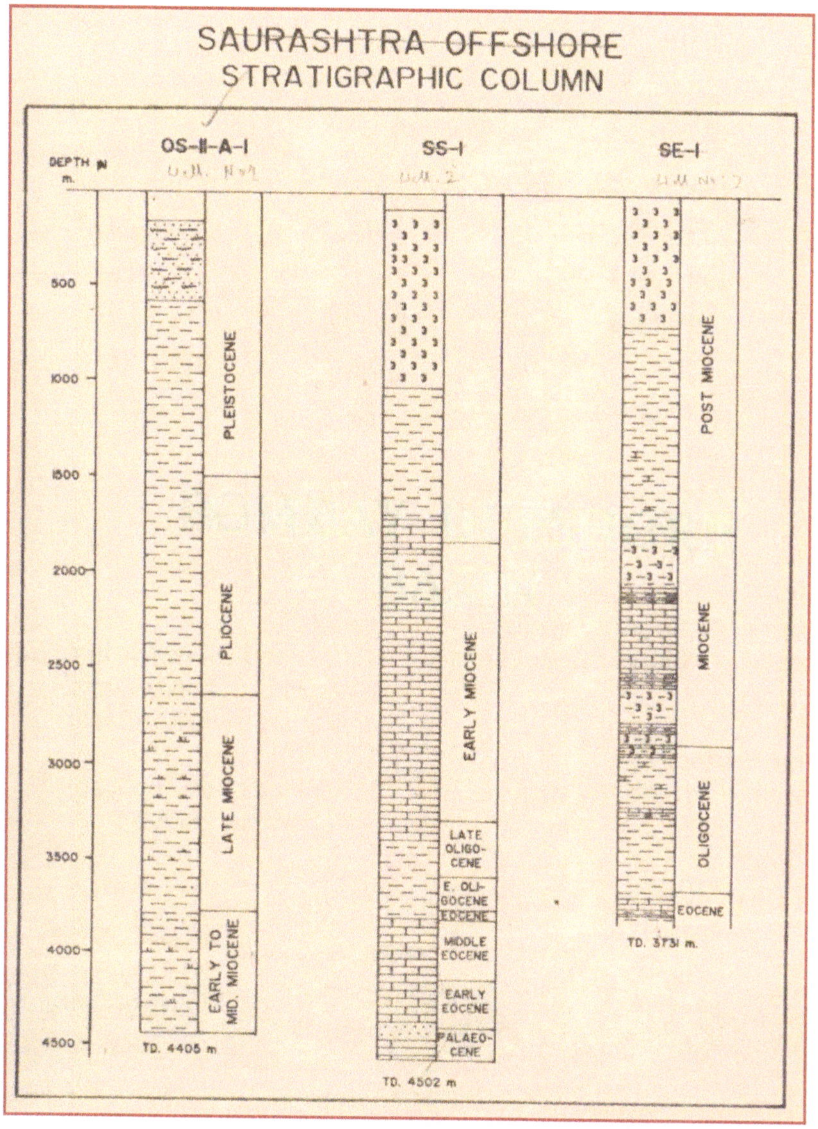

(Courtecy ONGC)

Fig. 5.1.b.2

Exploration and Production Status

Basin Area: 240,000 Sq. Km.

Offshore Seismic Data :

2D-45733 LKM

3D-7,517 Sq. Km.

Vintage	Type	Data acquired by	Quantum
2002-2003	2D	DG H	615 LKM
1981-2003	2D	ONGC	28,199 LKM
2001-2003	3D	ONGC	5904 Sq.Km
1997-2003	2D	Pvt./JV	8152 LKM
1997-2003	3D	Pvt./JV	1281 Sq.Km
1991	2D	OIL	8767 LKM
2002	3D	OIL	332 Sq.Km

Other Surveys
- Gravity Survey : 34,327 LKM
- Magnetic Surveys : 32,441 LKM
- Geochemical Sniffer Surveys : 4900 LKM
- satellite Gravity Survey

Well Data
- Number of Exploratory Wells : 22
- Hydrocarbon Shows / Discoveries : Gas and Oil shows in Middle Eocene to Miocene in 6 Wells

Basin Type : Pericratonic Rift Basin

Reserves and Production Status:

1. Hydrocarbon Resource Potential : Onland: Not Applicable; Offshore: 280MMT
2. Hydrocarbon Reserves : Nil
3. Production : Nil

Blocks on offer (NELP I to VIII)

NELP	Blocks Offered	Blocks Awarded	Awardees	Date of Signing/ Effective Date	Present area (sq.km)	Status
I						
II		GS-DWN-2000/1	ONGC (100%-OP)	17-Jul-2001/ 16-Aug-2001	Relinquished	
		GS-DWN-2000/2	ONGC(85%-OP), GAIL (15%)	17-Jul-2001/ 16-Aug-2001	Relinquished	
		GS-OSN-2000/1	RIL (90%-OP), HEPI (10%)	17-Jul-2001/ 16-Aug-2001	5890	
III	GS-OSN-2001/1	GS-OSN-2001/1	ONGC (100%-OP)	04-Feb-2003/ 12-Mar-2003	7101	
IV	GS-DWN-2002/1	GS-DWN-2002/1	ONGC (100%-OP)	06-Feb-2004/ 17-Mar-2004	Relinquished	

Contd...

NELP	Blocks Offered	Blocks Awarded	Awardees	Date of Signing/ Effective Date	Present area (sq.km)	Status
V	GS-OSN-2003/1	GS-OSN-2003/1	ONGC (51%-OP), CE7L (49%)	23-Sep-2005/ 15-Dec-2005	4477.5	
VI	GS-OSN-2004/1	GS-OSN-2004/1	ONGC (100%-OP)	02-Mar-2007/ 25-Apr-2007	6589	
VII	GS-DWN-2005/1					
	GS-DWN-2005/2					
	GS-DWN-2005/3					
	GS-DWN-2005/4					
VIII						

(Courtecy DDH)

The Blocks offered under production sharing contract from NELP I to VIII are shown in the Fig.5.1.b.3.

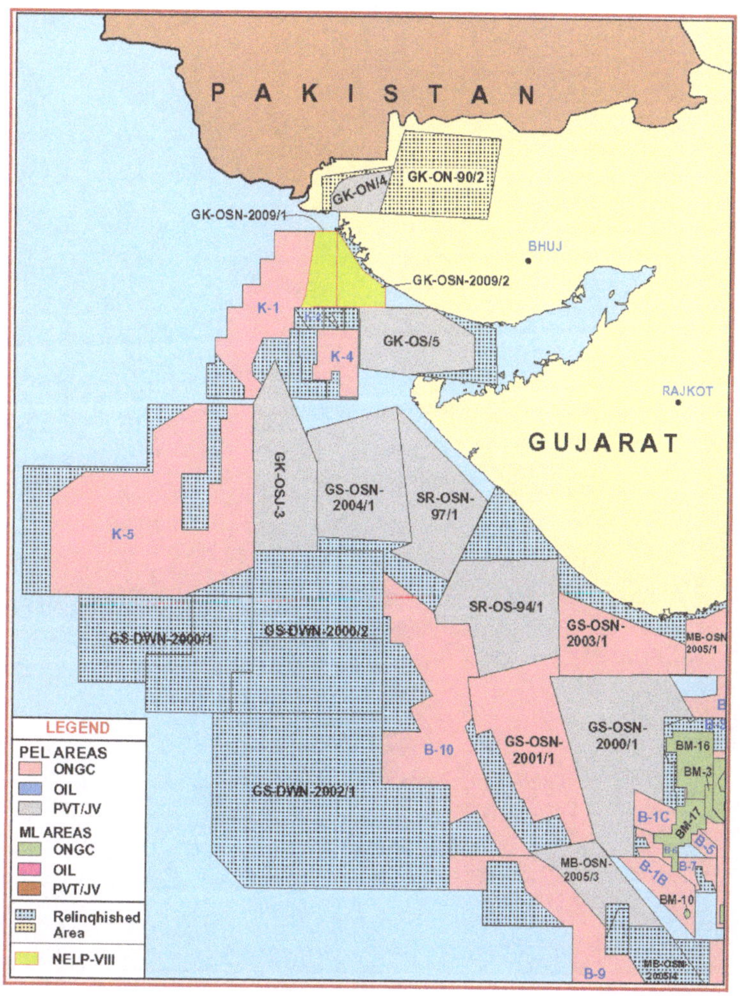

(Courtecy DGH)

Fig. 5.1.b.3

5.4 Bombay Offshore

Amongst the western offshore sedimentary basins of India, the Bombay offshore basin assumed the highest priority and importance due to the discovery of major oil and gas fields and also a number of marginal fields.

Bombay offshore basin covers an area of 63,000 Sq.Kms and the basin is divisible into two different regimes of sedimentation separated by an East-West trending regional line of dislocation namely KIM fault. To its north, lies the clastic basin called Surat depression while the carbonate platform lies to the south of the fault.

Bombay Offshore basin is further divided into six sub-blocks (Fig. 5.2.1) as below:

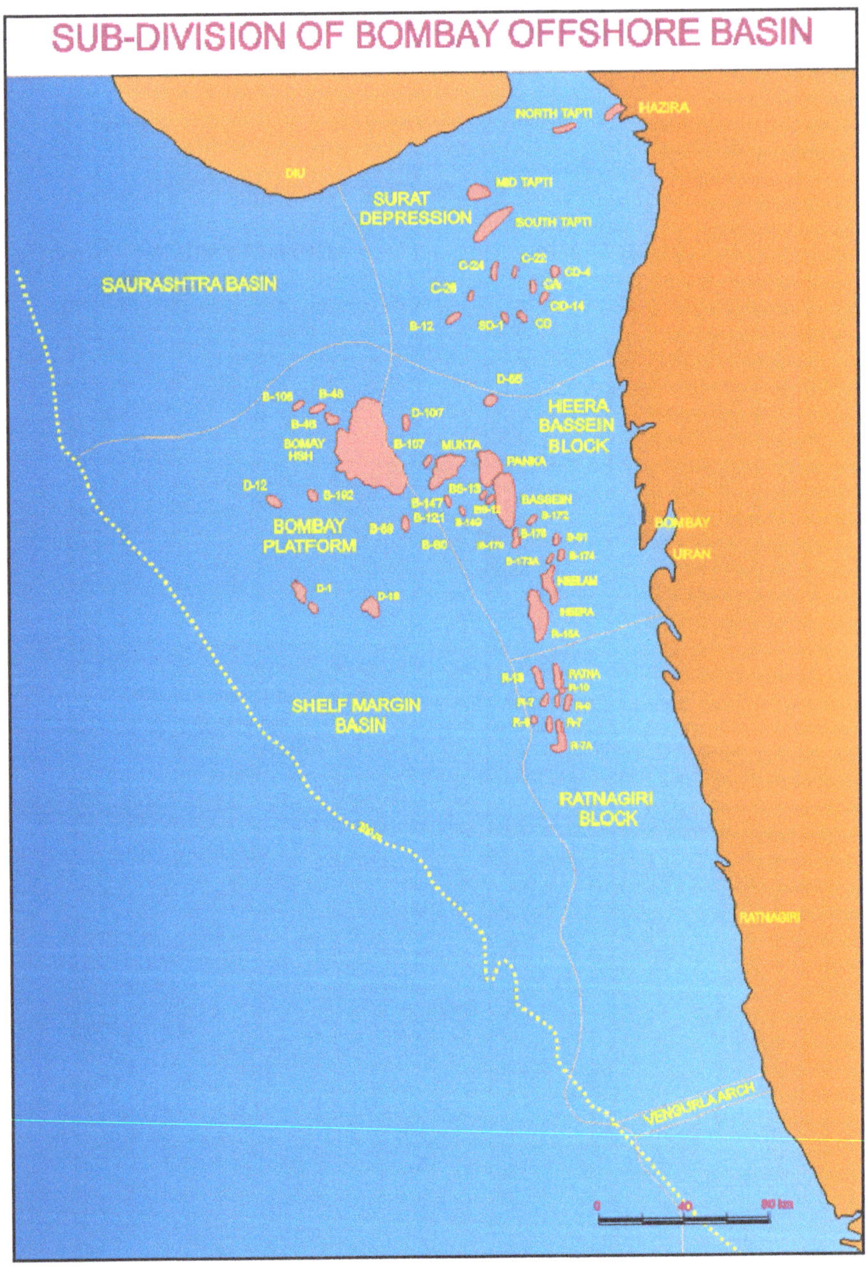

(Courtesy ONGC)

Fig. 5.2.1

1. Tapti-Daman-Dahanu area (Surat Depression to the north of Diu fault)
2. Bombay High platform including deep continental shelf area
3. Heera-Bassein-Panna-Mukta block
4. Eastern Homocline
5. Ratnagiri block
6. Shelf Margin block

Tectonic History

The tectonic element map of Bombay Offshore is placed at Fig. 5.2.2.

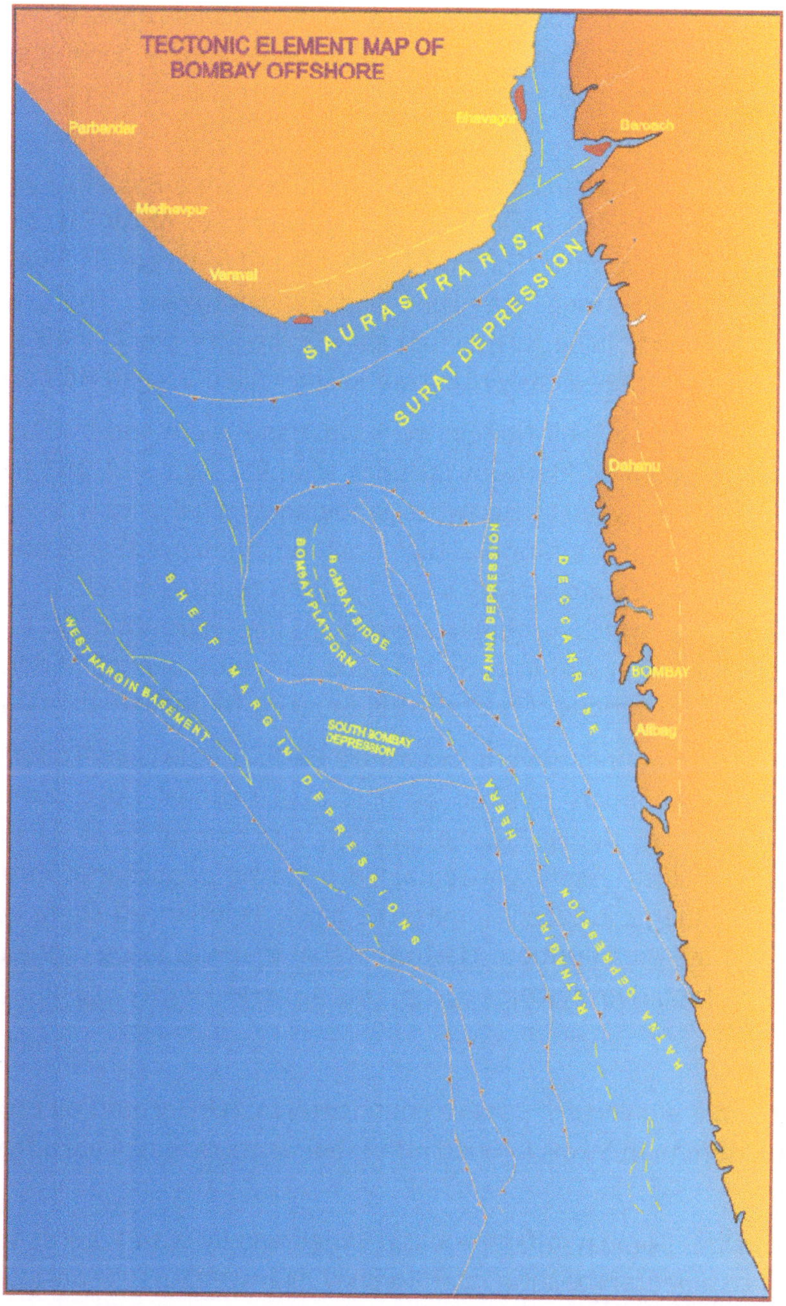

(Courtesy ONGC)

Fig. 5.2.2

(i) **Surat Depression:** Surat Depression is in Cambay Gulf and it forms the southward extension of the Cambay Basin and to the west it is separated from Saurashtra Basin by Diu Arch and Bombay High by KIM fault. An arm of this Depression extends far south into Panna-Bassein-Heera block (Central Graben) and further south into Ratnagiri block (Vijayadurg Graben). A 'high' feature interrupts the north-south continuity of these grabens. A few small-scale grabens radiating from these diastamise and circumscribe horsts in Ratnagiri block. The Surat Depression has numerous structural features of different origins like basement-controlled anticlines, differential compaction over sand bodies encompassed by shale, inversions and growth fault related roll over features.

Panna-Bassein-Heera Block: This block located east of Mumbai High/Platform and south of Surat Depression has three distinct N - S to NW-SE trending tectonic units which lose their identity in Miocene. The western block is a composite high block dissected by a number of small grabens. The Central graben is a syn sedimentary sink during Paleogene and Early Neogene. The eastern block is a gentle eastward rising homocline.

Ratnagiri Block: It is the southward continuation of the Panna-Bassein-Heera block. This block is differentiated into four distinct tectonic units by three sets of NNW-SSE trending enechelon fault systems. The western block is termed 'Shrivardhan Horst' and to its east is 'Vijayadurg Graben' which is also a syn sedimentary sink during Paleogene and Early Neogene. There is a general southward shallowing of this graben. Adjoining this is 'Ratnagiri Composite Block' with a number of 'highs' and lows' and further east, lies in the easterly rising homocline called 'Jaygad Homocline'.

(ii) **Bombay High:** It is bounded by Shelf Margin to its west and south and by Saurashtra Basin and Surat Depression to its north. Mumbai Platform includes Mumbai high and DCS (Deep Continental Shelf) area. The intervening area between these two is gentle homoclinal raise with a few structural 'highs' of different origins. Major part of the Mumbai High area remained positive almost up to Late Oligocene, missing much of the sedimentation activity. In comparison to other blocks in the basin, Mumbai block remained relatively stable which probably helped in the deposition of uniform carbonate-shale alternations over Mumbai High during early Miocene and early part of Middle Miocene, which later accommodated huge accumulations of hydrocarbons making Mumbai High, a Giant Oil Field.

(iii) **Shelf Margin:** Its northern boundary with Saurashtra Basin is indistinct. To its west lies the Deep Sea Basin with the western boundary marked by part of a regional ridge 'Kori High'. Except for the deposition of thin carbonates during Eocene (possibly due to paucity of clastic supply into the basin during this period) the block essentially remained a clastic depocenter throughout Oligocene and Neogene times. During the post Eocene times, the block experienced continuous sinking with varied intensity, to accommodate the enormous clastic material that was being brought into Surat Depression by proto Narmada and Tapti river systems and getting dispersed westward into this block.

Generalized Statrigraphy

The Bombay Offshore basin is a divergent passive continental margin located on broad continental shelf on the west coast of India. More than 5 kms (> 5000 m) of sediments ranging in age from Paloecene to Recent are present (Fig. 5.2.3).

The sedimentary fill of the Bombay offshore basin varied widely with regards to its age, thickness and facies. The thickness of sedimentary section over Bombay high structure is considerably less around 1800-2000m as compared to its thickness in depocenters, probably due to the absence of Eocene and Oligocene section over Bombay high. The sedimentary sequence in Bombay offshore basin can be classified as under:

1. The lower unit comprising of Basal sand and clay of Paleocene age
2. The Middle unit comprises of thick limestone in lower part and limestone shale in the upper part (Late Oligocene to Middle Miocene age).
3. In Bassein structure and on Panna-Bassein platform, Eocene and Lower Oligocene sequence is very well developed and occurs as thick limestone (Bassein limestone).
4. The overlying Late Oligocene to Middle Miocene is a sequence of limestone and shale alternations.
5. The post Middle Miocene sequence comprises of shale, claystone, and silt alternations.

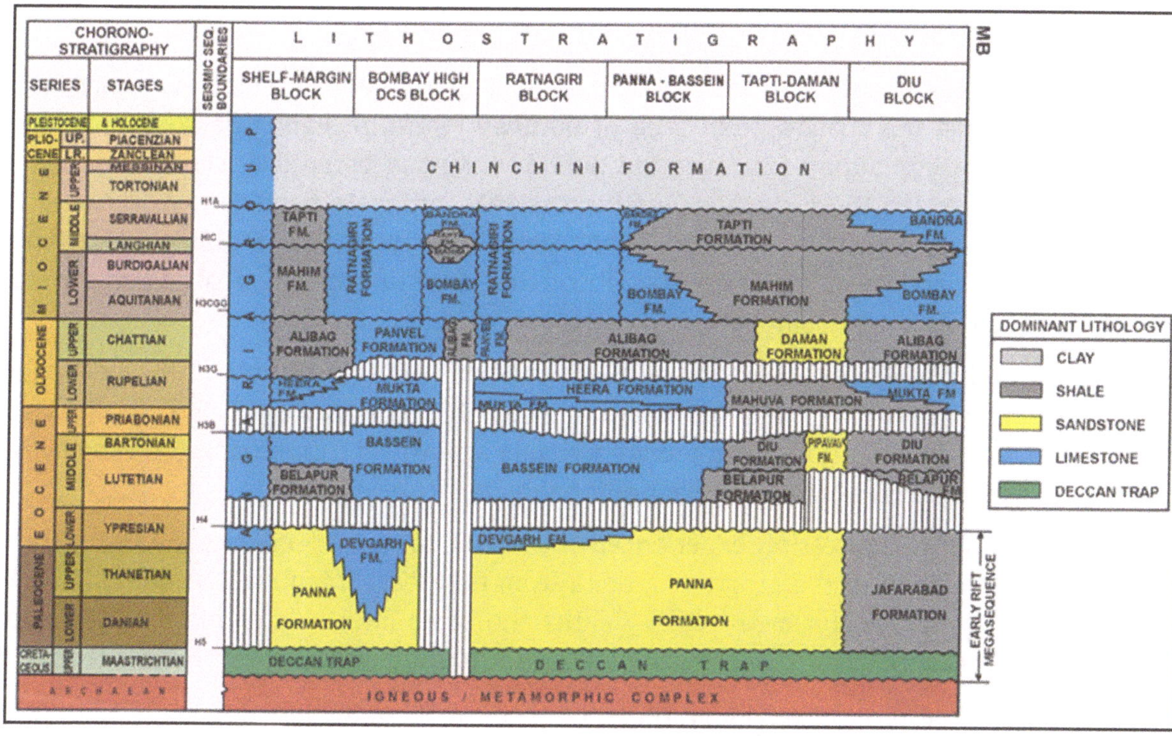

(Courtesy ONGC)

Fig. 5.2.3

To the North of Bombay High in the Surat depression, the Tertiary sequence is essentially an alternation of a shale/clay stone/ sandstone sequence and rests on volcanic rock which is considered as the technical basement. The Mesozoic section is expected below the trap which is not explored in the Surat depression.

Sedimentation History and Depositional Environment

This phase signifies the early syn-rift stage and is represented by trap-derived clastics contributed by the then existing paleo-highs essentially in continental to fluvial environment in its lower part (Panna Formation). It is overlain by grey to dark grey shales with subordinate sands possibly representing the first marine transgression into the basin. Presence of carbonaceous shale and coal at few places suggest localized restricted conditions.

Main clastic depocenters like **Surat Depression** and the contiguous southward lows like Central Graben **(Panna Bassien block)** and Vijayadurg Graben **(Ratnagiri Block)** received these sediments in considerable thickness, aided by syn-sedimentary activity of the bounding faults. A few localized depressions on Bombay Platform also received these sediments. Panna Formation is wide spread in the basin except over the crestal parts of prominent paleo-highs like Bombay High, Heera etc. Its thickness varies from almost nil to hundreds of meters in deep sinks.

Shelf Margin block, though under deep marine realm, seems to have received lesser quantities of sediments which were either derived from the Diu Arch (?) or from localized provenances. The facies developed in this block are mainly claystone and argillaceous carbonates with some amount of pelagic fauna.

Carbonate facies (Devgarh Formation) development is observed towards the southern edge of **Bombay High** in the form of muddy foraminiferal- algal banks; Deep Continental Shelf area and isolated off-shelf carbonate build-ups at a few places in **Shelf Margin** and **Ratnagiri**.

The syn-rift stage of Late Paleocene-Early Eocene period got terminated with a basin wide regression and development of an unconformity.

After a period of peneplanation, the basin witnessed a major transgression. Extensive carbonate sedimentation occurred in the shallow shelf area of **Bombay Platform, Panna-Bassein-Heera block** and **Ratnagiri block** (Bassein Formation). However the period witnessed essentially clastic sedimentation in **Surat Depression** (Belapur and Diu formations) with occasional carbonate bands and a few sand stringers and argillaceous carbonates and shales in Central and Vijayadurg Grabens (**Panna-Bassein-Heera block** and **Ratnagiri block**). Shelf margin was generally starved of clastics with deposition of minor claystone and carbonates of mixed middle shelf to bathyal origin (Belapur Formation)

Bassein Formation also indicates a wide range of environments – restricted platform, shelf lagoon with isolated shoals in Bassein area to open carbonate shelf in **DCS** and **Ratnagiri** and finally deep water carbonates in **Shelf Margin** area. It also formed wedge outs against the rising flanks of Bombay High and Heera, which can be considered as potential exploration targets.

During this period, **Surat Depression** experienced the maximum subsidence-accumulating thick under compacted claystones relating to the prograding delta from northeast (Mahuva Formation). The **Bombay platform** experienced generally shallower water depths and shale interbeds within limestone becoming more frequent. In **Shelf Margin** area thinner carbonates are deposited under relatively deeper conditions.

End of Early Oligocene also witnessed initiation of the westerly tilt of the basin.

Close of Early Oligocene is marked by a minor period of non-deposition except in **Shelf Margin** area. A few brief spells of transgression followed by continuous eustatic rise in sea level up to Early Miocene marked this period. Crestal part of Bombay High that hitherto remained a positive area also got submerged during this period.

Surat Depression witnessed a reduced subsidence resulting in a regressive coastline (Fig. 5.2.4). It is a package consisting of sand bodies deposited in distributary channels- coastal bars - tidal deltas- other transitional environments encased in marginal, marine, normally pressured silty and carbonaceous shale, overlying over pressured prodelta clay stone of Early Oligocene.

(Daman formation) The reservoir facies within this Formation have assumed great importance as they have been found to host significant amounts of hydrocarbons.

There was faster subsidence in **Shelf Margin** to accommodate the increased sediment load supplied by the westward prograding delta system. The finer clastics reaching the **Shelf Margin** block were mainly deposited in the depression between Kori High and the carbonate platform. (Alibag Formation)

Southward closure of Early Oligocene is marked by a minor period of non-deposition except in **Shelf Margin** area. A few brief spells of transgression followed by continuous eustatic rise in sea level up to Early Miocene marked this period. Crestal part of Bombay High that hitherto remained a positive area also got submerged during this period.

Southward from Surat Depression, clays got dispersed over Panna-Bassein-Heera block, including the crestal areas and the northern part of Ratnagiri block and as well as Bombay Platform. In Bombay High-DCS area and southern part of Ratnagiri, the unit is termed as Panvel Formation, whereas in Panna-Bassein-Heera and northern part of Ratnagiri, it is named as Alibag Formation.

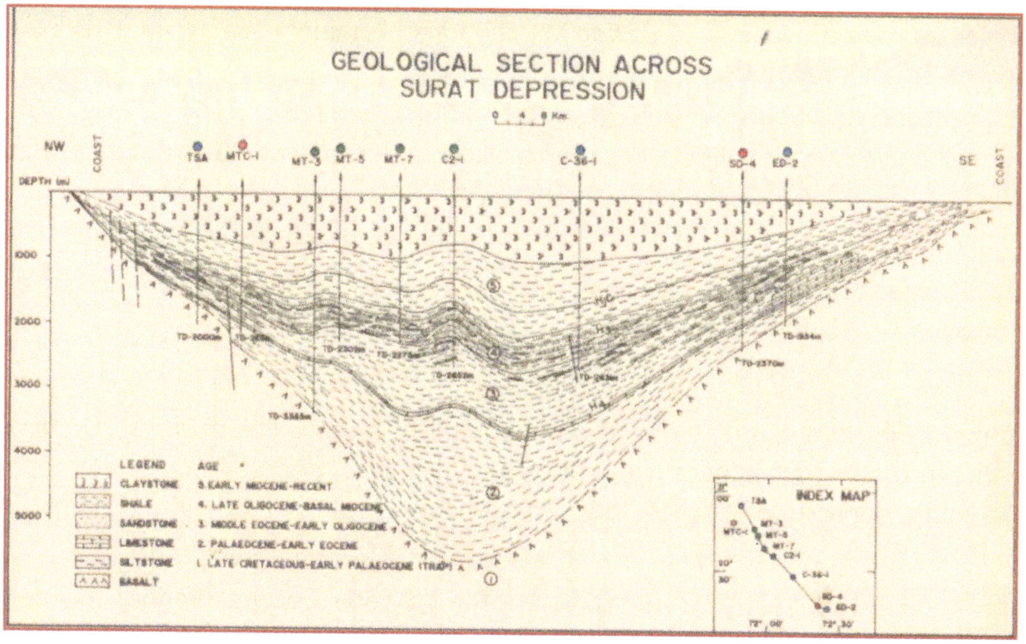

(Courtesy ONGC)

Fig. 5.2.4

It was a period of eustatic rise in sea level punctuated by a brief spell. The finer clastics entering into **Surat Depression** got mostly dispersed westward into Saurashtra basin and **Shelf Margin area**. Limited quantity of clastics got dispersed southward and entered **Bombay Platform** and its southeast and also up to Heera area. In response to the rising sea level, the delta being formed in **Surat Depression** in Late Oligocene shifted eastward.

Bassein and the area to its south that experienced shoaling conditions during Eocene was the site for fine clastic deposition during Early Miocene. **Bombay High** and its western part **(DCS)** underwent fairly thick carbonate sedimentation. In fact the major reservoir of **Bombay High** that hosts major part of the Country's hydrocarbon reserves belongs to this unit. While over the **Bombay High** area the facies are low energy, very fine grained to clayey carbonate reservoirs, the **DCS** area represents high-energy bio-clastic grainstone facies along with minor mudstone and wackestone.

The sea level continued to rise during this period. Clastic supply also continued into the basin. However much of the clastic material got dispersed westward into Saurashtra and **Shelf Margin** areas. Considerable quantity of clastics got dispersed southward also covering the entire **Panna-Bassein** area and also the **Bombay High** and its immediate surroundings to the west and south. This clastic unit over **Bombay High** includes sheet like sand, which has also been found to be hydrocarbon bearing. Carbonate sedimentation continued in **Ratnagiri** and **DCS** areas. Toward the later part of Middle Miocene, clastic deposition almost came to a halt in **Bombay High** and other areas and consequently carbonates got deposited over many areas. Uppermost part of the Middle Miocene Limestone in Heera field has been found to be hydrocarbon bearing. Close of Middle Miocene was marked by a very pronounced unconformity.

Post Middle Miocene witnessed a major transgression covering the entire basin coupled with spectacular increase in clastic supply. The earlier initiated westerly tilt of the basin also became more pronounced. All these events brought the carbonate sedimentation to a total halt. The increased clastic supply also resulted in a significant progradation of Miocene shelf at places up to 80 km (Chinchni Formation).

Petroleum system and Habitat of Oil and Gas

In Bombay offshore basin, oil and gas accumulated in three different types of reservoirs.

(a) Sandstone reservoirs (Oligocene to Lower Miocene age) – mainly gas accumulations in North Tapti, Mid Tapti and South Tapti fields.

(b) Miocene limestone reservoirs – oil with gas cap as a Bombay High.

(c) Bassein limestone reservoir (Middle Eocene to Oligocene) – oil and gas as in giant Bassein gas field, Panna, Neelam, Heera and Mukta fields.

(d) Basal clastics sandstone reservoir – oil and gas like in B-121, B-129, B-126 and Ratna fields.

The Bombay oils are paraffinic and very light with very high API gravity. Oil would have migrated from Surat depression and also Southern depression to the South of Bombay high.

Source Rock

Hydrocarbon generation potential is found to be abundant in mature Paleocene to Upper Oligocene fine grained sediments in the various generative depocenters like Surat depression, Mahim depression, Murud depression, Vijaydurg depression, Daman low, Navsari lows and shelf margin troughs to the west of Bombay High.

Geochemical studies indicate that syn rift Paleocene to Early Eocene 'Panna' formation (below H4 seismic marker) is the main source rock which has contributed to a large extent to Hydrocarbon generation and accumulations. The Panna formation is spread over the entire Bombay Offshore basin. In Panna formation, the TOC ranges between 0.5 and 20.5%, hydrogen index (HI) ranges between 50-300 mg HC/g. and the VRo is around 0.60. The maturity curves indicate that the hydrocarbon generation is around 20 Ma and migration around 10 Ma.

Reservoir Rocks

In the basin, both limestone and sandstone are reservoirs besides the fractured basement and Basal clastics. However, the limestones are the main reservoirs. In the Miocene sequence, L-III limestone reservoir is the main pay zone in Bombay High. Both primary (intergranular) and secondary porosities are present in the limestone. The other limestone reservoirs in Miocene are L-I & L-II.

The sandstone reservoirs are generally thin having local significance as S1 and basal sands in Bombay High.

Bassein limestone of Middle Eocene to Oligocene is an important reservoir in structures like Bassein, Heera, and Ratna etc. The limestones developed both primary (intergranular) and secondary porosity (Vugs).

In some wells, fractured basement rocks are also acting as reservoirs.

Cap-Rocks

The post Middle Miocene shales are the principal cap rocks throughout the basin. Upper Oligocene shales provide the caprock for Bassein limestone. Hard and unfractured tight carbonates (limestone / Dolomite) also act as cap rocks at some places.

Traps and Entrapments

Mostly, anticlinal reversals and fault closures are the main trapping mechanisms for oil and gas in the Bombay offshore area. The anticlinal structures are mostly developed by the drape of sediments over

Paleo-highs and tilted fault blocks. This type of entrapments is mostly structural type. The accumulations in the wedge outs and the accumulations in the lenticular sands and sand bodies are having structural cum stratigraphic plays.

Geochemistry, Hydrocarbon Generation

Geochemical studies in Bombay Offshore Basin throw light on the conditions of hydrocarbon generation, probable source rock, their development and nature.

Geochemical studies in Bombay High, Panna, Heera and South Bassein areas had indicated Paleogene shale sequence to be the potential source rock.

The Miocene prodelta muds and Palaeogene clastics deposited in the Dahanu depression may be considered as the main source sediments for the generation of hydrocarbons in the basin. Also, the northern and central depression of shelf margin where fine clastics of large thickness are expected, can be the locales for the generation of hydrocarbons.

Geochemical work carried out in sub-basins like Tapti-Daman, Ratna, DCS and Shelf Margin is summarized below:

Tapti-Daman Block

Geochemical studies on the sedimentary sequence ranging from Eocene to Post Miocene form nine wells (Mehrota et al 1988) have indicated Oligocene and Eocene sequences as containing potential source rock sequences. However, Eocene sequence is identified as the dominant source rock for oil and gas occurrence in this area (Fig. 5.2.5 to 5.2.9).

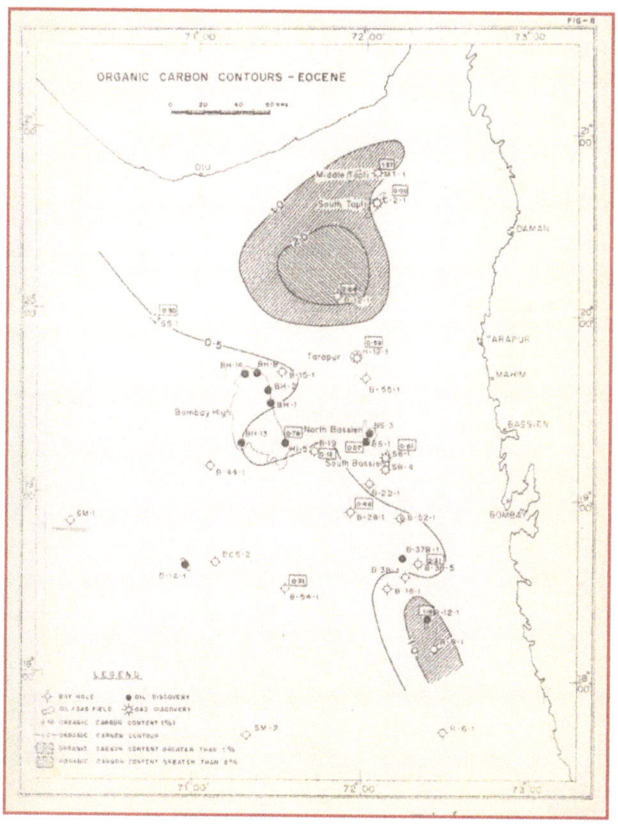

(Courtesy ONGC)

Fig. 5.2.5

(Courtesy ONGC)

Fig. 5.2.6

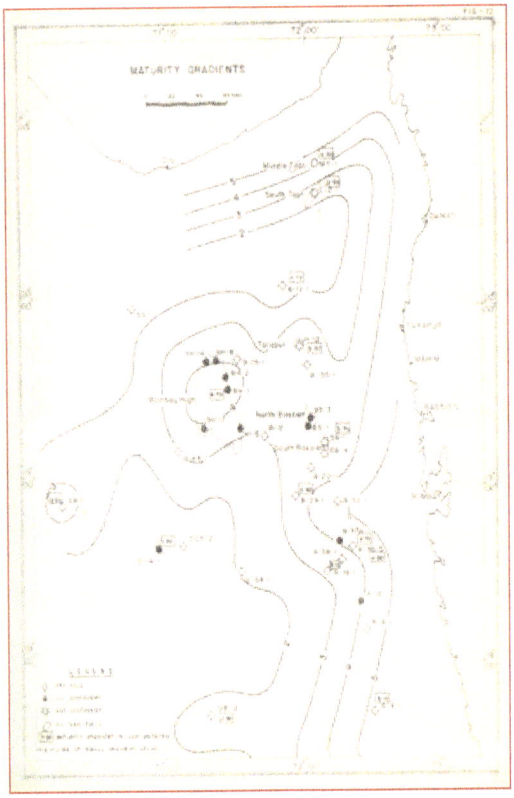

(Courtesy ONGC)

Fig. 5.2.7

(Courtesy ONGC)

Fig. 5.2.8

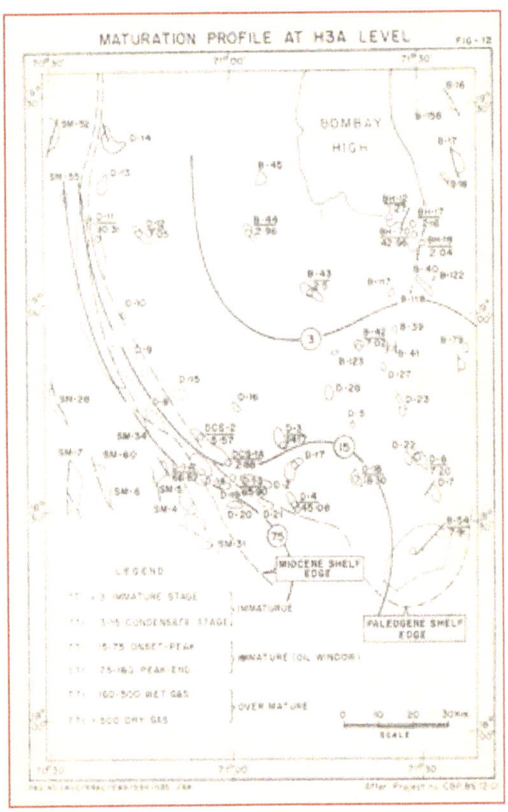

(Courtesy ONGC)

Fig. 5.2.9

DCS Area

Geochemical studies (Mangal Sain et al 1988) in DCS area suggest that organic matter is dominantly Humic (Type-III) derived from terrestial sources with evidence of mixed input in some sections particularly in Eocene sediments from wells D-11-1, D-12-1 and D-17-1.

Whole of the Palaeogene section is indicative of adequate maturation for hydrocarbon generation in DCS area. General organic matter distribution pattern in Oligocene/Eocene sediments indicate better hydrocarbon generation potential for South Western part in comparison to North Eastern part of DCS. Organic matter quality i.e. hydrogen richness for Oligocene sediments is better in southern part (around D-17-1 area) as compared to northern part. Eocene sediments in the northern DCS part (around well D-11-1) have indicated better quality organic matter as compared to southern part (Fig. 12).

In this area under study, burial history, and time temperature relationship as proposed by Lapotin (1971) and later modified by Waples (1980) indicate the maturation, the time of generation of hydrocarbon and possible paths for the migration of oil from the deeper parts of the basin, towards the Bombay High upwrap.

As per the Lapotin's TTI charts, the following maturation profiles/indices are obtained:-

3TTI	:	Immature stage	Oil window (Mature stage)
15TTI	:	Onset of oil generation	
75TTI	:	Peak oil generation	
160TTI	:	End of oil generation	
500TTI	:	40° API Preservations	Gas Zone
1000TTI	:	50° API Preservations	Over Mature stage
1500 TTI	:	Wet gas	

With the above indices as controlling factors, it is worked by chugh et al. that the Horizon H4A and H4 in the DCS chain of structure fall in the oil window zone, within the threshold of oil generation with TTI 15 and end of oil generation with TTI 160. The H3B & H3A levels will fall in the threshold on set of oil generation in the DCS structures (Fig.13) (i.e.TTI between 15 to 75). A zone with oil window can be extrapolated to the SW of D-1 structure. The onset to peak of oil generation is confined around the Palaeogene-Miocene shelf edge. The H3(CGG) & H2 levels fall within the immature stage of oil generation with possible stage of condensate formation with TTI values 3 to 15 particularly to the SW of D-4 structure.

Migration and Accumulation

According to Lapotin, the peak oil generation (TTI 75) corresponds to the initiation of oil migration. In other words, the migration might have started at about 6.38 million years ago (as observed in D-4-1) and continued as late as 1.72 million in D-18 (i.e. Oligocene to Middle Miocene).

The migratory trend of hydrocarbon from clastic basin appears to be mostly confined up to the Paleocene shelf edge and hence structures situated nearer to the Paleocene and Miocene Shelf edge are of interest. It is also brought out that the structures between the DCS and Bombay High upwarp might be in a different migratory trend, which may migrate from NE in Surat depression and also from the paleosink situated between DCS and Bombay High upwarp.

Shelf Margin

It has been established, that the limits of organic matter for the source rock potential is 0.5% in case of shales and 0.3% for carbonates.

Kuldeep Chandra & Misra et al (1984) indicated that the samples from the interval 1980-3415 m in SM-2-1 have 1.10 to 2.30% organic matter (Type III/II), which is a very good concentration for source rock potential. SM-1 well (780-2650m) yielded 0.6 to 1.27% of fairly rich organic matter of type III (Vitrinite Type). The B-54-1 has fair a concentration of organic matter in the Miocene section around 2850m ranging between 0.15 to 1.07% of Type III, vitrinite. Thomas had evaluated the geochemical data of shelf margin basin, calculating the TTI for thermal maturity and subsidence history and indicated that Basal Miocene and Paleogene sequences buried below 2900 m in Shelf Margin area are thermally matured to generate hydrocarbons. These studies indicate that a good thickness of organic rich sediments fall within the oil generating window, making the area attractive for the exploration of hydrocarbons. Two main depressions, one in the northern part and the other in the central part of the area, appear to be the locales with over 5000 to 7000m of organic rich pelagic sediments, capable of generating hydrocarbons. Structures situated near the peripheries of these depressions, deserve attention.

Ratnagiri Area

The total sedimentary section has been divided into 6 seismic sequences.

Geochemical studies were carried out for the bottom two sequences viz. sequence I and sequence II. While the sequence I comprises the section between a reflector close to the acoustic basement and the base of Bassein Limestone, the sequence II corresponds to the lower part of Bassein Limestone i.e. between top of Ratnagiri Limestone II (RP-IV) and base of Bassein Limestone.

Bottom part of sequence I comprises Lateritic clay with lignite and coal bands deposited in deltaic to marginal marine environment. The upper part consists of dark grey marine shales with a good source potential. These shales are considered as the main contributors for the hydrocarbon accumulations in the area. Sequence II is dominated by limestone interbedded with thin shales.

Source rock studies suggested better source potential for sequence I than sequence II (Fig. 5.2.10). However, the studies were restricted to the wells drilled in the north Ratnagiri Shelf. Details of the studies for sequence I are briefly discussed below.

(Courtesy ONGC)

Fig. 5.2.10

Two wells over Ratnagiri Composite Block (R-9-4 & R-14-1) and six well on Shrivardan Horst (R-7-1, R-8-1, R-13-1, R-43-1A, R-66-1 & R-71-1) have been chosen for the study. Total organic carbon (%) type, vitrinite reflectance, hydrocarbon generation potential (mg. H.C/gm of rock), quality are determined for the samples from these wells.

While good quantity of organic matter (1-2%) is observed in R-9-4, it is fair (0.5-1%) in R-14-1. The organic matter is predominantly of type-III with significant contribution from liquid rich material. The contribution is from both terrestrial and marine algal material. Relative contribution of each type is estimated and the organic matter is inferred to have a fair potential for generation of liquid hydrocarbons in addition to gaseous one. Based on the vitrinite reflectance studies, the oil window in R-14-1 is worked out to be 1940-4040m and 2660-5900m in R-9-4 well. By these studies, it is observed that the sequence I is immature in R-9-4 and just mature In R-14-1.

As regards the wells on Shrivardan Horst, the sequence I in these wells, have good to very good (1-5%) organic matter and the wells closer to the Vijayadurg graben on the eastern part of Shrivardan Horst have better predominantly type III kerogens with some contribution from type-II. They have a liquid hydrocarbon generating capacity. Predomanance of type-III kerogene in northern wells (R-13-1, R-66-1 & R-71-1) suggests that the original kerogen may have had greater contribution of type-II kerogen.

Vitrinite reflectance studies suggest that oil window starts from 2260m onwards and extends up to 5000m in different wells of Ratnagiri area.

Thus, these studies had brought out that, the potential source rich clastic sequence I in North Ratnagiri has type III kerogence in fair to good quantities and is expected to have matured in the deeper parts of Vijaydurg Graben and has a liquid hydrocarbon generating capacity. Thus, the prospects close to the Vijaydurg graben are good exploration targets.

In south Ratnagiri area, source rock studies were carried out on sediments from two wells R-2-1 & R-61-1. Presence of marginally organic rich sediments in Eocene age, coupled with the samples showing hydrogen index over 100, indicate good source rock sequence in R-2-1. Similarly, the samples of R-61-1 indicate a very good concentration of organic matter and good S2 peak in rock eval analysis. The sequence I has a better quality and type of organic matter.

Moreover, it is seen from the time interval maps (H5 & H4) that the burial depths and type of sedimentation of Devagarh depression in the south and Vijaydurg depression in the north are comparable. Drilling results confirmed the generation of hydrocarbon in the northern Vijaydurg graben.

Based on the above studies, generation of hydrocarbons can be envisaged in the southern depression (Devagarh & Malwan), as well. Hence, the peripheral structures of these depressions need attention for the exploration.

Petroleum Plays

Major Identified play types

1. Paleogene Synrift clastics(Paleocene-Lr. Eocene, Panna Fm)
2. Eocene Carbonate Platform (Bassein formation)
3. Lr.Oligocene Carbonate plays (Mukta and Heera formations)
4. Oligocene-Lr. Miocene deltaic Play (Mahuva &Daman formations)
5. Up. Oligocene carbonates (Panvel and Ratna formations)
6. Lr. Miocene carbonates (L-III and L-IV reservoirs, Bombay / Ratnagiri formation)
7. Lr-Mid. Miocene clastics (S1 sands),
8. Mid. Miocene carbonates (L-I and L-II reservoirs, Bandra Formation)

1. Paleogene Synrift clastics (Paleocene-Lr. Eocene, Panna Fm)

- Area: Western and southeastern flank of Bombay High, western flank of central graben, Heera-Panna Block
- Reservoir rock: sandstone
- Depositional environment: Continental (parallic) to coastal
- Trap: structural /stratistructural(updip pinch outs)
- Source rocks: Paleocene-Eocene (Panna Formation)

2. Eocene Carbonate Platform (Bassein Formation)

- Area: Heera-Panna Composite Block, part of MH-DCS block (NW and SW flank of Mumbai High), Ratnagiri
- Reservoir rock: Limestone
- Depositional environment: shallow shelf /Shelf-lagoon carbonates
- Trap: structural /strati-structural (wedge outs)/diagenetic traps (?)
- Source rocks: Paleocene-Eocene (Panna Formation)

Commercial production from several medium sized and marginal Fields-

- Mukta-Panna
- Bassein
- Heera
- Neelam
- B-55
- B-173A
- South Heera
- NW of Mumbai High
- SW of Mumbai High
- West and SW of Bassein
- Bassein East

3. LR. Oligocene Carbonates (MUKTA AND HEERA FORMATIONS)

- Area: Heera -Panna composite block, MH-DCS platform
- Reservoir rock: Limestone
- Trap: Structural
- Source rocks: Paleocene-Eocene (Panna Formation)
- Fields: Heera,panna,neelam,Basseim,B-55,D-18 etc

4. Oligocene - LR. Miocene Distal Deltaic-Coastal Play (DAMAN AND MAHUVA FORMATIONS)

- Area: Tapti-daman Block
- Reservoir rock: Sandstone

- Depositional environment: Deltaic to coastal
- Trap: Structural /strati-structural
- Source rocks: Paleocene-Eocene (Panna Formation)
- Fields: South Tapti/Mid Tapti
- Areas identified for development/pre-development studies: C-series structures, North Tapti.

5. Up. Oligocene carbonates (Panvel and Ratnagiri formations)

- Area: MH-DCS Block
- Reservoir rock: limestone
- Trap: structural
- Source rocks: Paleocene-Eocene (Panna Formation)
- Fields: B-121/119
- Tested commercial potential from wells located in the MH-DCS Blocks Area identified for Dev/pre-development studies: B-46, B-48 (NW of Mumbai High), B-192, B-45, and WO-24 etc.

6. Lr. Miocene carbonates (L-III and L-IV reservoirs)

- Area: Bombay High
- Reservoir rock: Limestone
- Trap: Structural
- Depositional environment/facies: Deposited under cyclic sedimentation with each cycle represented by lagoonal, algal mound, foraminiferal mound and coastal marsh facies
- Source rocks: Paleocene-Eocene (Panna Formation)
- Fields: Mumbai High, D-1 Prospects identified for development WO-24, B-45 along with other pays

7. Lr-Mid. Miocene clastics (S1 sands)

- Area: Bombay High and adjoining area
- Reservoir rock: Sandstone/siltstone
- Trap: Strati-structural
- Source rocks: Paleocene-Eocene (Panna Formation)
- Fields: Bombay High, recent discovery on BNP prospect

8. Mid. Miocene carbonates (L-I and L-II, Bandra Formation)

- Area: Bombay High, DCS and adjoining area (L-I and L-II)
- Reservoir rock: Limestone
- Trap: Structural
- Source rocks: Paleocene-Eocene (Panna Formation)
- Fields: Bombay High,Heera,D-1

Exploration History & Hydrocarbon Finds

Till date geological, geophysical and geochemical surveys in Western Offshore lead to identification of about 600 prospects of which around 250 prospects have been drilled, and this resulted in establishing around 70 oil and gas discoveries/strikes. 13 of these are large and medium size fields which have been developed either by ONGC or through joint venture participation and are put on commercial production. The remaining are small fields individually having hydrocarbon reserves of less than 16 MMT and some are having insignificant reserves. In the marginal fields, about 132 MMT of oil and OEG reserves are locked up. These marginal prospects could not be developed due to limitations of economic viability through conventional methods, even though efforts are on hand to integrate a few of them, together for development. To bridge the widening gap between demand and production to the extent possible, it is important to develop these fields with innovative and cost effective methods.

The oil and gas fields / discoveries so far made is listed below (Fig. 5.2.11)

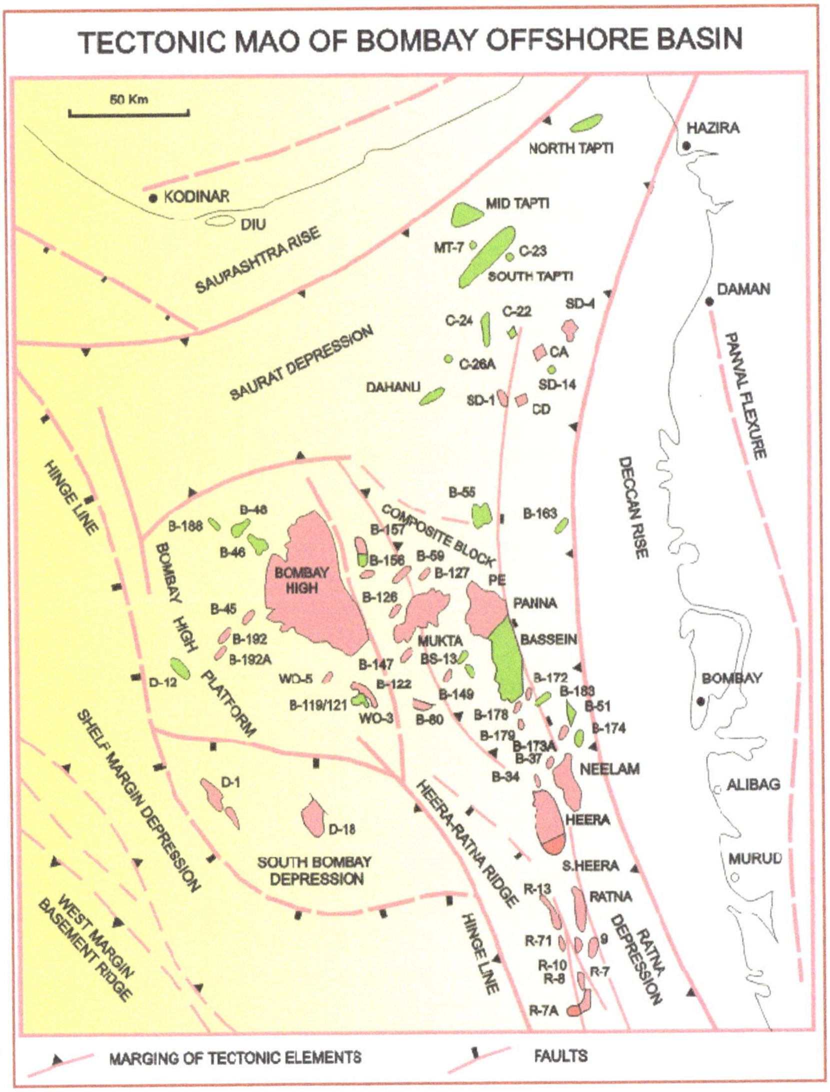

(Courtesy ONGC)

Fig. 5.2.11

A. Giant, Large and Medium Sized Fields

S.No.	Oil	Gas
1.	Bombay High (Giant Field)	Bassein
2.	Heera	Mid Tapti
3.	South Heera (R15A)	South Tapti
4.	Ratna	Hazeera
5.	Neelam	Gauri
6.	Panna	Laxmi
7.	Mukta (B-57, B-19)	
8.	B-126	
9.	D-1	

B. Marginal Fields

S.No.	Oil	Gas
1.	R-7	B-55
2.	D-173A	C-24
3.	R-9	B-157
4.	D-18	DS-13
5.	B-178	North Tapti
6.	B-119 / B-121	C-26
7.	B-192	B-149
8.	R-13	B-48
9.	B-HE	C-22
10.	R-10	B-46
11.	C-A	D-12
12.	B-134	B-188
13.	R-7A	GK-29
14.	B-179	B-12
15.	SD-4	B-174
16.	CD	BS-12
17.	D-80	B-51
18.	KD	C-23
19.	B-45	GK-22c
20.	B-172	SD-14
21.	B-37	B-163
22.	R-8	B-183
23.	B-180	
24.	WO-5	
25.	B-59	
26.	SD-1	
27.	R-71	
28.	B-127	
29.	B-192A	
30	B-147	

Brief on Few Important Fields

(a) Tapti Daman offshore block (Surat depression): Surat depression is congruous with onland Cambay basin but forms a part of Bombay offshore basin. It covers an area of about 27000 Sq.Kms. which is limited by Saurashtra coast in the North-west, Saurashtra basin to the west, and west coast of India in the East, and Heera Bassein block to the South. The basin extends into Gulf of Cambay towards north-East. Average water depth varies from 10-40 m. Seismic surveys in Surat depression were initiated in 1964. Since then continuous exploration efforts are being carried out in the basin.

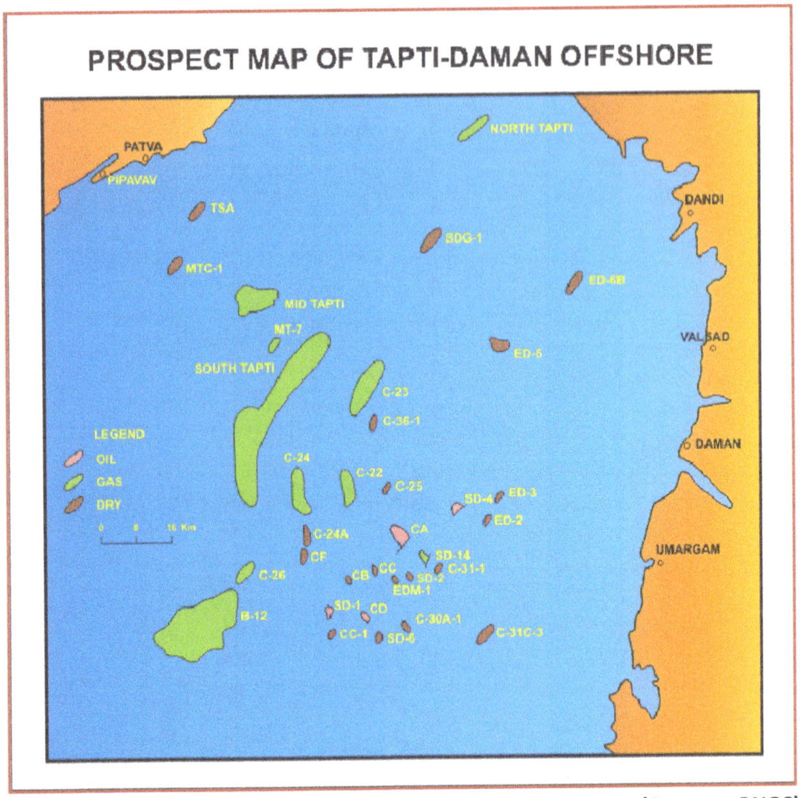

(Courtesy ONGC)

Fig. 5.2.a.1

The seismic data indicated the presence of over 6 Kms. of sedimentary fill in the basin. More than 50 prospects have been identified and mapped (Fig. 5.2.a.1). Exploratory drilling commenced in 1977 leading to the discovery of South Tapti gas field in 1978. Subsequently, Mid Tapti field was discovered. Till date around 36 prospects have been drilled resulting in 11 gas and four oil strikes.

The total prognosticated resources of the block are 1120 MMt of (O+OEG) of which around 20% has been upgraded into geological reserves. Both Mid Tapti and South Tapti fields have been delineated, and integrated technological and development schemes have been prepared. Presently, the fields are under development by Joint Venture partners -ONGC, Reliance, & BG.

(b) Mid Tapti and South Tapti blocks: These are situated in the North western part of Surat depression, 10 Kms from Daman town. Water depth varies from 13 to 14 m. South Tapti and Mid Tapti fields were discovered in 1978 and 1980 respectively. The fields have been well delineated, integrated technological schemes for development have been drawn and the fields are on production. Mid Tapti field is an anticlinal structure trending NE-SW direction with no major faulting. South Tapti field is a doubly plunging anticlinal structure trending NE-SW with steep South-East flank. It is dissected by number of NW-SE trending faults with vertical displacements across the faults, less than 10 m. Structural contour map on top of pay zone (Late Oligocene) to Basal Miocene is shown in Fig 5.2.b.1.

Reservoir

Late Oligocene-Basal Miocene is the main hydrocarbon bearing section in Mid. and South Tapti fields. Sandstones and occasionally siltstones form the reservoirs. The sandstones are loose, non-compact and usually pose problems of sand cutting during production testing.

15 sand reservoirs have been identified in the Mid Tapti and South Tapti fields, of which around 7 are productive in Mid Tapti and 11 in South Tapti. Generally, sands are 3-5 m thick.

Maximum gross sand thickness varies up to 25 m with an average porosity of 18-20%.

The field is under production for gas and it is producing @ 40 MM standard CFT/day.

Reserves

Reserves of Mid Tapti and South Tapti are as follows

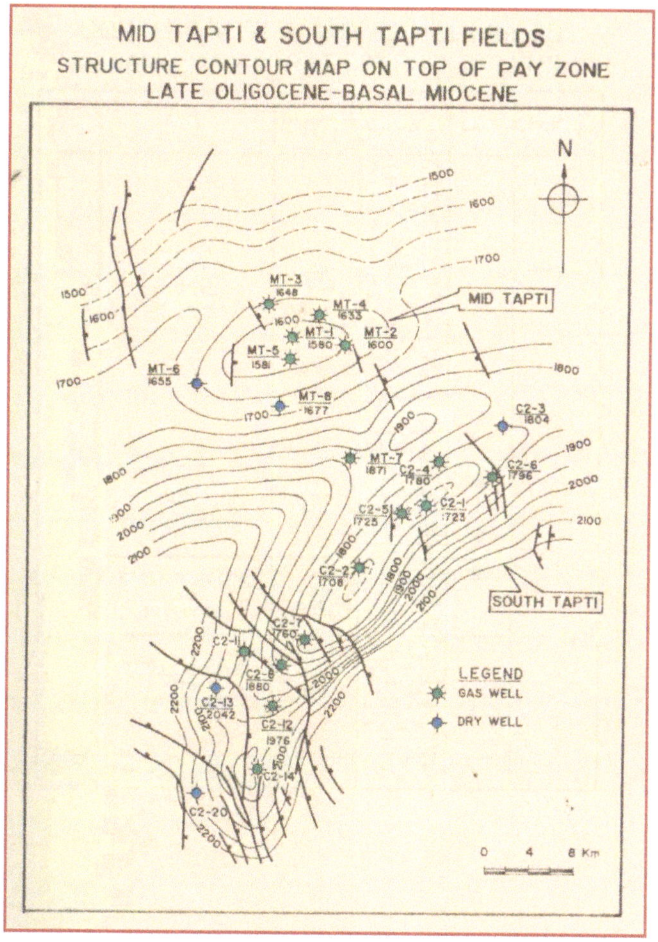

(Courtesy ONGC)

Fig. 5.2.b.1

	Free Gas Geological Reserves (MMm^3)	Free Gas Recoverable Reserves (MMm^3)
Mid Tapti	22407.8	10657.3
South Tapti	45219.8	20730.6

The general gas composition of the Tapti field is as below:

Methane	-	91.04%
Ethane	-	4.06%
Propane	-	1.53%
I-Butane	-	0.39%
N-Butane	-	0.27%
N-Pentane	-	0.09 – 0.07%
Hexane	-	0.07%
Heptane	-	0.02%
Carbondioxide	-	1.22%
Nitrogen	-	1.18%
Total	-	**100%**

(c) CA Prospect: This prospect is located 150 kms NNW of Bombay city in a water depth of 33 m. In the exploratory well CA drilled in Sept 1986, oil was discovered in a lenticular sand belonging to

Oligocene age, deposited in tidal deltaic environment. Isochron map of horizon H3A is shown in the Fig. 5.2.c.1.

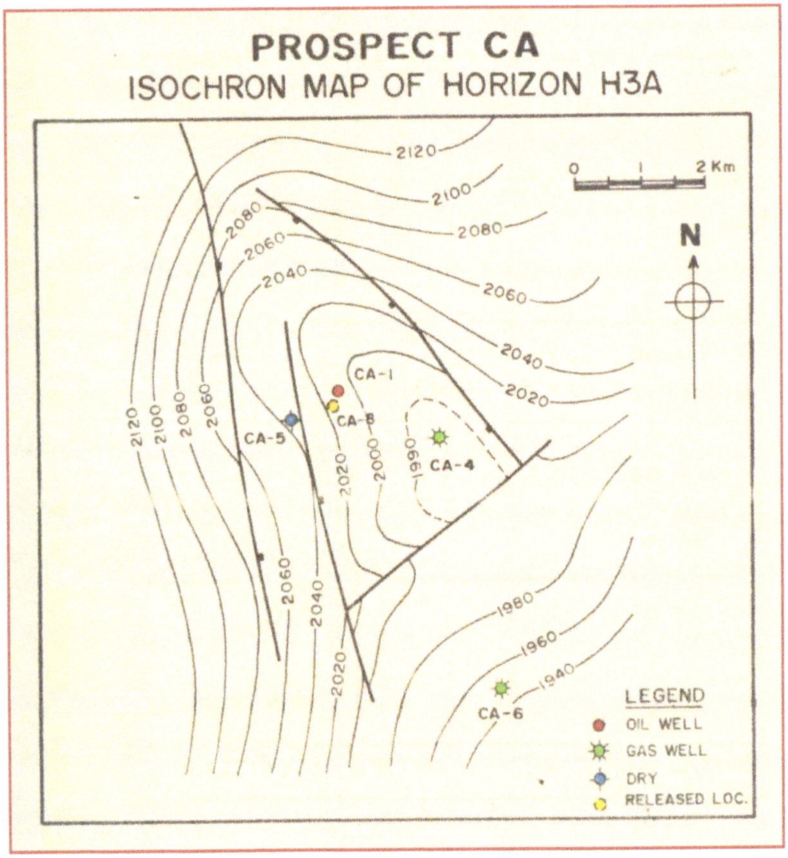

(Courtesy ONGC)

Fig. 5.2.c.1

Test Results

The prospect is planned for production on EPS.

Horizon	Bean	Oil	Gas	FTHP	Remarks
1990-1992m (Early Oligocene)	½ "	6010 bbls/day	Nil	2117	GOR 33 API 34^0 PP 36^0

Reserves

	Geological Reserves	Recoverable Reserves
Oil (MMT)	2.05	0.03
Condensate (MMT)	-	-
Solution Gas (MMm^3)	132.5	71.6
Free Gas (MMm^3)	197.7	118.6

(d) CD Prospects

Oil was struck in May 1986 in the exploratory well CD-1, in the lenticular sands of Early Oligocene age deposited in tidal environment. The prospect is 140 Kms NNW of Bombay city. Water depth is 33m. Isochron map at H3A is shown in the Fig. 5.2.d.1.

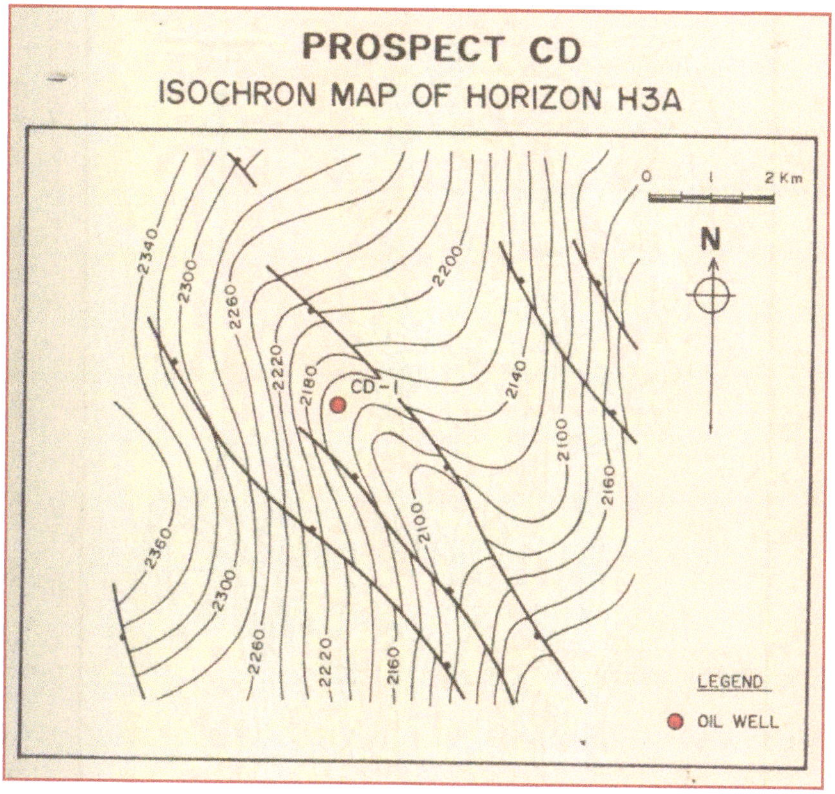

(Courtesy ONGC)

Fig. 5.2.d.1

Test Results

Horizon	Bean	Oil	Gas	FTHP	Remarks
2029.5-2035 m **2041.5-2047.5 m**	½ "	2164 bbls/day	Nil	2552 psi	GOR 319 API 45^0 PP 33^0

Reserves

	Geological Reserves	Recoverable Reserves
Oil (MMT)	0.05	0.01
Condensate (MMT)	-	-
Solution Gas (MMm^3)	25.0	15.0
Free Gas (MMm^3)	20.70	14.5

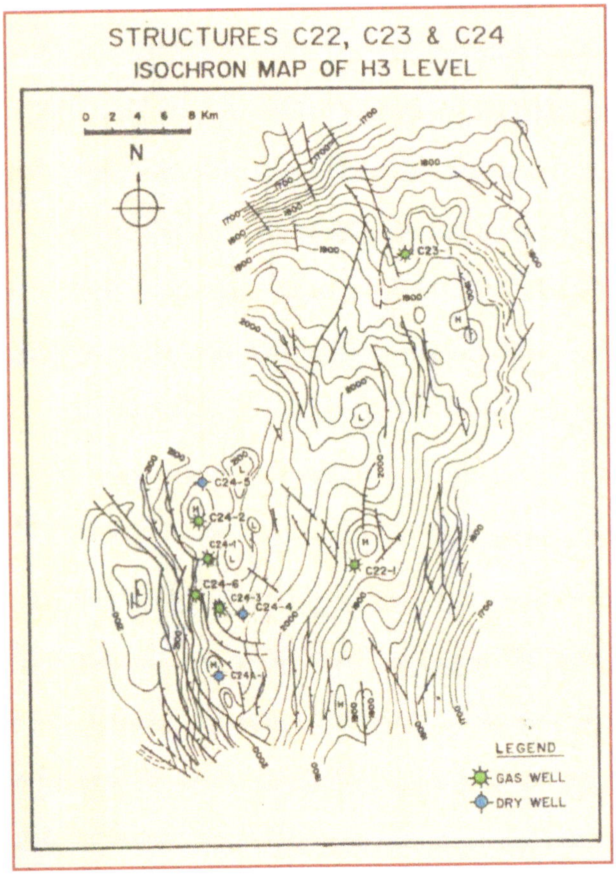

Fig. 5.2.e.1

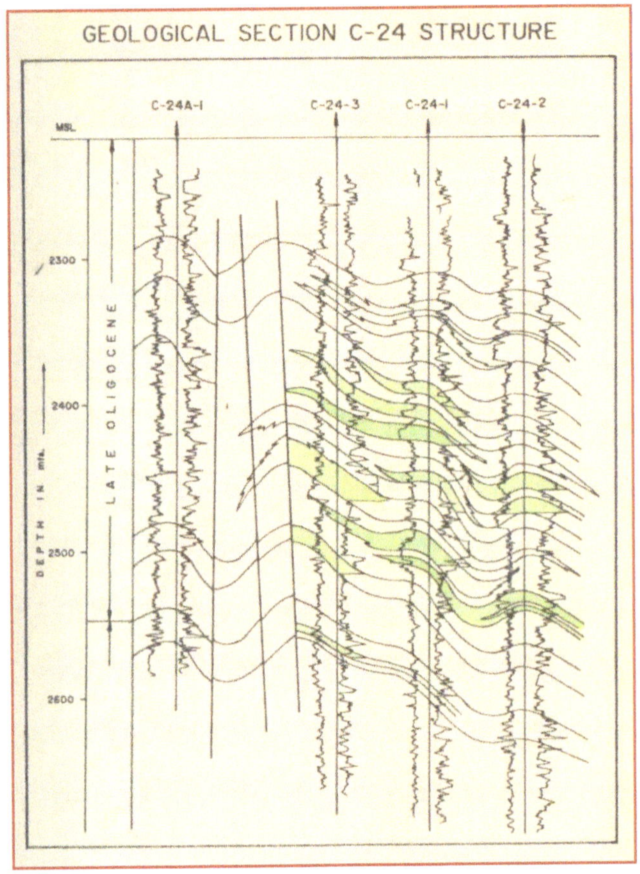

(Courtesy ONGC)

Fig. 5.2.e.2

(e) Integrated development of C-22, C-23 and C-24 prospects: C-22 was drilled in 1987. Gas has been struck in the sandstone reservoir of Oligocene age. The prospect is located 70 kms WSW of Daman town. C-23 prospect is located 7 Kms East of Mid Tapti in a water depth of 15 m. It is a NNE-SSW running anticlinal feature having an aerial extent of about 20 Sq.Kms. The well was drilled in 1992, which produced gas from Early Oligocene sand reservoirs. C-24 prospect, located about 81 Kms west of Daman, in a water depth of 24 m, was drilled in 1987. Gas has been found in Oligocene sandstone reservoirs. Isochron map of H3 level is at Fig. 5.2.e.1 and geological section with sand geometry at Fig. 5.2.e.2.

Test Results

Prospect	Horizon	Bean	Oil (bbls/d)	Gas (m^3/d)	FTHP (psi)	Remarks
C-22	2276-2284 m	½"	-	3,26,387	2552	Condensate 199 bbls/d API 54^0
C-23	2345-2347 m (E.Oligocene)	¼"	-	2,02,129		Condensate 96 bbls/d
	2469-2472 m (E.Oligocene)	$^3/_8$"	-	4,00,805		Condensate 252 bbls/d
C-24	2720-2722 m	½"	360	74,160	630	GOR 1295
	2518-2530 m	$^3/_8$"	1,060	2,15,466	2750	Condensate 192 bbls/d
	2472-2476 m	$^5/_8$"	-	2,25,556	2750	Condensate 130 bbls/d
	2442-2450 m	$^3/_8$"	-	2,18,150	2740	Condensate 197 bbls/d
	2430-2435 m	$^3/_8$"		2,04,148	2650	Condensate 162 bbls/d

Reserves

	Geological Reserves			Recoverable Reserves		
	C-22	C-23	C-24	C-22	C-23	C-24
Oil (MMT)	-	-	0.10	-	-	-
Condensate (MMT)	0.13	-	-	0.08	-	-
Solution Gas (MMm^3)	-	248.10	6180	-	51.10	3708
Free Gas (MMm^3)	987.30	-	-	592.40	-	-

Gas Analysis

Constituent		Compostiion (% Vol)
Nitrogen	-	3.16
Carbondioxide	-	1.73
Methane	-	87.24
Ethane	-	4.17
Propane	-	2.31
I-Butane	-	0.59
N-Butane	-	0.45
I-Pentane	-	0.13
N-Pentane	-	0.08
Hexane	-	0.09
Heptane	-	0.05
Calorific Value ($Kcal/m^3$)	-	8577
Specific Gravity @60^0F & 14.73 gsi	-	0.6501

(f) B-12 Structure (Dahanu) – C-26 Prospect: Dahanu structure is located about 160 kms North-west of Bombay city and about 50 Kms SSW of South Tapti field in a water depth of 33 m. Well B-12-1 was drilled in the year 1978 and gas was discovered in Oligocene sandstone reservoir. The Ariel extent of the structure is about 100 Sq.kms. Isochron map of horizon H3 is in Fig. 5.2.f.1.

Test Results

Horizon	Bean	Oil	Gas	FTHP	Remarks
2225-2229 m 2243-2249 m 2253-2254.5 m	$^5/_8$ "	-	1,46,448 m^3/d	2850 psi	-

Reserves

	Geological Reserves	Recoverable Reserves
Free Gas MMm^3	56.50 (C-1 Category) 396.20 (2 Category)	33.90 -
Total :	**452.70**	**33.90**

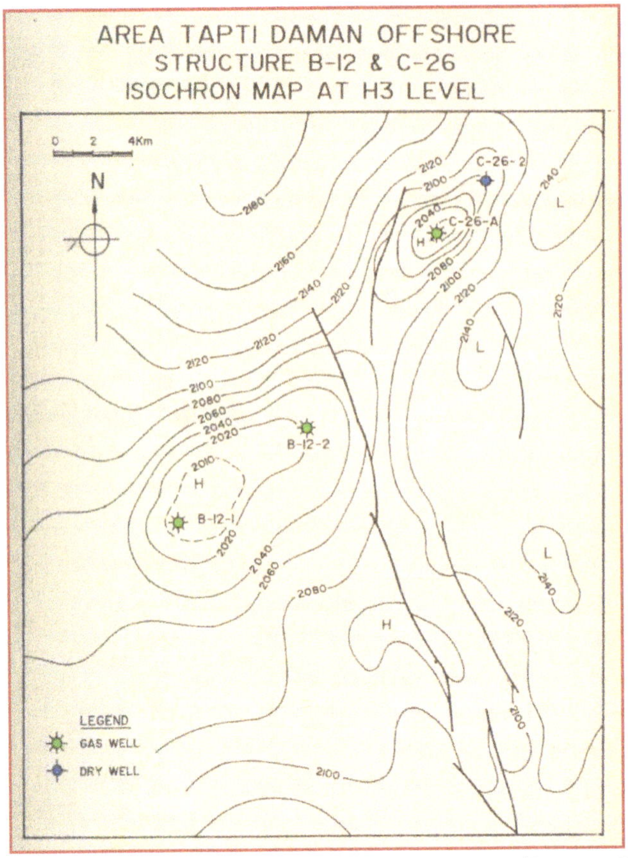

(Courtesy ONGC)

Fig. 5.2.f.1

C-26 is a stratigraphic cum structural prospect about 15 kms North-east of Dahanu structure in a water depth of 27 m. The well was drilled in 1991 and gas was found in Upper Oligocene sandstone reservoir.

Production Test Results

Horizon	Bean	Oil	Gas	FTHP	Remarks
2307-2309 m 2312-2316 m	$^{3}/_{8}$ "	-	1,71,990 m^3/d	2420 psi	-
2241-2243 m	$^{3}/_{8}$ "	-	12.776 m^3/d	500 psi	-
2155-2157 m	¼ "	-	54,517 m^3/d	2675 psi	Condensate 102 bbls/d

Reserves

	Geological Reserves	Recoverable Reserves
Free Gas MMm^3	1196.50	708.80

(g) Laxmi and Gauri Fields: Cairn India discovered Laxmi gas field in the Gulf of Cambay during May 2000. The oil field was added in April 2002 (Pre NELP, CB-OS/2). The production is governed by PSC between Government of India and a Consortium consisting of ONGC, Tata Petrodyne Ltd. and Cairn India Ltd. The mining lease areas of Laxmi and Gauri fields are 121.06 Sq.Kms and 52.7 Sq.Kms respectively. The water depths are 6 to 30 m. The gas reservoirs of the Laxmi and Gauri fields are of Mid Miocene age; approximately at a depth of 735 and 1150 m (Fig. 5.2.g.1). Gauri field is adjacent and connected in parts to the Hazira field. Laxmi and Gauri fields have also crude oil bearing reservoirs which are of Early Miocene age at depths of 1175 m and 3025 m. In LA-2 discovery well of Laxmi field 1050 bopd was produced. These sands are considered for development in Laxmi field. The average gross production was 13,799 bopd (gas production 33.3 million standard Cft/d; Oil / condensate production 8254 bopd).

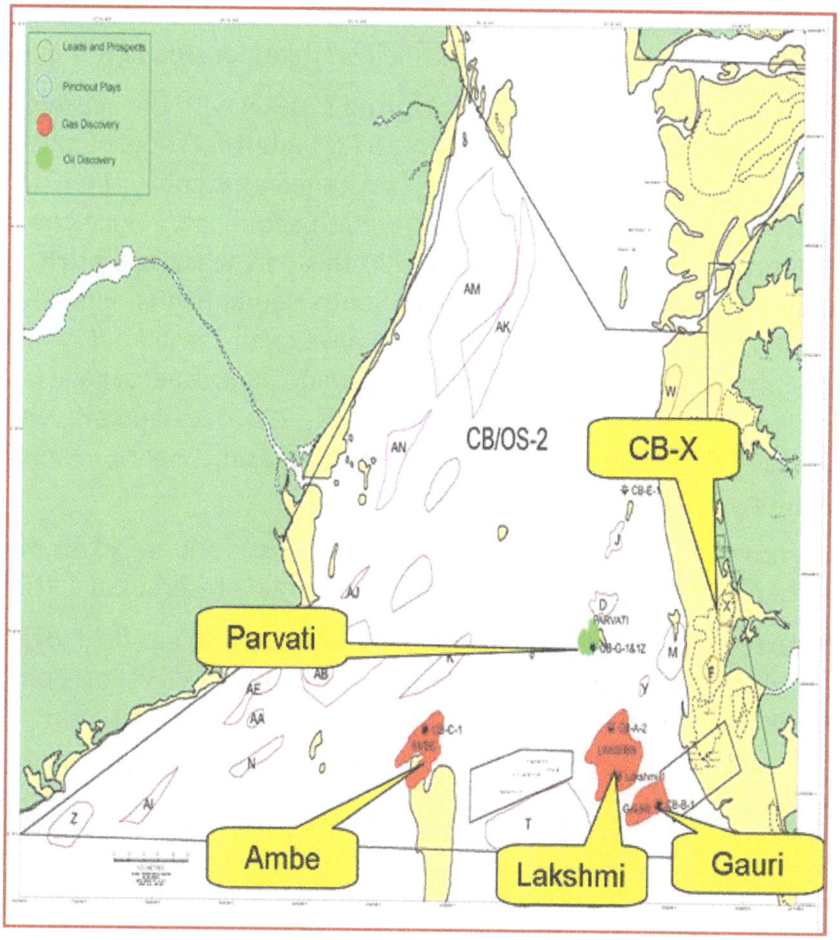

(Courtesy ONGC)

Fig. 5.2.g.1

(h) CB-X Field: Cairn India in Feb 2004 discovered CB-X gas field (CB-OS/2). The mining lease area of CB-X-1 is 33.30 Sq.Kms. Cairn commenced natural gas production from this field @ 325 million standard Cft/d.

(i) Ambe Field: Cairn India discovered Ambe field in Nov 2000 (CB-OS/2). The mining lease area is 107.47 Sq.Kms and is located in a water depth of 5-13 m. Free gas is obtained from Miocene reservoirs. The field has the potential of GIIP of the order of 43 BCF of gas with potential delivery rate of 3.2 million standard CFt/d. The DST-4 & 5 carried in Babaguru sandstone flowed gas @ 20.4 million standard CM^3/d and 20.8 million standard CM^3/d respectively. In the well DST-3, the Tarakeshwar formation produced oil at 643 bopd.

(j) **Bombay High-Heera Panna/Bassein/Neelam Block:** Bombay Offshore basin covers an area of about 63,000 Sq.Kms. Bombay platform, Heera-Panna-Bassein-Neelam block are important for exploration because of several oil and gas fields/strikes located in this part of the basin.

Exploration activities in the area started in 1964 with the seismic survey by a Sovietship and subsequent detailed mapping by CGG, France in 1972. Interpretation of this data resulted in mapping a NNW-SSE trending anticlinal feature which was christened as Bombay High. First well BH-1drilled on this structure in 1974 produced oil from Miocene carbonate sequence. Since the discovery of Bombay high field major exploratory efforts were concentrated in this part which paid rich dividends by discovery of a number of other oil and gas fields such as Heera, South Heera, Panna, Bassein, Neelam, Mukta, B-126, D-1 and D-18. More than 2 lakh LKM of seismic surveys have been carried out, which include 3D surveys also. Around 250 prospects have been mapped of which around 130 prospects have been drilled, which resulted in 28 oil / 14 gas strikes. Around 300 exploratory wells, 750 development wells have been drilled. Of the above discovered fields, Bombay High, Bassein, Neelam, Heera, Panna and Mukta are under commercial production.

Prognosticated resources of oil + OEG of this block are 7160 MMT, of which 3301.5 MMT have been converted into geological reserves with recoverable reserves of 1172.5 MMT.

1. **Bombay High Field:** Bombay high is a 'Giant' oil and gas field located in the western offshore of India. It was discovered in 1974 and locate**d** 160 Kms WNW of Bombay city in a water depth of 75 m (Fig. 5.2.h.1.1 and 5.2.h.1.2).

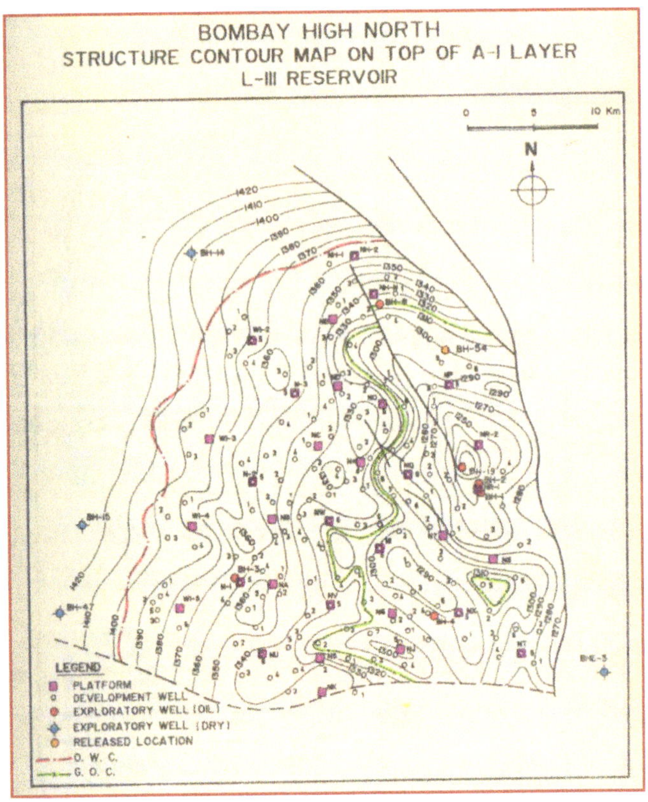

Fig. 5.2.h.1.1

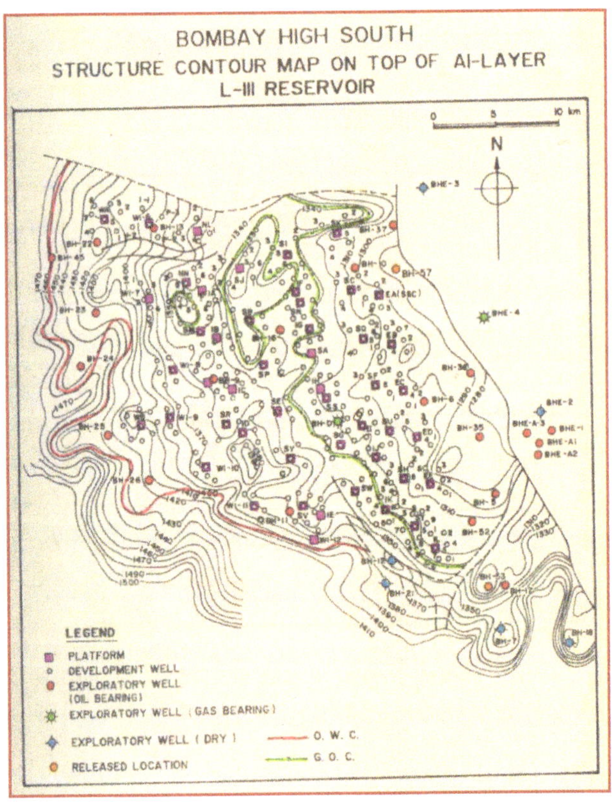

(Courtesy ONGC)

Fig. 5.2.h.1.2

Bombay high is a NNW-SSE trending doubly plunging asymmetric anticline and has an Ariel extent of about 1300 Sq.Kms. A large number of reservoirs belonging to various stratigraphic units from Pre-

Cambrian to Miocene age have been proved to be hydrocarbon bearing. L-III, L-II and S-1 reservoirs are presently under production. L-III and L-II are multi heterogeneous carbonate reservoirs. AT the level of L-III reservoir, the field is divided into three parts as:

1. North Bombay High
2. South Bombay High
3. Central Graben (BH-13)

L-II reservoir is developed and confined to Bombay High North only. Still exploration is in progress to explore / delineate the Fractured Basement, Basal clastics and sub L-III carbonate reservoirs. The various stratigraphic reservoirs are as below:

Reservoir	**Type of Hydrocarbon**
Limestone L-I	- Gas bearing
Limestone L-II	- Oil and Gas bearing
Sandstone/Siltstone S-1	- Gas bearing
Limestone L-III	- Oil and Gas bearing
Limestone L-IV	- Oil and Gas bearing
Limestone L-V	- Oil bearing
Limestone L-VI	- Oil bearing
Basal Sand	- Oil and gas bearing
Fractured Basement	- Oil and Gas bearing

The generalized stratigraphy column for Bombay high is shown in (Fig. 5.2.h.1.3).

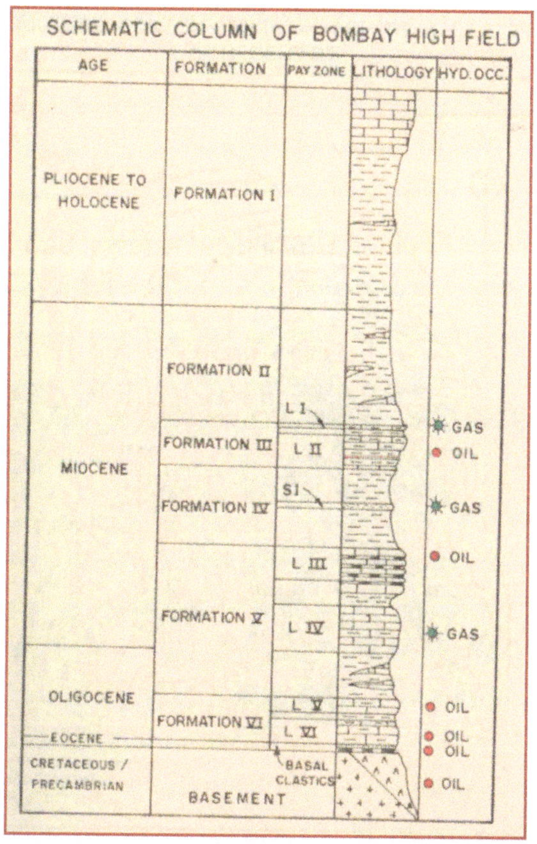

(Courtesy ONGC)

Fig. 5.2.h.1.3

	Geological Reserves	Recoverable Reserves
Oil (MMT)	1,590.72	460.37
Condensate (MMT)	8.99	5.70
Solution Gas (MMm^3)	2,03,926.50	1,05,977.10
Free Gas (MMm^3)	1,84,284-80	1,13,861.10

Reserves

Development schemes have been prepared for three reservoirs L-III, L-II & S-1. Bulk of the oil deposits are confined in L-III carbonate reservoir. Both Bombay High North and South are put on commercial production in May 1976 and September 1980 respectively. Geological cross section of L-III reservoir of Bombay high is shown in the Fig. 5.2.h.1.4. L-II reservoir of Bombay high was put on stream in the year 1981. Peripheral water injection in three rows is being carried out in L-III and L-II reservoirs to obtain the ultimate recovery. The reservoir and fluid properties of various pay zones is as per the Table 5.2.h.1.1.

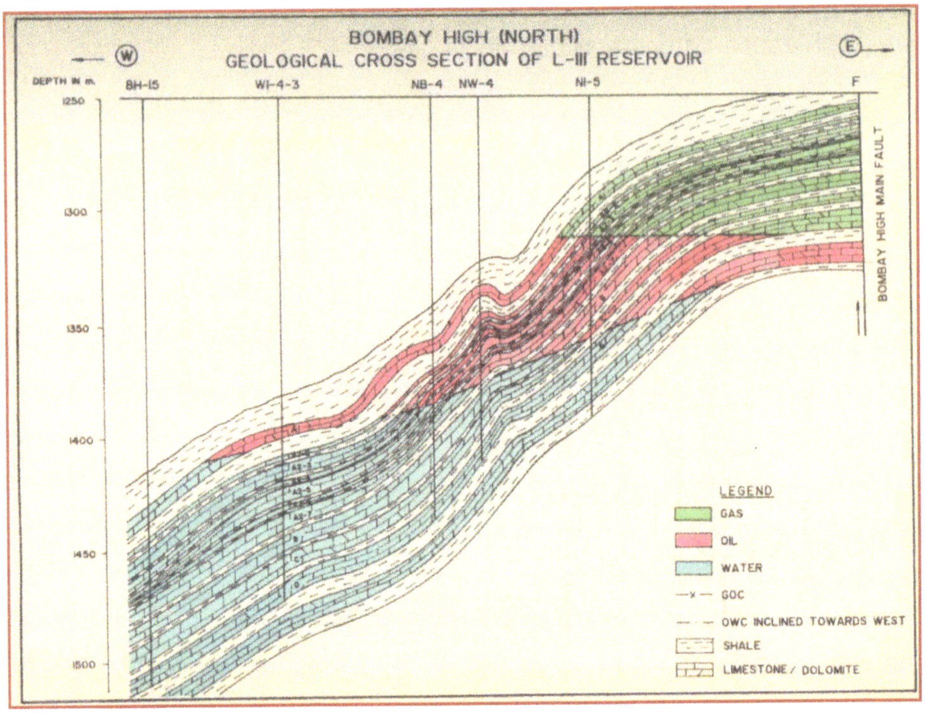

(Courtesy ONGC)

Fig. 5.2.h.1.4

Table 5.2.h.1.1

	Pay Zones								
Properties	**L-I**	**L-I**	**L-I**	**L-I**	**L-I**	**L-I**	**L-I**	**Basal Sand**	**Basement**
Top (m)	905	905	1130	1270	1350	1704	1730	1850	-
Fluid nature	Gas	Oil & Gas	Gas	Oil & Gas	Oil & Gas	Oil & Gas	Oil & Gas	Oil & Gas	Oil
GOC / OWC (m)	-	970	GSC	1315 – 1332 (GOC) 1340 – 1440 (OWC)	-	-	-	-	-
Pay Thickness (m)	-	-	10 - 25	40	-	3.5	2.2	1.64 – 4.5	6.5 – 31.25

Table 5.2.h.1.1 ***Contd...***

Properties	L-I	L-I	L-I	L-I	L-I	L-I	L-I	Basal Sand	Basement
Porosity (%)	8 - 20	5 - 20	35	20 - 22	10 - 24	15	15	16 - 22	1.5 – 2.0
Saturation (%)	-	30 - 60	30 - 90	60	40 - 50	45	56	48 - 65	-
FVF / Z	-	-	-	1.39	1.39	1.37	1.37	1.43	1.35
Sp.Gr.	-	-							
Pb (kg/cm^2)	-	-	-	155.6 – 164.7	-	-	-	-	-
P (kg/cm^2)	-	-	-	156.82	-	-	-	-	

There are about 126 platforms positioned as per the development scheme, of which 107 are well platforms, four well cum process platforms and 15 process platforms.

Water injection in BWPD

Bombay High North L-III Reservoir

	Planned	Actual
Up-dip	85,000	77,720
Down-dip	1,45,000	1,46,750
WJ-6 sector	10,000	4,640
Total :	**2,40,000**	**2,29,110**

Bombay High South L-III Reservoir

	Planned	Actual
Peripheral	**1,50,000**	**1,37,630**
Middle	**3,20,000**	**2,44,350**
Up-dip	**1,30,000**	**1,18,630**
Total :	**6,00,000**	**5,00,610**

Bombay High South L-II Reservoir

Water injection: Planned – 45,000 BWPD and Actual – 46,090 BWPD.

A number of horizontal wells and few drain holes have been drilled to obtain optimum production. Production from Bombay High field was around 19 million tons per Annum. It peaked to 21 million tons. The field is producing with natural decline. GOR in some wells are high.

L-II reservoir is divided into three sub units as L-II-A, L-II-B & L-II-C (total thickness 30-60 m), porosity 15-26%, permeability 10-60 md, initial reservoir pressure around 116.5 kg, temperature

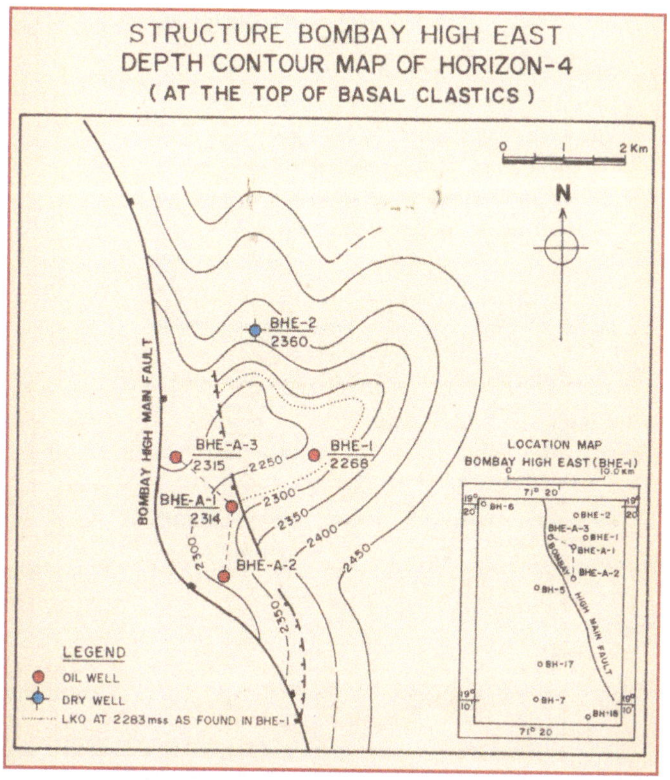

(Courtesy ONGC)

Fig. 5.2.h.1.5

105^{0}C, initial GOR 10 v/v and formation volume factor 1.365. The GOC and OWC are at 937 and 978 respectively. L-III reservoir is heterogeneous with 11 layers (A2-1 to A2-7, B, C & D) separated by thin shales. L-III reservoir is present both in Bombay High North and South forming separate pools. The reservoir is 100-160 m thick, porosity 20-30%, permeability 50-300md and initial reservoir pressure is around 156.7 Kg/cm^2. The initial GOC & OWC in Bombay high North are 1346 m & 1408 m and in Bombay high South they are at 1353 and 1448 m. The reservoir covers around 750 Sq.Kms in Bombay high South (BHS) and 360 Sq.Kms in Bombay High North (BHN). The cumulative production worked out to over 265 MMT by March 1997. The crude oil is of 39^0 API gravity, 0.44 CP viscosity and 30^{0}C pour point and it contains 10-12% wax, 0.18-21% Sulphur and 0.83-1.60% Asaphaltene.

The structure Bombay High East (BHE) is shown in Fig. 5.2.h.1.5.

2. **Bassein Field:** Bassein field is located in the central part of Heera Bassein block, 75 kms Northwest of Bombay city and 70 Kms East of Bombay High field. The average water depth is 55 m. Bassein structure is a North-south trending elongated asymmetric anticline with an areal extent of around 225 Sq.Kms at B zone level with a vertical closure of around 120 m (Fig. 5.2.h.2.1). The discovery well SB-1 was drilled in March 1976 which produced gas from limestone reservoir of Middle Eocene and Early Oligocene age. Presence of oil column was also established in these wells. Development plan of Bassein field was drawn in 1986 for exploitation of free gas @ 20 MMm^3/day, to be processed at two process platforms BPA & BPB. Production of oil from oil rim was also planned. Horizontal well was also attempted in the oil zone. After establishing the platforms as per the development plan, the envisaged gas production was around 30 MMm^3/day from 5 platforms. Structural contour map on top of the zone with platform locations is shown in the (Fig 5.2.h.2.1). The stratigraphy established in Bassein field is shown in the (Fig 5.2.h.2.2). In Bassein field, hydrocarbons have been established in Basal clastics, B&A zones of Bassein limestone. The reservoir facies of Bassein limestone is mainly represented by thick monotonous unit of limestone.

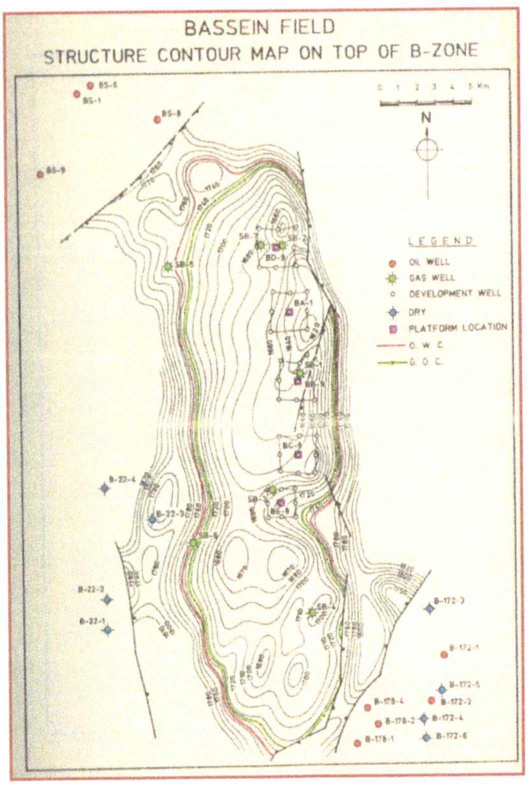

(Courtesy ONGC)

Fig. 5.2.h.2.1

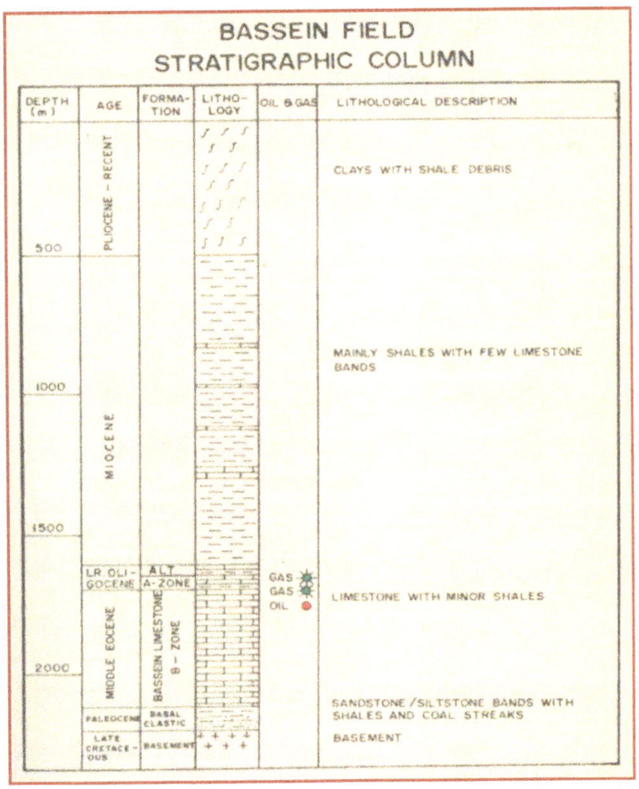

(Courtesy ONGC)

Fig. 5.2.h.2.2

Basal Clastics

This is a sandstone reservoir underlying the Bassein limestone which indicated the presence of oil and gas in well BC-9 having good porosity which ranges from 15-20%.

B-Zone

This is a thick carbonate sequence of about 425 m. In this zone, the primary inter-granular and secondary solution channels, vugs, fractures and moldic porosities are present. The hydrocarbon accumulations are within the upper part of B Zone of Bassein limestone. The porosity ranges between 12-35%, in the upper part of the Bassein limestone. However, there is a rapid degradation in *porosity with depth in B-zone. The hard, tight and compact layers of Bassein limestone act* as seals for the lower Bassein clastics.

A-Zone

This is a 45 m thick zone and represents a heterogenous reservoir, comprising both porous and tight limestones with few thin shales. Main porosity type is inter-granular and ranges from 10-25%.

Oil and Gas Pools

The oil and gas accumulations are found in B&A zones of Bassein limestone and the zones have common gas-oil and oil-water contacts. The gas-oil and oil-water contacts are 1733 m and 1743 m (MSL) respectively.

Reserves

	Geological Reserves	Recoverable Reserves
Oil (MMT)	113.81	0.09
Condensate (MMT)	40.59	24.35
Solution Gas (MMm^3)	19,448	60.20
Free Gas (MMm^3)	2,73,255	2,04,941

Bassein field commercially developed with four platforms (36 wells) to produce gas @ 25 MMm^3/day. Subsequently, one more platform (BE) was added in the southern part with 9 wells and it is producing @ 30 MMm3/day. Oil production is continued from two wells on BA platform. As on April 2002, 6.15 MMT oil and 108 BCM gas had been produced. Gas Gravity is 0.7 and contains H_2S gas, up to 250 PPM. The condensate is of 55^0 API gravity.

3. **Mukta and B-126 Fields:** Mukta and B-126 fields are located in the Northwestern part of Heera-Bassein block 25 Kms east of Bombay high, 25 Kms west of Panna field and about 120 Kms WNW of Bombay city in an average water depth of 65 m.

   The structures B-57 and B-59 were identified in 1973. B-57-1 drilled on the structure in Aug 1981, produced oil and gas from various zones of Mid Eocene to Oligocene carbonates called as Bassein limestone. The B-57 structure was delineated.

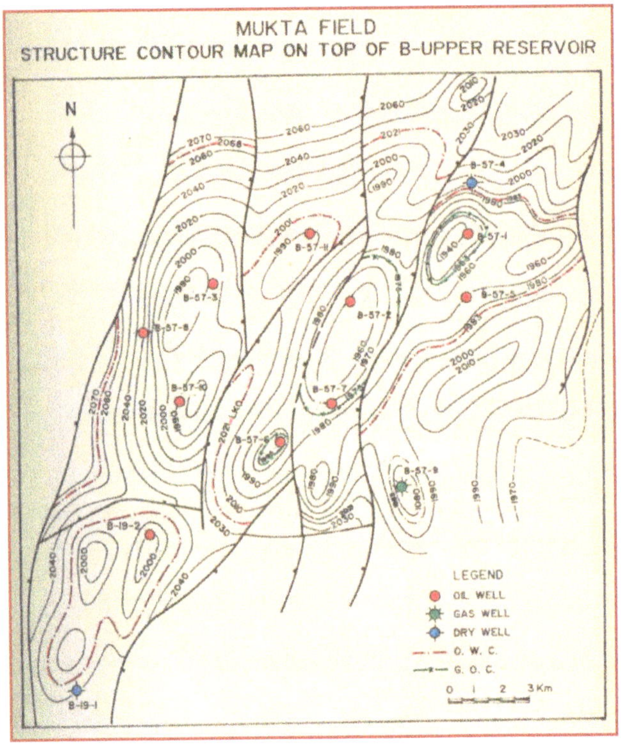

Fig. 5.2.h.3.1

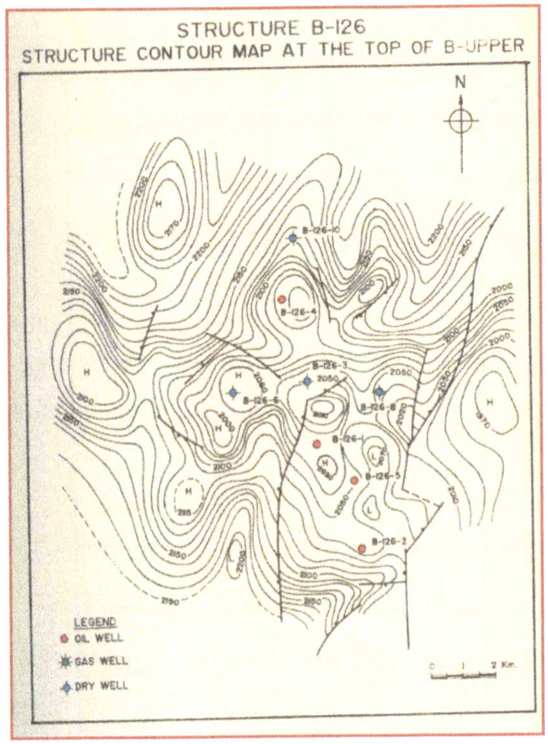

(Courtesy ONGC)

Fig. 5.2.h.3.2

Subsequently, oil was struck in an adjoining structure B-19. Mukta field covers B-57 & B-19 structures (Fig 5.2.h.3.1). To the NE, another anticlinal culmination, B-126 is situated (Fig 5.2.h.3.2). This is also clubbed with Mukta field. Generalised stratigraphy obtained from Mukta wells is indicated in the (Fig 5.2.h.3.3).

The field was delineated and put on production. In Mukta and B-126, hydrocarbon reserves are located in the Basal clastics, Bassein limestone, alternation zones and in the fractured basement, in both oil and gas have been discovered (Fig 5.2.h.3.3).

Development of the Field

Technological scheme for the development of Mukta field was prepared by ONGC during 1988. The scheme envisaged 5 well platforms and a process platform with peak oil production rate of about 1.33 MMTA. Development plan of B-126 had been clubbed with Mukta for economic reasons. Development of A zone for free gas production was also included in the development plan.

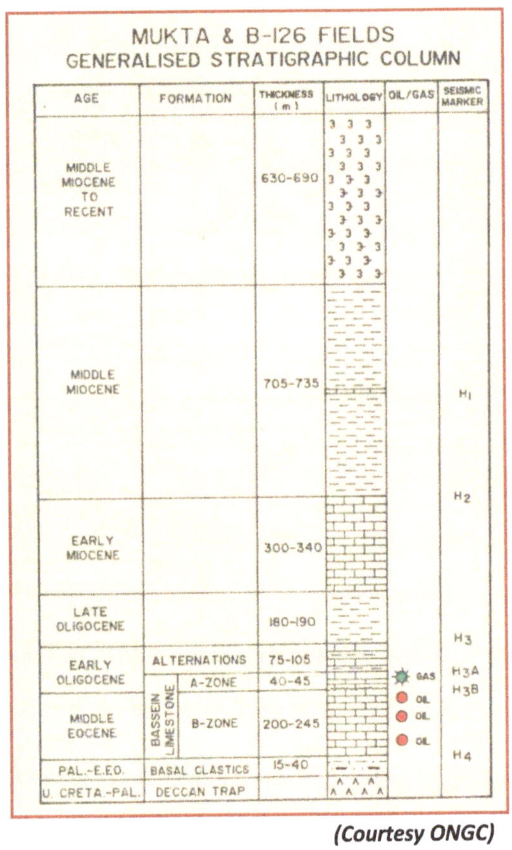

(Courtesy ONGC)

Fig. 5.2.h.3.3

Reserves

	Geological Reserves	Recoverable Reserves
Oil (MMT)	97.06+6.59	12.26+0.78
Condensate (MMT)	0.99+0.17	0.57+0.10
Solution Gas (MMm^3)	11259.4+35937	6025.4+2153.70
Free Gas (MMm^3)	6506.3+663.8	3769.1+398.3

The field was awarded to consortium of JV partners ONGC, BG, Reliance and Vidoecon. Water injection was planned from the beginning, for pressure maintenance in the Mukta field.

4. **Neelam Field:** Neelam Field was discovered in Jan 1987. Oil was discovered in Bassein limestone belonging to Paleocene sequence. It is located 45 Kms southwest of Bombay city and has an areal extent of about 10 Sq.Kms. Water depth is about 50-60 m. A total of 14 exploratory wells have been drilled to delineate the field Fig. 5.2.h.4.1. The generalized stratigraphy of the Neelam field is shown in the Fig. 5.2.h.4.2.

Reserves

	Geological Reserves	Recoverable Reserves
Oil (MMT)	167.97	70.55
Condensate (MMT)	0.26	0.21
Solution Gas (MMm^3)	21403.40	8989.4
Free Gas (MMm^3)	6521.8	5217.4

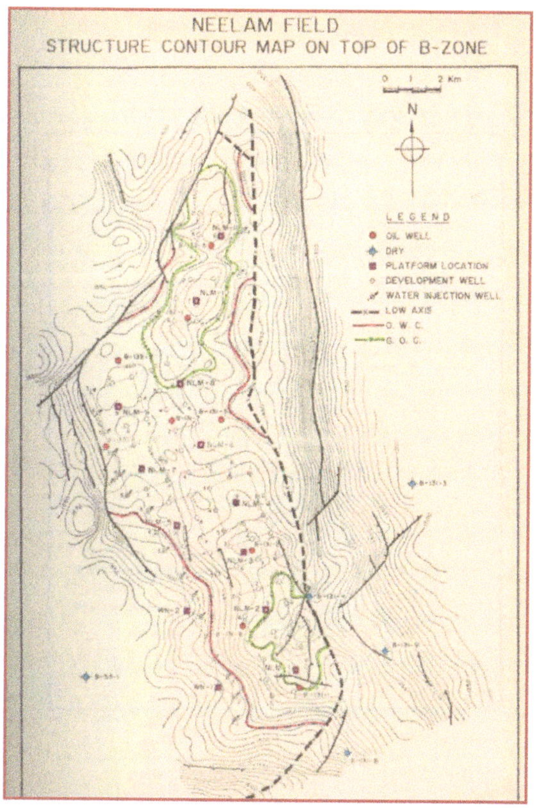

(Courtesy ONGC)

Fig. 5.2.h.4.1

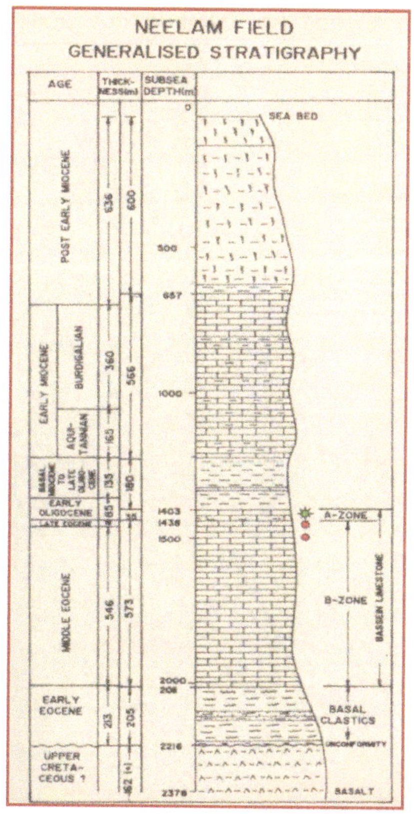

(Courtesy ONGC)

Fig. 5.2.h.4.2

Neelam field comprises of two culminations B-132 and B-131. B-132 is the northern culmination trends NNE-SSW. B-131 southern culmination trends NNW-SSE (Fig 5.2.j.43). Limestones of B zone and A zone of Paleogene age constitute the reservoirs. B zone is the main oil producer (Fig 5.2.j.4.2). The GOC of northern culmination is at 1456 m (MSL) and southern culmination has GOC @ 1402 m MSL. The regional OWC of the field is at 1474 m. The porosity is good with 20-22%, permeability 100-500 md and the reservoir pressure is around 151 Kg/cm2.

Development Plan

The Neelam field was planned for development through 11 platforms including the process platform. A plateau production @13615 m^3/d of oil for 5 years and gradually declining to 5700 m3/day. In some wells gas lift is also planned. Till 2002, the field has produced 20MMT oil and 7420 MMm3 gas. The crude oil is having 39.2^0 API gravity.

5. Heera Field

Heera field was discovered in 1977 and is located at a distance of about 70 Kms southwest of Bombay city and has an areal extent of 10 Sq. Kms. The field has been delineated by drilling 17 exploratory wells of which 14 wells produced oil and gas.

Stratigraphy

Generalised stratigraphy of Heera field is shown in the Fig. 5.2.h.5.2.

Structural contour map on top of H3B zone (Fig 5.2.h.5.1) and a geological section depicting OWC are shown in the Fig 5.2.h.5.2.

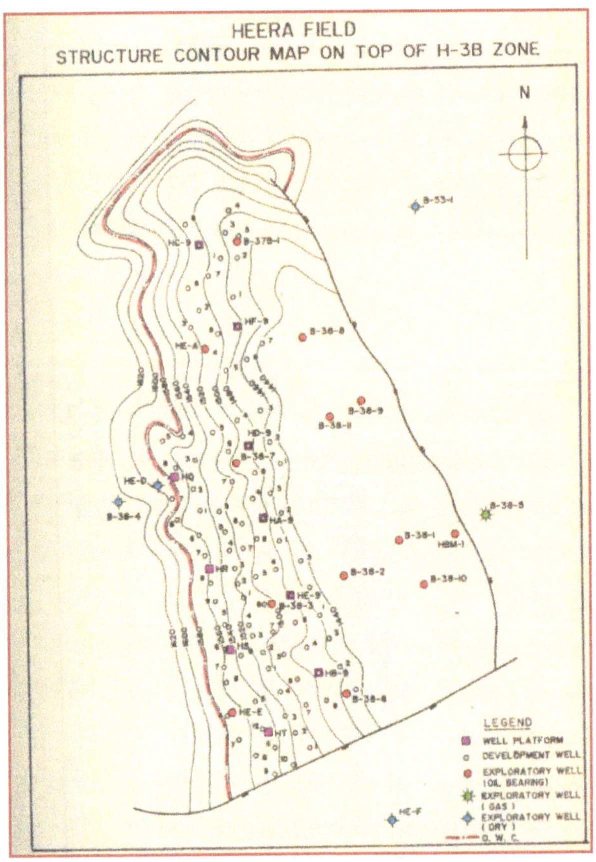

(Courtesy ONGC)

Fig. 5.2.h.5.1

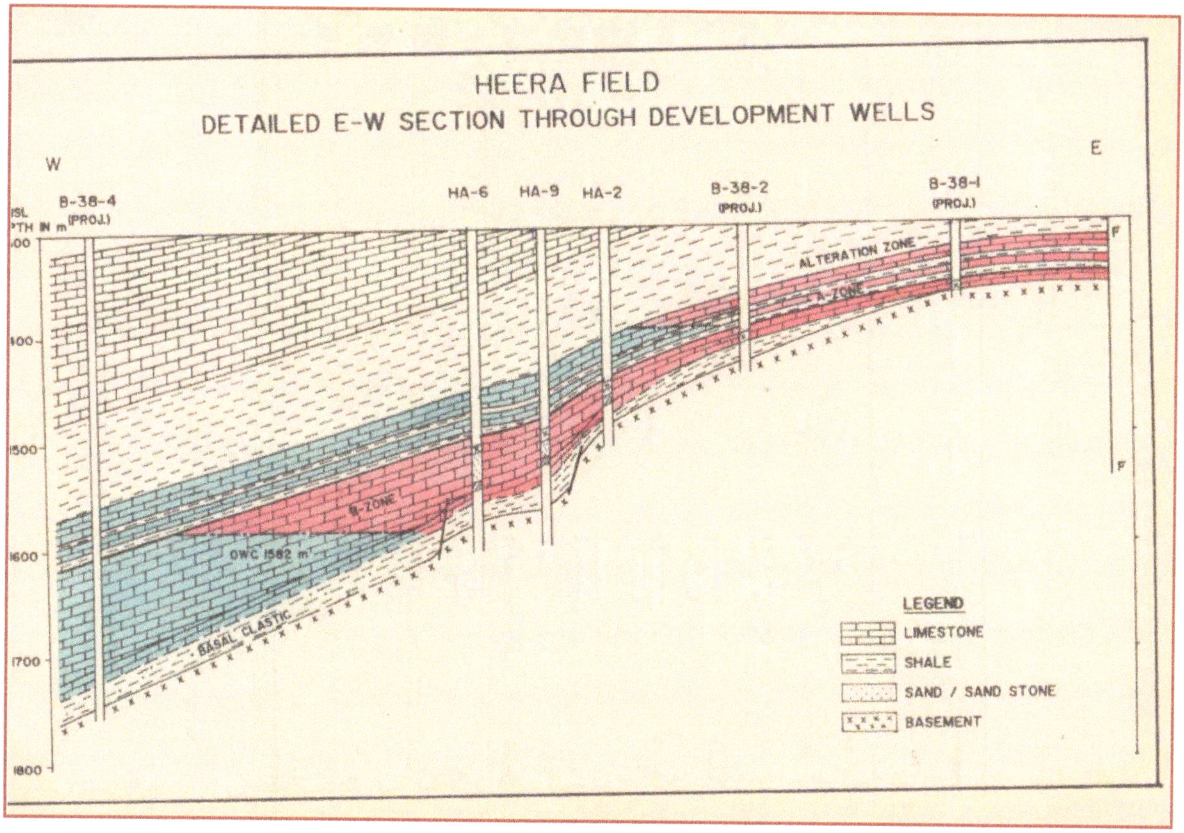

(Courtesy ONGC)

Fig. 5.2.h.5.2

Reserves

	Geological Reserves	Recoverable Reserves
Oil (MMT)	178.83	49.89
Solution Gas (MMm^3)	22780	11938.7
Free Gas (MMm^3)	7659.7	4595.80

Development Plan

Development scheme had been drawn for oil production from Bassein limestone. 14 platforms had been planned in the development. The field had produced so far around 29 MMT oil and 7460 MMm^3 gas. The crude oil is of 37-39^0 API gravity with a pour point of 30-33^{0}C.

Peripheral water injection was planned to maintain the reservoir pressure.

South Heera

R-15A is hydrocarbon bearing structure immediately south of Heera field and is named as South Heera field. It covers an area of 23 Sq.Kms and the field was discovered during Dec 1990, proved to be oil bearing in Basal clastics B&A zones of Bassein limestone and alternations. Total 7 wells have been drilled to delineate the field (Fig 5.2.h.5.3).

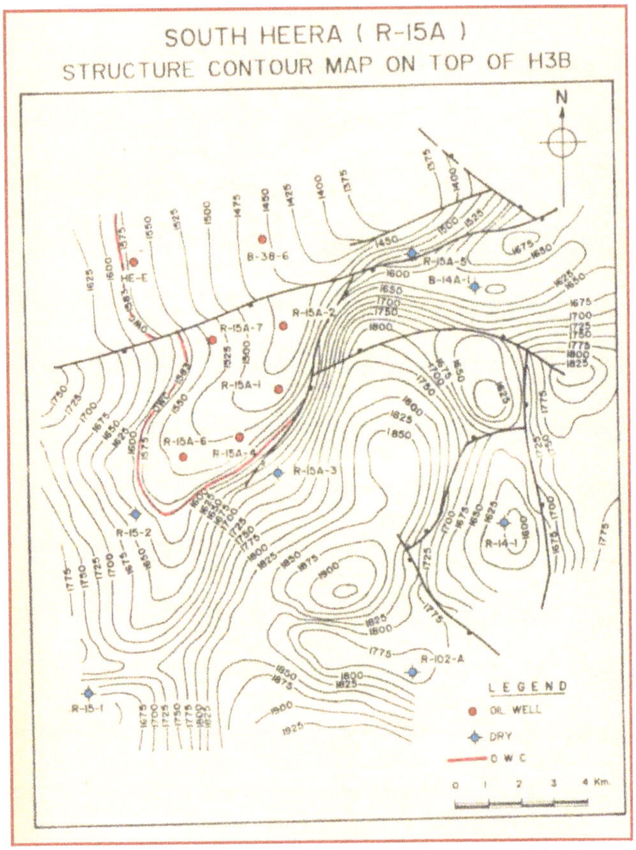

(Courtesy ONGC)

Fig. 5.2.h.5.3

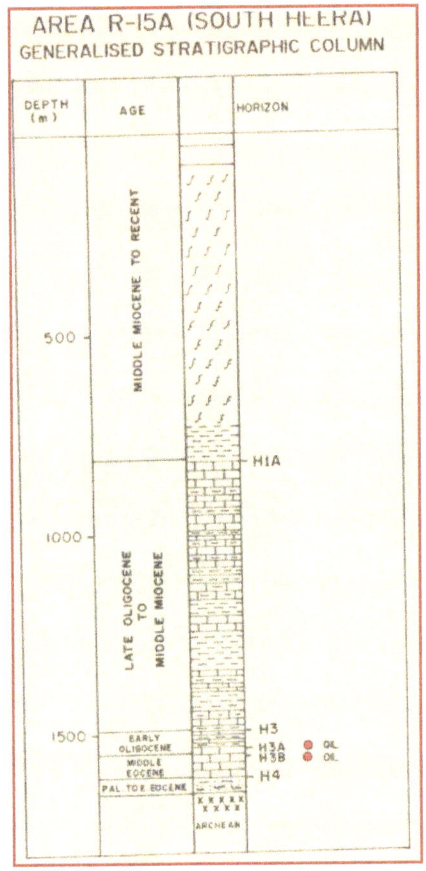

(Courtesy ONGC)

Fig. 5.2.h.5.4

Stratigraphy

The generalized stratigraphy column encountered in this field is shown in the (Fig 5.2.h.5.4).

South Heera is a southern continuation of Heera field which plunges towards southwest. South Heera and Heera fields are separated by ENE-WSW trending fault.

Reserves

	Geological Reserves	Recoverable Reserves
Oil (MMT)	41.16	9.35
Solution Gas (MMm^3)	5518.20	1444.80

Development

South Heera development scheme had been drawn in two phases, with 1 platform in Phase-I and two platforms in Phase-II. Water injection was planned from the beginning to maintain the reservoir pressure.

6. Panna Field

Panna field is located at a distance of 95 Kms Northwest of Bombay city in an average water depth of 45 m. The discovery well BS-1 was drilled in February 1996 which produced oil and gas in Middle Eocene and Lower Oligocene limestone reservoirs. Five wells BS-2, BS-3, BS-4, BS-5 and BS-7 were drilled to delineate the field.

Structure

Panna structure is a NNW-SSE trending elongated asymmetrical anticlinal feature. The eastern limb of which is faulted and down thrown towards east. The structural contour maps drawn on top of pay sequence bring out 3 culminations within the structure. The areal extent of Panna field is about 135 Sq.Kms with a vertical closure of 60 m. Structure contour map on top of B&A zones of Bassein limestone are depicted in the Fig 5.2.h.6.1

Stratigraphy

The generalized stratigraphy column of Panna field is shown in the Fig 5.2.h.6.2.

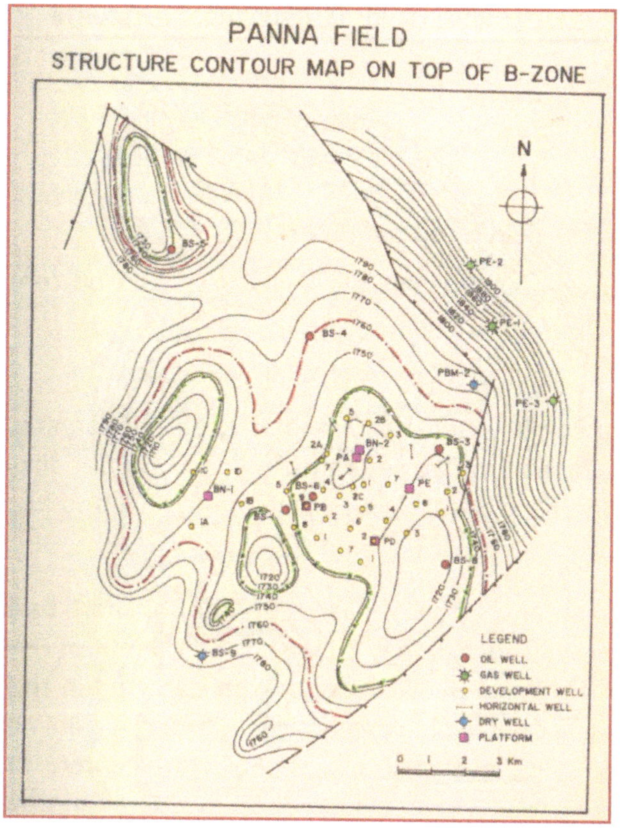

(Courtesy ONGC)

Fig. 5.2.h.6.1

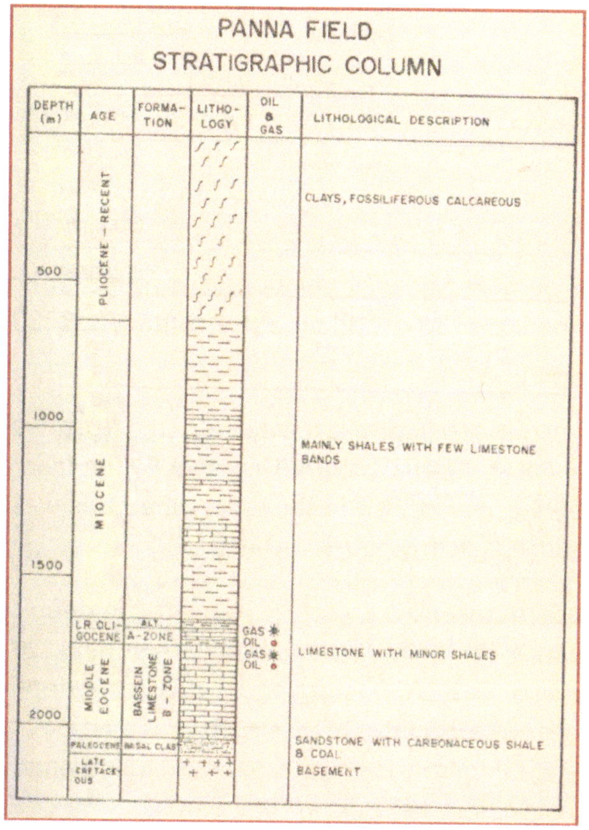

(Courtesy ONGC)

Fig. 5.2.h.6.2

The field had been delineated by 12 exploratory wells. One well PBM-2 was drilled to explore hydrocarbon prospects in Basement. In this well Basal clastics produced oil and gas.

Development

In Panna field, 38 development wells including seven horizontal wells have been drilled so far from two templates and four well platforms (PA, PB, PC and PD). Subsequently, one more platform PE was added with nine wells.

Reservoir Geology

In Panna field, hydrocarbon reserves have been established in A&B zones of Bassein limestone and Basal clastics. The sandstone reservoirs of Basal clastics indicate reserves of oil and gas in wells PB-2 and produced gas in well PE-1 and PE-2. B zone (Bassein limestone) is monotonous limestone having thickness of about 325 m. Primary inter-granular porosity and secondary porosity in the form of solution channels, solution vugs, micro fractures and molds are present. Development of the porosity in the upper part of the zone is better and ranges from 12-30% while the lower part is heterogeneous and tight. A zone is a 50 m thick heterogeneous reservoir having porous and tight layers. Main porosity type is inter-granular, ranging from 5-15%.

Reserves

	Geological Reserves	Recoverable Reserves
Oil (MMT)	92.05	16.41
Condensate (MMT)	2.14	1.23
Solution Gas (MMm^3)	15740	2805.7
Free Gas (MMm^3)	21895	12570

Oil produced @ 562 bopd and gas 0.137 MMSCMD. The field is offered under JVC.

The oil and gas accumulations are found in both A&B zones of Bassein limestone. Gas oil contact is defined at 1738 m and oil water contact at 1760 m.

Panna East

Panna East structure is located 90 Kms Northwest of Bombay city and immediately east of Panna oil field in a water depth of 39m. The main Panna field and Panna East are separated by a NE-SW trending fault. The PE-1 was drilled in the year 1985. Gas was found from Basal clastics section. In PE-2 and PE-3 also produced gas from BASAL clastics.

7. **D-1 Structure:** D-1 structure is located in deep continental shelf (DCS) part of Bombay offshore basin at a distance of 70 Kms from Bombay High field, 20 Kms WSW of Bombay city in a water depth of 85-90 m. The well D-1-1A was drilled in 1976 to a depth of 3245 m. Hydrocarbons were detected in the Early Miocene carbonates. D-1 structure is a NE-SW trending doubly plunging anticline, having 4 culminations. The areal extent is about 180 Sq.Kms and amplitude at H2 level is 60 msecs. Structure contour map on top of Middle pay is as in Fig 5.2.h.7.1.

Stratigraphy

The generalized stratigraphy of the field is in the Fig 5.2.h.7.2

Early to Middle Miocene carbonates proved to be oil producers. 10 wells have been drilled to delineate the D-1 structure. The Mid-Miocene limestone reservoir is having number of layers which are oil and gas bearing. These zones have been grouped as Lower, Middle and Upper pays. Middle pay is good producer. The average oil production is_around 4000 bbls/day - gas 9288 m3/day-FTHP 720 psi with a very low GOR- API gravity of oil is around 38^0.

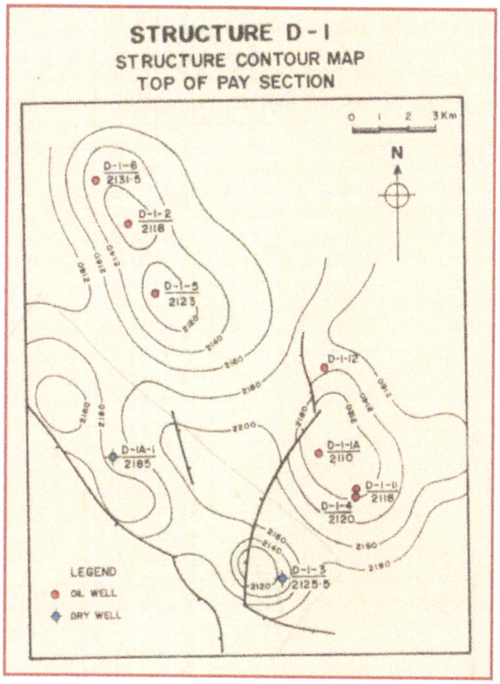

(Courtesy ONGC)

Fig. 5.2.h.7.1

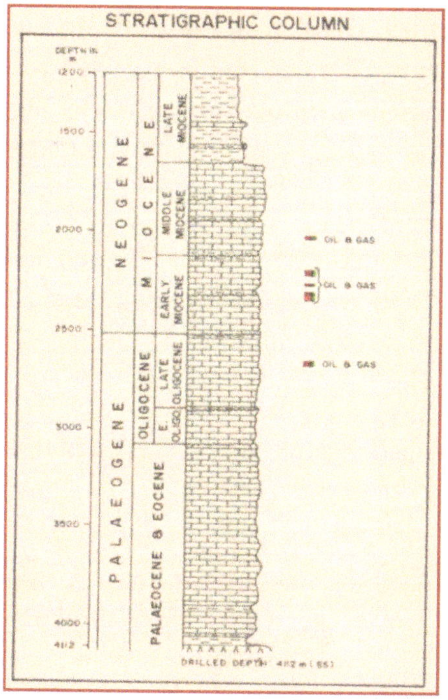

(Courtesy ONGC)

Fig. 5.2.h.7.2

Development Plan

Development plan is drawn and the field is offered for exploitation under JVC.

Reserves

	Geological Reserves	Recoverable Reserves
Oil (MMT)	39.55	4.62
Condensate (MMT)	1062.30	618.20

Crude oil analysis is included in the Table 5.2.h.7.1.

Table 5.2.h.7.1

Properties	Well # D-1-4		Well # D-1-5	Well # D-1-6
Zones (m)	2213–2322 2325-2330	2219-2212 2218-2227 2245-2247	2234-2237	2130-2132
Density at 15°C (gm / ml)	0.8309	0.8386	0.8366	0.8474
Sp.Gr at 60°F	0.8313	0.8390	0.8370	0.8478
API Gr. at 60°F	38.71	37.15	37.55	35.40
Pour Point at 60°C	+24	+27	+24	+33
Kinematic Visc. at 37.8°C	4.6	4.6	2.92	7.2
Sulphur Total	-	-	-	0.16
Aspheltenes (% wt.)	-	0.6	7	1.10
Resins (% wt.)	-	5.2	4.7	6.7
Wax (% wt.)	-	9.8	10	16
Flasti Point (°C)	<+5	<+5	<+5	<+5
Molecular Wt.	235	230	225	240
Base of Crude	Intermediate	Intermediate	Intermediate	Intermediate

8. **D-18 Structure:** D-18 structure is located in Bombay offshore basin, 160 Kms WSW of Bombay city in a water depth of 85 m. D-18-1 was drilled to 4106 m with in Early Eocene carbonates. The field proved to be a commercial producer of oil from Oligocene carbonates. 6 exploratory wells have been drilled to delineate the structure, of which 4 are oil bearing.

Structure

It is a broad structural terrace at H3 and H3B levels having NW-SE alignments. The area of the structure is around 32 Sq.Kms at H3 level with a closure of 35m.secs. Structural map at the level of Upper pay is shown in the Fig. 5.2.h.8.1.

Stratigraphy

The generalized stratigraphy is shown in the Fig. 5.2.h.8.2.

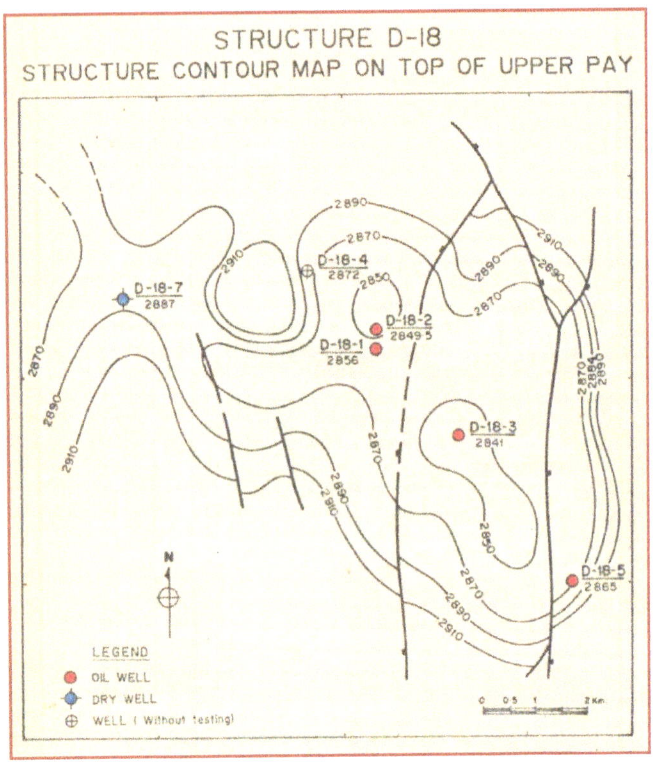

(Courtesy ONGC)

Fig. 5.2.h.8.1

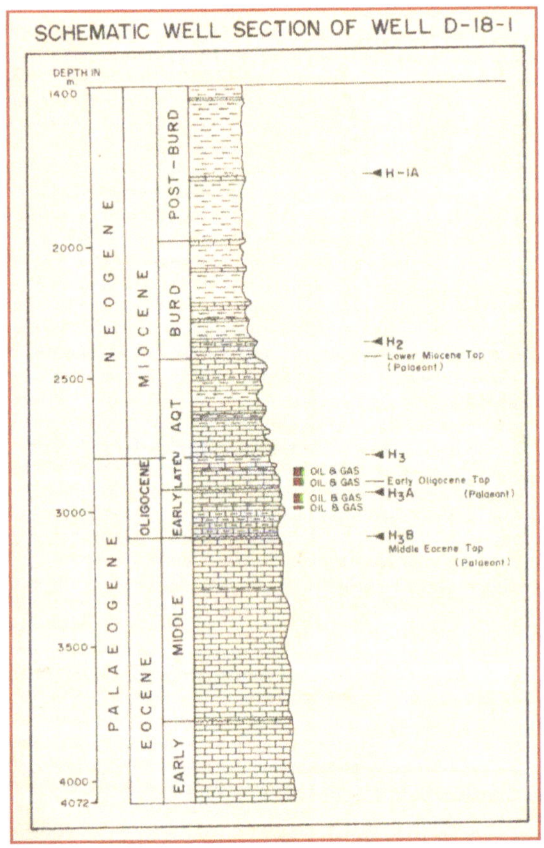

(Courtesy ONGC)

Fig. 5.2.h.8.2

The field was put on production through an EPS – Tahara for the production from Lower and Middle pays. Initially it produced through EPS about 180 bbls/day of oil but the EPS had to be de-hired due to flow line choking by wax deposition.

A number of exploratory wells have been drilled around Bombay High and on its flanks covering both structural and stratigraphic prospects like wedge outs. Oil and gas are found in a number of wells ranging in age from Eocene to Younger.

Bombay High East Structure

Bombay High East Structure is located immediately East of main Bombay High field about 140 Kms WNW of Bombay city. Fig. 5.2.h.8.3. Oil was found in the first well BHE-1 drilled in Sept 1976 in Basal clastics and B zone of Bassein limestone. Subsequently, 4 exploratory wells BHE-A1, BHE-A2, BHE-A3 and BHE-2 were drilled of which three wells proved to be oil bearing. BHE-1 produced around 120 bbls/day oil from Basal sandstone (2296-2308) with ½" choke with FTHP at 388 PSI and GOR at 65. The B zone Bassein limestone in the interval 2151-2156 m also produced around 1200 bbls/day of oil with a GOR of 64. Subsequent wells BHE-2, BHE-3, BHE-4, BHE-5 and B-18-1 proved to be dry.

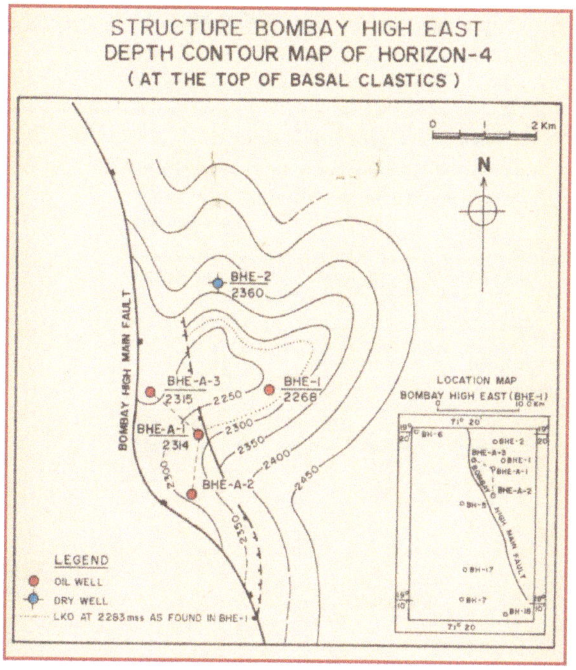

(Courtesy ONGC)

Fig. 5.2.h.8.3

9. Ratna and R-Series of Structures: Ratna and R-Series of structures are on the Ratnagiri Paleo shelf which is the southern extension of Bombay offshore basin, from Heera in the North to Vengurla arch in the South. Ratnagiri Paleo shelf covers an area of about 30,000 Sq.Kms. Out of 37 structures drilled so far in Ratnagiri area, 8 proved to be oil / gas bearing and another 4 structures have oil/gas indications. R-12 structure known as Ratna field is situated south of Heera field and on commercial production since November 1982. The remaining oil strikes namely R-7, R-7A, R-8, R-9, R-10, R-13 and R-71 grouped as a cluster south of Ratna field and are referred as R-series of structures. The prognosticated resources of Ratnagiri paleo shelf are estimated to be 660 MMT of which 57.61 MMT of oil and OEG have been converted into geological reserves so far.

Ratna and R-Series of structures are located in an average water depth of 45 m at a distance of about 130 Kms southwest of Bombay city.

Structure

Ratnagiri Paleo shelf is divisable into three NNW-SSE trending tectonic blocks. These blocks from East to West are referred as Jayagarh Homocline, Vijaydurg graben and Srivardhan horst. R-12 structure (Ratna) is situated in the composite block boardering Jayagarh Homocline in the east and Vijaydurg graben in the West. All the structures have North-south to NW-SE trend in general.

Ratna Field

Ratna field is situated 90 Kms. SSW of Bombay city. The discovery well R-12-1 was drilled in 1979. Oil was discovered in Ratna pay of Bassein limestone. In addition to Ratna pay non-commercial flow of oil was found in Basal clastic section in the northern part of the field. Another one well in the southern part of Ratna also proved presence of oil in Ratna pay. Central part of Ratna field is developed and is under production since 1983.

Structure

Ratna field is a doubly plunging anticline having NNW-SSE trend situated in the northern part of Ratnagiri shelf. The structure is a long, narrow anticline intersected by number of NNW-SSE trending faults (Fig. 5.2.h.9.1).

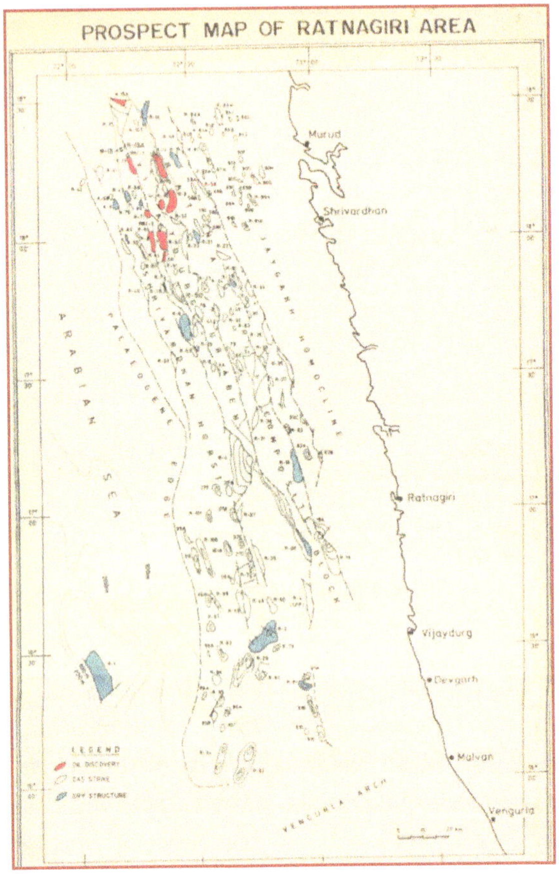

(Courtesy ONGC)

Fig. 5.2.h.9.1

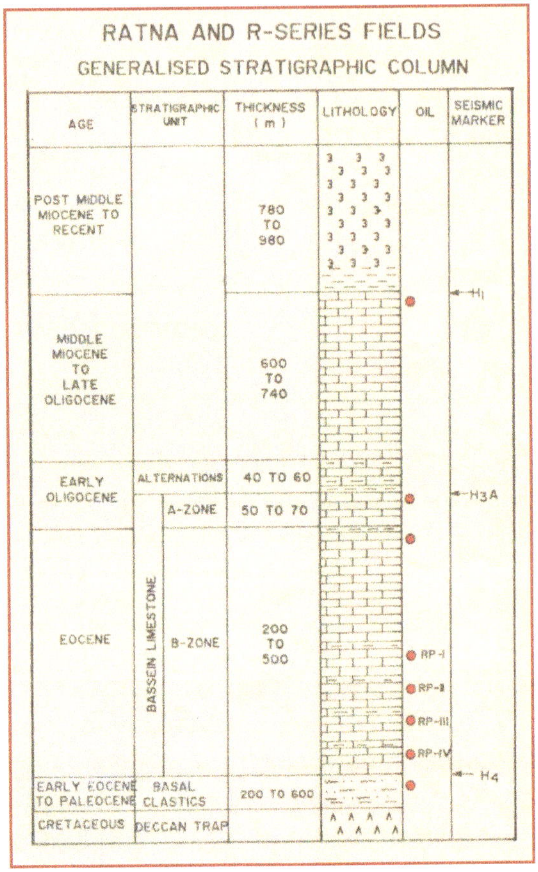

(Courtesy ONGC)

Fig. 5.2.h.9.2

Stratigraphy

The generalized stratigraphy is as shown in the Fig. 5.2.j.9.2.

Reserves

	Geological Reserves	Recoverable Reserves
Oil (MMT)	19.13	3.82
Solution Gas (MMm^3)	2223.6	1070.1
Free Gas (MMm^3)	404.7	-

Status of Development

The field is under production through well cum process platform installed in 1993. Development field is confined to the central part of the structure. Ratna pay, in the lower part of the Bassein Limestone, is the main pay zone of the field. Presence of commercial oil from the same pay was observed in the southern part R-12-2. Ten development wells have been drilled, of which five are oil bearing.

R-Series of Structures

The R-series of structures are located nearer to the Ratna field. They are R-7, R-7A, R-8, R-9, R-1-, R-13 & R-71.

R-7 Structure

In R-7 structure oil was found in A zone of Bassein limestone. An interval at 1577-1587m produced around 1700 bbls/day and 5000 m3/day of gas (GOR 18.2, H_2S-300-400 PPM, API gravity 34.07).

Reserves

	Geological Reserves	Recoverable Reserves
Oil (MMT)	11.68	3.82
Solution Gas (MMm^3)	245.3	80.2

Average porosity is 17.55%, ISBHP 575 psi, bottom hole temperature 188^{0}F, OWC 1580 m, reservoir pressure 157.7 Kg/cm2, and saturation pressure 56 Kg/cm2.

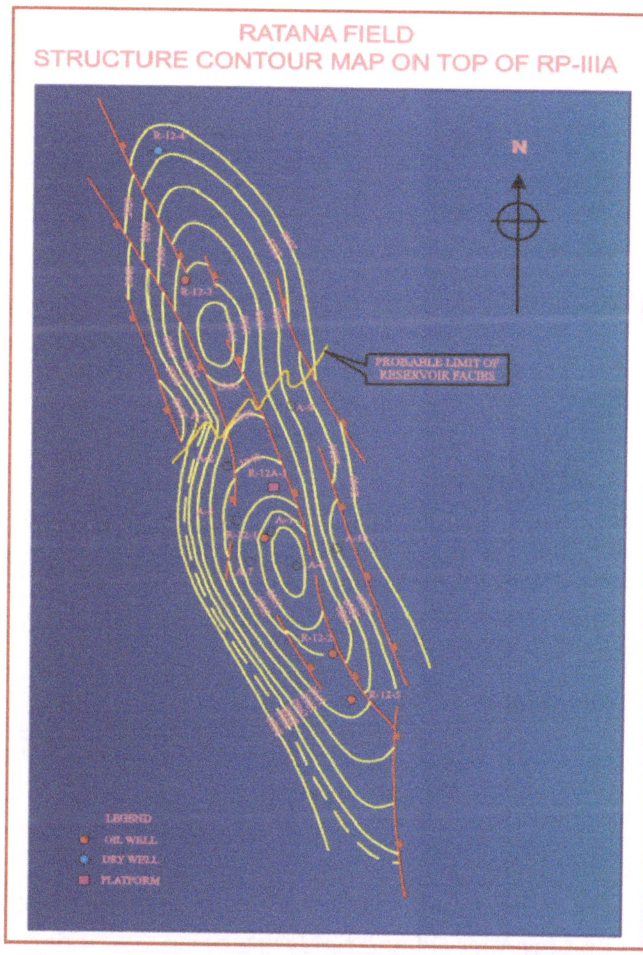

(Courtesy ONGC)

Fig. 5.2.h.9.3

R-8 Structure

R-8 Structure was drilled in 1982, oil was found in Miocene Carbonate reservoir.

R-7A Structure

R-7A structure is a North-South trending anticline located in North Ratnagiri block in a water depth of 48 m. Oil was discovered in A zone of Bassein limestone belonging to Early Oligocene age. An interval at 1592-1600 m produced around 2864 bbls/day, with FTHP 410 psi, GOR 24, H_2S 20 PPM, and API Gravity 39^0. Two more delineation wells have been drilled which proved to be oil bearing.

Reserves

	Geological Reserves	Recoverable Reserves
Oil (MMT)	1.59	0.19
Solution Gas (MMm^3)	38.3	23

R-9 Structure

R-9 structure is located in a water depth of 42 m

in North Ratnagiri shelf. R-9-1 drilled in 1979, produced oil from A&B zones of Bassein Limestone of Paleogene age. Total four exploratory wells have been drilled on this structure. Oil bearing area is around 14 Sq.Kms. An interval at 1543-1553 m produced oil @ 1088 bbls/day and gas 12975m3/day through 5/8" choke. (GOR 75, API Gravity 36.4, H_2S 30 PPM).

Reserves

	Geological Reserves	Recoverable Reserves
Oil (MMT)	12.02	2.56
Solution Gas (MMm^3)	841.4	188.5

R-10 Structure

R-10 structure is located 15 Kms SSW of Ratna field in a water depth of 42 m. R-10-1 drilled in 1988, produced oil and gas from A-zone of Bassein Limestone. Three more delineation wells have been drilled on this structure. An interval between 1679-1704 m produced oil @ 3763 bbls/day and gas 48000 m3/day through 3/4" choke (GOR 80).

Reserves

	Geological Reserves	Recoverable Reserves
Oil (MMT)	**3.05**	**0.46**
Solution Gas (MMm^3)	**268.8**	**161.3**

R-13 Structure

R-13 structure is located 10 kms west of Ratna field. R-13-1 drilled in 1983, discovered oil in A&B zones of Bassein Limestone. Five exploratory wells have been drilled to delineate the field. An interval at 1950-1955 m produced oil @ 2413 bbls/day and gas 330m3/day through 1/2"choke. (GOR 86, H_2S 11000 PPM).

Reserves

	Geological Reserves	Recoverable Reserves
Oil (MMT)	3.83	0.63
Solution Gas (MMm^3)	589	525

R-71 Structure

R-71 structure is situated to the west of R-10 structure in a water depth of 57 m. R-71 was drilled in 1986 and oil was found in Basal clastics. The areal extent of the structure at H4 level is 15 Sq.Kms. An interval at 2839-2843 m produced oil @ 412 bbls/day and gas 4500 m3/day through 1/2" choke. (GOR 69, FTHP 102 psi).

Reserves

	Geological Reserves	Recoverable Reserves
Oil (MMT)	0.56	-
Solution Gas (MMm^3)	39.6	-

Ratna field and R-Series of Structures have been awarded for production to the Consortium of Essar Oil and Premier Oil and fields are being developed.

Exploration & Production status of Western Offshore

Basin area : About 236,000 km

Offshore Seismic Data : 2D: 266,127 LKM 3D : 44036 Sq. Km.

Data acquired by	2D (LKM)	3D (Sq. Km.)
O NG C	264543	39052
Pvt. /JV	1584	4984

Other Surveys : Gravity – 75,118 LK: Magnetic – 123,387 LKM

: Satellite Gravity

Well Data Number of wells : 774(Exploratory), 1369 (Development)

Hydrocarbon shows / Oil & gas fields : Established petroleum province with a number of oil & gas fields including two giants viz BombayHigh Oil Field & Bassein G as Field

Basin Type : Passive Margin

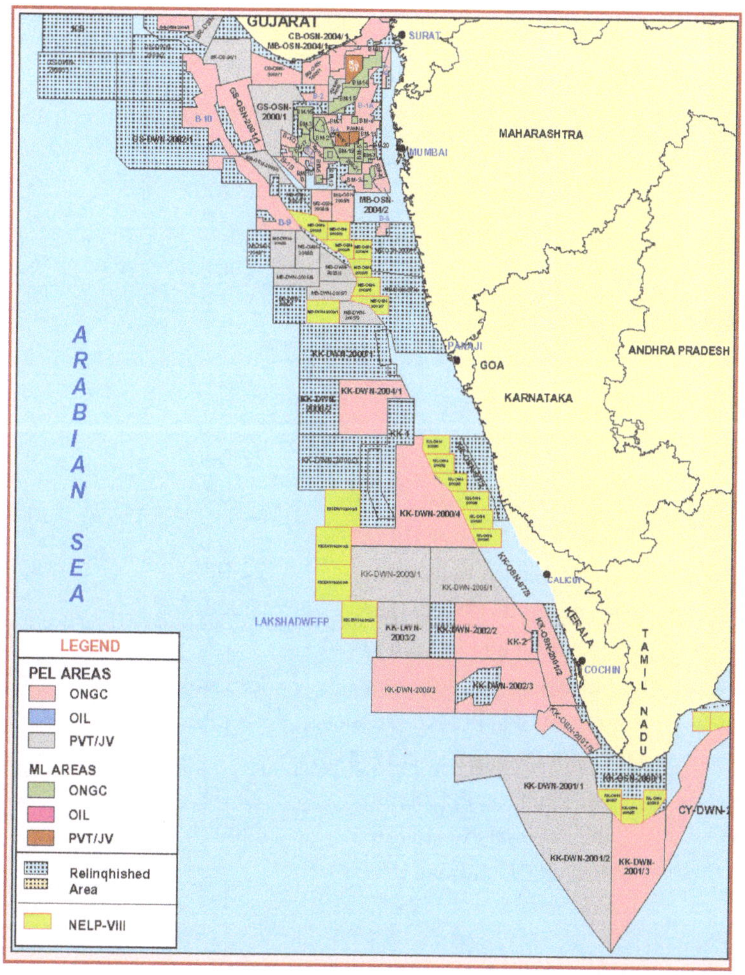

(Courtesy DGH)

Fig. 5.2.h.9.4

Blocks On Offer (NELP I to VIII)

The blocks offered and awarded under production sharing contract from NELP I to VIII are shown in the Fig. 5.2.h.9.4.

NELP	Blocks Offered	Blocks Awarded	Awardees	Date of Signing/ Effective Date	Present area (sq.km)	Status
I		MB-OSN-97/2	RIL(100%-OP)		Relinquished	
		MB-OSN-97/3	RIL(100%-OP)		Relinquished	
		MB-OSN-97/4	ONGC(70%-OP), IOC (30%)	12-Apr-2000/	Relinquished	
II		MB-DWN-2000/1	ONGC (85%-OP), IOC (15%),	17-Jul-2001/	Relinquished	
		MB-DWN-2000/2	ONGC (50%-OP), GAIL (15%), IOC (15%), OIL (10%), GSPC (10%)	17-Jul-2001/	Relinquished	
		MB-OSN-2000/1	ONGC (75%-OP), IOC (15%), GSPC (10%)	17-Jul-2001/ 16-Aug-2001	Relinquished	
III						
IV						
V						
VI	MB-OSN-2004/1	MB-OSN-2004/1	GSPC (20%-OP), IOC-20%, GAIL-20%, HPCL-20%, PETROGAS-20%	02-Mar-2007/ 12-Jun-2007	1520	
	MB-OSN-2004/2	MB-OSN-2004/2	PETROGAS(20%-OP), GAIL (20%), IOC-(20%), GSPC (20%), HPCL (20%)	02-Mar-2007	741	
VII	MB-DWN-2005/1					
	MB-DWN-2005/2	MB-DWN-2005/2	BHP Billiton Petroleum International Pty.Ltd (26% - OP), GVK Oil and Gas Ltd (74%)	22-Dec-2008/	3660	
VII	MB-DWN-2005/3	MB-DWN-2005/3	BHP Billiton Petroleum International Pty.Ltd (26% - OP), GVK Oil and Gas Ltd (74%)	22-Dec-2008/	3097	
	MB-DWN-2005/4	MB-DWN-2005/4	BHP Billiton Petroleum International Pty.Ltd (26% - OP), GVK Oil and Gas Ltd (74%)	22-Dec-2008/	3408	
	MB-DWN-2005/5	MB-DWN-2005/5	BHP Billiton Petroleum International Pty.Ltd (26% - OP), GVK Oil and Gas Ltd (74)	22-Dec-2008/	3169	
	MB-DWN-2005/6					
	MB-DWN-2005/7	MB-DWN-2005/7	BHP Billiton Petroleum International Pty.Ltd (26% - OP), GVK Oil and Gas Ltd (74%)	22-Dec-2008/	3324	

Contd...

NELP	Blocks Offered	Blocks Awarded	Awardees	Date of Signing/ Effective Date	Present area (sq.km)	Status
	MB-DWN-2005/8					
	MB-DWN-2005/9	MB-DWN-2005/9	BHP Billiton Petroleum International Pty.Ltd (26% -OP), GVK Oil and Gas Ltd (74%)	22-Dec-2008/	3138	
	MB-OSN-2005/1	MB-OSN-2005/1	ONGC (80%-OP), GSPC (20%)	22-Dec-2008/	2811	
	MB-OSN-2005/2	MB-OSN-2005/2	Adaani Welspun Exploration Ltd (100%-OP)	22-Dec-2008/	1191	
	MB-OSN-2005/3	MB-OSN-2005/3	Essar Exploration & Production Ltd (50%-OP), Nobel Energy International Ltd (50%)	22-Dec-2008/ 02-Feb-2009	2810	
	MB-OSN-2005/4					
VII	MB-OSN-2005/5	MB-OSN-2005/5	ONGC (70%-OP), GSPC (30%)	22-Dec-2008/	2402	
	MB-OSN-2005/6	MB-OSN-2005/6	ONGC (80%-OP), GSPC (20%)	22-Dec-2008/	2820	
VIII	M B-DW N-2009/1	M B-DW N-2009/1	Cairn Energy India Pty Limited (10%-OP), Cairn India Ltd (90%)	30-Jun-2010/	2961	
	M B-OSN-2009/1					
	M B-OSN-2009/2					
	M B-OSN-2009/3	M B-OSN-2009/3	BHP Billiton Petroleum International Exploration Pty Ltd (100%-OP)	30-Jun-2010/	1492	
	M B-OSN-2009/4					
	M B-OSN-2009/5					
	M B-OSN-2009/6	M B-OSN-2009/6	BHP Billiton Petroleum International Exploration Pty Ltd(100%-OP)	30-Jun-2010/	1876	
	M B-OSN-2009/7	M B-OSN-2009/7	BHP Billiton Petroleum International Exploration Pty Ltd(100%-OP)	30-Jun-2010/	1865	

(Courtesy DDH)

5.5 Kerala-Konkan-Lakshadweep Basin

The Kerala-Konkan basin covers an area of 77000 Sq.Kms up to 400 Isobath and another 5,00,000 Sq.Kms of basinal area in the deep waters beyond 400 m depth. The Kerala-Konkan offshore basin is a pericratonic rift basin located in the Southern part of the west coast of India. The structural frame work is generally North-South and can be divided into following (Fig. 5.5.1)

1. Shelf with Horst and Graben Blocks
2. Kori-Comorin Ridge
3. Kori-Comorin Depression
4. Laxmi Laccadive Ridge
5. Laxmi Laccadive Depression

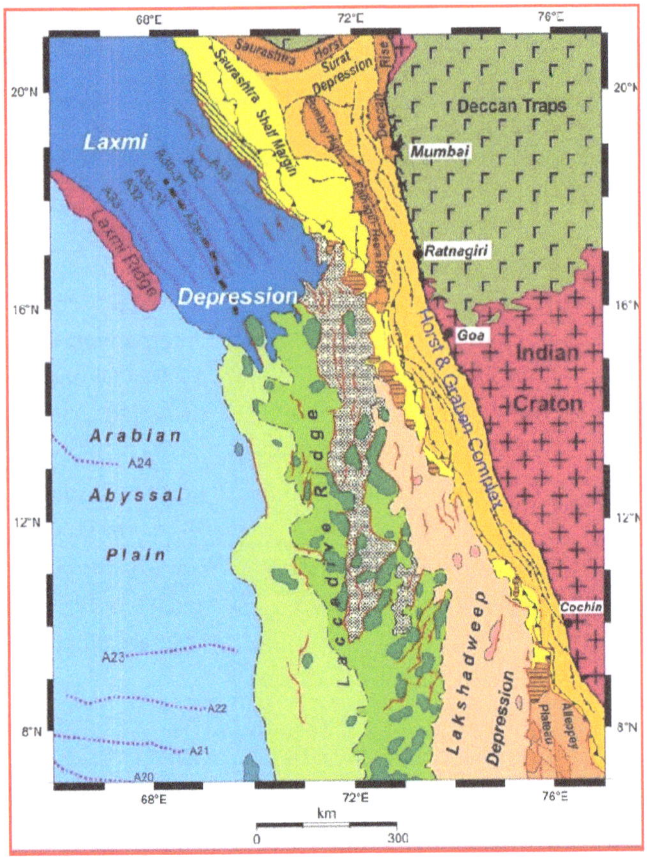

(Courtesy ONGC)

Fig. 5.5.1

Sediments ranging from Cretaceous to Recent are present in the block. Further, the Kerala-Konkan-Lakshadweep basin can be divided into the following sub-basins. They are (Fig. 5.5.2)

1. Konkan Basin
2. Kasargod Basin
3. Kerala Basin
4. Lakshadweep Basin

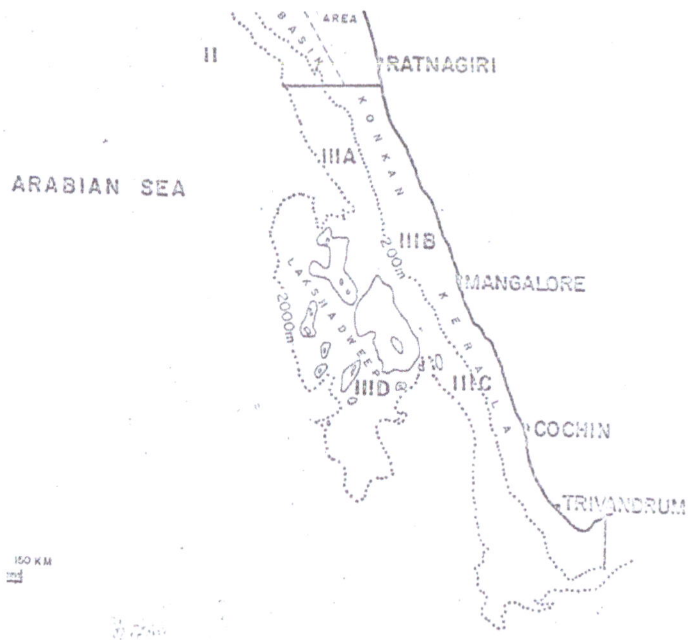

(Courtesy ONGC)

Fig. 5.5.2

The continental shelf of the Kerala coast is about 75 kms wide. The basin is delineated to the east by Pre-Cambrian igneous metamorphic rocks of the shield area and to the west by Laccadive ridge.

The northern limit of the basin is arbitrarily taken as 14^0 N Latitude, i.e. to the south of Vengurla arch (south of Ratnagiri area). This basin opens up to the south into the Indian Ocean. The shelf width varies from 110 Kms near Karwar to 50 Kms near Trivandrum. Around Laccadive group of Islands, the shelf is practically absent and the islands are surrounded by the deep ocean varying from 1000-3000 m water depths. Scanty information is available from Deep Sea Drilling Project (DSDP). DSDP well number 219 of JOIDES (Joint Oceanographic Institution for Deep Earth Sampling) near Laccadive ridge was drilled by the drill ship Glomor Challenger. In the Karwar-Kasargod area, 3 wells have been drilled by ONGC. The DSDP well number 219 of International Ocean Expedition was drilled on the eastern flank of Laccadive ridge in a water depth of 1755 m. (Fig.5.5.3)

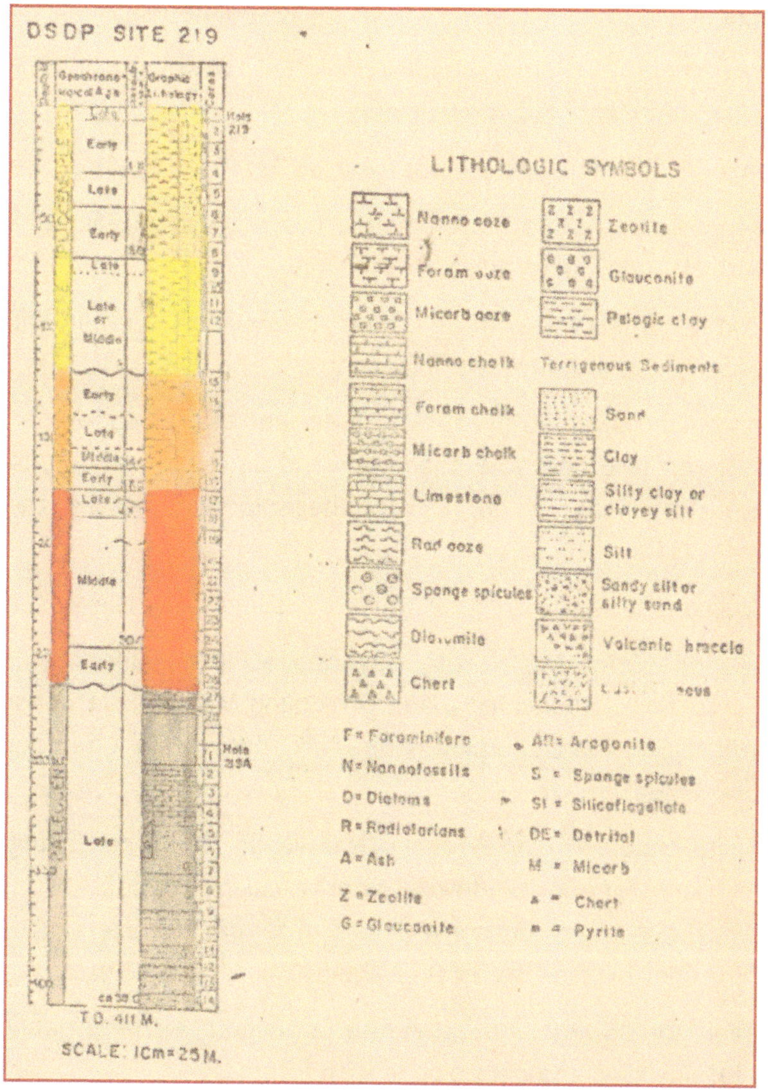

(Courtesy ONGC)

Fig. 5.5.3

A. Stratigraphy

1. **Basement:** Igneous and metamorphic rocks of Precambrian age form the basement on the land part of the basin. These include gneisses, granites, charnokites and schist.

2. **Upper Paleocene:** Sediments of upper Paleocene age are not exposed on the land part of the basin.The oldest sediments met in DSDP 219 include grayish, green limestones containing gluconite and tuffaceous material. Layers rich in shell fragements, pelecypods, forams, ostracods, red algae, Bryozoas and Belemnites and containing volcanic material are present. The Benthoric foraminiferal assemblages dominate.

3. **Eocene:** Sediments of Eocene age are not exposed in the land part of the basin. The Eocene sediments met in DSDP 219 unconformably overlie those of Paleocene age. These include white radiolarian rich micarb ooze, grayish yellow to yellow chert, brown chalk and ooze, limestone, chert

and clay. Angular tuffaceous material and some Glauconite are present in the basal part with Benthonic foraminifera. The rest of the sequence contains rich Planktonic foraminifera.

The intrabasalt sediments met in K-1-1 are found to be barren of fossils. However, palynological studies indicate a Paleocene to early Eocene age to these sediments.

In Kasargod well No.1, about 890 m of limestone (micritic to biomicritic, buff coloured, hard, compact, and vuggy) of early Eocene to Late Eocene was met.

4. **Oligocene:** Sediments of Oligocene age are not exposed in the land part of the basin.

   Sediments of Oligocene age unconformably overlie those of Eocene age at DSDP 219; consisting of white ooze, chalk with Nannos and forams.

   About 378m of dolomitic limestone with minimum shale was met within Karwar-1. Undifferentiated Late Eocene to Oligocene was also met within Kasargod-1 and Cochin-1.

5. **Miocene:** The oldest onland sedimentary rock include sand stones, sandy clay with lenses of clay and peat, which are met with in a borehole near Mayyand and dated as Aquitanian(?) and reffered as Mayyand Formation (Desikachar et al, 1960). These directly overlie the igneous/metamorphic basement.

   The Quilon formation consisting of dirty grey argillaceous limestone is exposed near Quilon.

   The Palynofossils include Strombus Fortisi, Cassidea Sculpta, Subrinora, Stylophora Puldierrima, leptocyathus Epithicate, Nocula Cancellate, and Orbiculine Malbarica. This assemblage is referable to Early Miocene (Burdigallian). The Quilon limestones are met with in a number of shallow wells drilled in the Coastal tract of Kerala.

   The Warakali beds are exposed near Warkala (Narayanan,1958), consisting of chiefly buff to pink coloured, weathered sandstone, grits, detrital laterite current bedded sandstone, peaty clays, with thin seams of lignite. The Warakali beds are devoid of fauna. The age of these beds cannot be fixed with certainity but considered to be Burdigalian age (Desikachar, 1970).

   The Miocene sequence of DSDP 219 overlies the Oligocene unconformably, comprising of grey to greenish gray, detrital silty clay, and nanno ooze with forams.

   About 680m of Dolomite limestone and minor shales in Karwar-1, 560m of micritic, fossiliferous limestome in Kasargod-1 and 390m of dominating coarse sandstone and thin band of clay and limestone in K-1-1 were met with.

6. **Pliocene, Pleistocene & Recent:** Sediments of Pliocene/Pleistocene age are met at DSDP 219, comprising of light grey to greenish grey silty clay, nanno ooze with forams. On the land part, it is represented by alluvial deposits (sand, clay and sandy clays).

   Litho-stratigraphy is shown in the Fig. 5.5.4 & 5.5.5

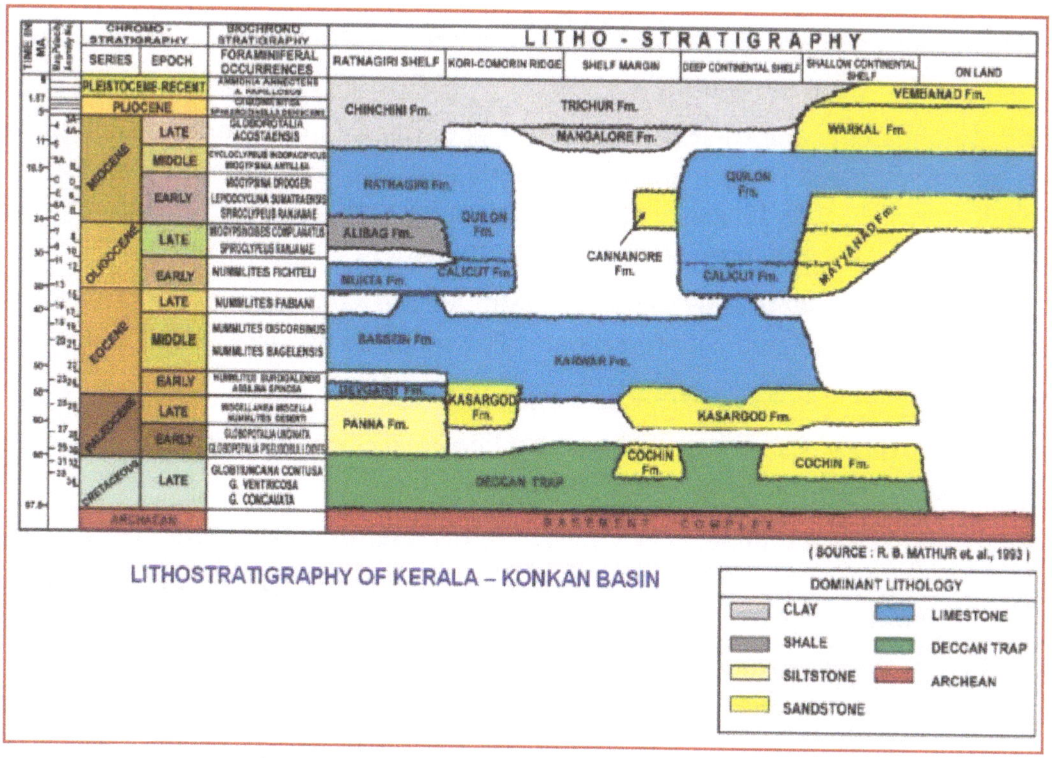

(Courtesy ONGC)

Fig. 5.5.4

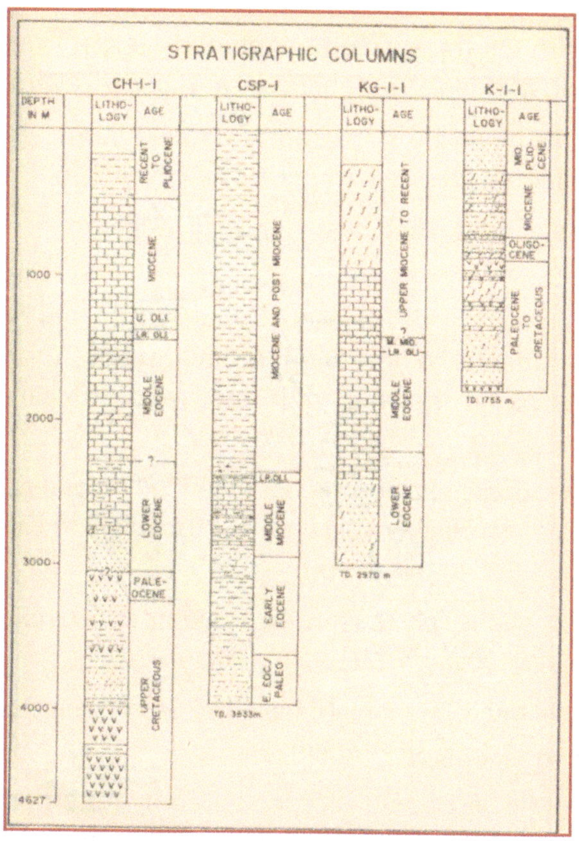

(Courtesy ONGC)

Fig. 5.5.5

(a) **Sediments and Provinance:** Over a large part of the basin, the Neogene sequence is claystone, particularly in the Northern part. Towards south, the lithology becomes more arenaceous. Neogene beds drilled in K-1-1 are sandstones.

Miocene sediments over the shelf part in Karwar and Mangalore sector are likely to be carbonates, where as near Cochin and further south, these sediments are represented by sandstones with minor limestone. Beyond Miocene hingeline in the northern part micritic limestone predominates, whereas in the southern part, there is more argillaceous component in the section.

Early Tertiary is represented by limestones in Karwar and Mangalore area which becomes progressively clastic dominated with few Basaltic flows in Cochin-Quilon sector and farther south.

Provenance

The west coast offshore basins under reference might have come into existence only in Paleocene and no Pretertiary sediments can be expected in this part.

Early Tertiary arenaceous sediments are present in the southern part of the basin, which might have been transported towards north, by long shore currents, and being deposited as a narrow clastic belt on the outer shelf. Neogene over a large part of the basin is a claystone.

The peculiarity of Kerala basin is that, there is no present day river, large enough to transport the sediments to the basin from the nearby land mass. Indus cone sedimentation is restricted to the northern part of the Arabian sea to the west of Laccadiv ridge, and no influence on the sedimentation to the east of Laccadiv ridge. The sedimentation brought by Narmada and Tapti rivers into Gulf of Cambay filled the DAHANU & SURAT depression and could not reach this basin. Therefore, the Kerala offshore basin must have received the sediments, derived from the peninsular Indian Archean gneisses, schists and granites transported by the early tertiary river system. It was only in the Late Miocene the western coast of India was uplifted, causing the reversal of drainage to the east.

(b) **Geological Evolution:** The available stratigraphic data, regional tectonic setting of the basin and the seismic data had been utilized for visualizing the basin evolution.

The N-S aligned Laccadiv ridge zone and the shelf edge zone were prominently involved in the evolution of the basin.

1. The upper cretaceous through Paleocene represent the period of continental uplift, followed by volcanism and continental to near shore sedimentation. The K-1-1 well indicates that there are several Basaltic flows on the shelf with intertrappean coarse sandstones and reddish brown clays.

   In DSDP-219, the upper Paleocene is essentially composed of greenish grey glauconitic limestone with greenish black rhyolitic tuffs. Probably, the Laccadiv ridge might be nearer to the western coastal belt during this period.

2. Early to Middle Eocene is represented by rapid down warping along N-S lineaments leading to basinal condition of deposition.

   The Kerala-Laccadiv depression might have formed during this period, along the N-S lineaments of Laccadiv ridge and Kerala shelf edge.

   In DSDP-219, the Middle Eocene is represented by micarbooze, chalky limestone, chert bands and clay, indicating an increase in the bathymetry of deposition during this period.

3. Late Eocene-Early Oligocene represents a period of uplift, leading to regression in parts of the basin.

   In DSDP-219, the interval consists of white foram rich micrab and the nanno ooze and chalk with radiolarian.

   Reefoidal development is envisaged in the more uplifted parts of the basin, like Laccadiv ridge, as well as the shelf edge.

4. Late Oligocene to Middle Miocene represents a period of seaward tilt of the continental margin towards the Arabian Sea. In the well K-1-1, the section is represented mainly by continental to lagoonal sands with lignite intercalations.

5. Middle Miocene to recent is represented mainly by Terriginous influx, coarser towards southern part but finer towards the northern part of the basin. The shelf had prograded significantly during this period.

(c) **Geochemical Studies:** Geochemical data is available only for the wells K-1-1 and Karwar-1.

The Geochemical source rock evaluation has taken into consideration the aspects of (a) organic carbon concentration and (b) the nature and maturation of organic matter.

Organic carbon concentration

In general the quantity of organic matter in most of the analysed samples of K-1-1 is fairly high. In the Burdigalian clays and carbonates the organic carbon content is about 0.85 to 1.82%. The upper Palaeocene and Eocene consist of sandstones with intervening lignite bands and the organic carbon content lies between 16 to 51%. In the lower Paleocene sequence the organic matter is 0.5 to 1.5%. In the interval 840-1150m the organic matter from the intertrappean section varies between 0.39 - 6.72%

In Karwar well-1 the Late Miocene clays are having organic matter 0.9-1.40%, and the Mid Miocene Limestones from 0.35% to traces.

Nature of organic matter

Two samples each from Burdiglian and Eocene of K-1-1 are found moderately rich in lignites and vitrinites. H/C ratio from elemental analysis varies from 1.12 to 1.14.

Nature of Bitumen

The Bitumen in the samples of K-1-1 contains 30-60% resins and 10-30% asphaltines which are characteristic of humic-sapropelic to sapropelic organic matter.

Maturity

Vitirnite reflectances (Ro) of some samples were measured. The Ro values range from 0.31 to 0.46 indicating that the entire drilled sequence in K-1-1 is immature. A maturation profile based on this data indicates that reflectance of 0.50 would be reached at a depth of 2000 m. (In Kerala offshore shelf) with the oil window extending up to 3900m.

The visual kerogen analysis of the samples of Karwar well-1 also shows that the entire drilled sequence is immature.

As per the thermal maturation studies of the sediments in the Kasargod-Calicut area of the Kerala offshore basin, worked out using Lopatin's time temperature index (TTI), the Neogene and Oligocene sediments are immature. The Palaeocene sediments have gone through the full phase of oil generation in the deepest portions of the Kasergod-Calicut area, and are in the stage of wet gas generation. The Paleocene sediments buried in the depth range of 2450-3650m are located within the oil generation window.

(d) Hydrocarbon Prospect

1. The hydrocarbon prospect zone is demarcated in the offshore part of the basin, comprising the deeper parts of the shelf, and the shelf edge zone, and the zone between the Miocene hinge line and the present day shelf (i.e Konkon and Kasargod areas) towards east and the Laccadiv uplift zone on the west.
2. The Kerala-Laccadiv depression contains sediments up to 1500-2000m and is likely to serve as the source area. The hydrocarbons generated, might migrate towards east and west and get entrapped in the structures over the shelf edge zone and deeper parts of the shelf on the east and Laccadiv uplift on the west.
3. The area between the present day shelf and Paleo-Shelf may contain carbonates in the Eocene through Early Miocene and coarse clastics in Middle Miocene, may act as good reservoir rocks.
4. Hydrocarbon prospects of Laccadiv uplift zone appears to be good. Good thickness of sediment of the order of 2000-2500m is expected in the area outlying the Eikalponi and Kalponi islands.
5. Vengrula arch with the associated fracture zone tectonically controls/separates the hydro-dynamics and the hydrocarbon migration to its north (Bombay High, Ratnagiri area) and south (Kerala-Laccadiv basin).
6. Indus cone sedimentation is restricted to the west of Laccadiv Ridge and a large pile of Neogene can be expected in deep water basin to the west of Laccadiv Ridge, where the stratigraphic play for hydrocarbons can be expected in the submarine complex / turbidite fan complex regimes.

(e) Konkan Basin (Block-III A): Twenty one structures are mapped in this area, of which five structures are confirmed. As mentioned earlier, drilling had been carried out in only one structure, i.e Karwar-1. This well had to be terminated at 1678m due to drilling problems. The closure indicated was within Eocene and the well could not reach to that level (Fig. 5.5.2).

(a) The well was drilled under total mud loss conditions. Hydrocarbons cannot show up under such huge loss conditions, even if present in the formations.

(b) For such wells, proper casing policy has to be evolved (i.e the casing has to be set just above the loss zones) and each loss zone should be tested immediately after the formation is opened, by conducting open- hole DST keeping the packer in the casing. If the well is to be tested, after reaching the target depth, to activate the loss formation and get anything out of it, is next to impossible.

(c) The zone between present day shelf and the Miocene hinge line in Karwar is likely to contain good mature source sequence in the Tertiary rocks. Reservoir development is likely in the porous and vuggy limestone/reefs. The broad shelf, lying on either side of the fringing shelf ridge also attracts attention.

(f) Kasargod Block (Block IIIB)

(a) In this block, on the basis of seismic interpretation about 22 structures are identified. On one of the structures, Kasargod-1 well was drilled to 2970m against a projected depth of 3200m.

(b) This well also encountered total mud loss conditions between 940-1304m. Several cement plugs were to be placed for continuing drilling. Inspite of this, minor gas shows were observed in the section 944-1304m. As mentioned earlier, under total mud loss conditions hydrocarbons cannot show up, even if they are present. Proper casing policy and open -hole DST programmes have to be chalked out carefully, for such wells.

(c) Geochemical analysis of several samples showed the presence of signigicant amount of organic matter, but lacking in maturity. The studies also indicated the presence of migratory bitumen, suggesting the presence of oil prone source rocks in the deeper parts of the basin.

(d) The zone between the present day shelf and the Miocene hinge and also the broad shelf of Kasargod area is likely to create interest for hydrocarbon prospect. Reservoir development is likely in porous vuggy limestone/reefoidal limestones of early to late Tertiary period.

(g) Kerala offshore block (Block III C)

(a) About 15 structural features have been identified in this block based on the seismic interpretation.

(b) One deep well, Cochin-1 (K-1-1) was drilled to a depth of 1755m bottomed in basalt layer. In Paleocene, this well drilled clastic sequence, as against a carbonate section in other two wells.

In the Cochin prospect area, a prominent unconformity between Paleogene and Neogene sequence is mapped on seismic sections, with the development of pinch out features. So, information of primary stratigraphic and unconformity traps can be expected in this shelf Margin zone.

A significant seismic reversal is obtained on the profile TM-4 in the offshore area around Alleppy and the area may contain comparatively thicker sedimentary column of 3000m. The shelf margin in this area is also quite wide.

(h) Lakshadweep Block (Block III D)

(a) Lakshadweep islands are situated at about 100 to 200 miles away from the Kerala Coast. There are altogether 19 islands. Androth is the nearest and Agati is the farthest from the Kerala Coast (situated within the latitudes 8° to 12° 30′ N and longitudes 70° to 74° E) (Fig. 5.5.6 & 5.5.7)

(b) The most prospective zone/area for exploration is that of the Laccadiv regional high.

(c) This is a tectonically active zone/area, with which oil and gas pools may be associated.

(d) In the east, it is bounded by Laccadiv-Kerala inter cratonic graben, where in about

(e) 1500-2000m thick sediments are expected. The hydrocarbons generated might migrate towards east and west and get entrapped in the structures over the shelf edge zone to the east and Laccadiv uplift on the west.

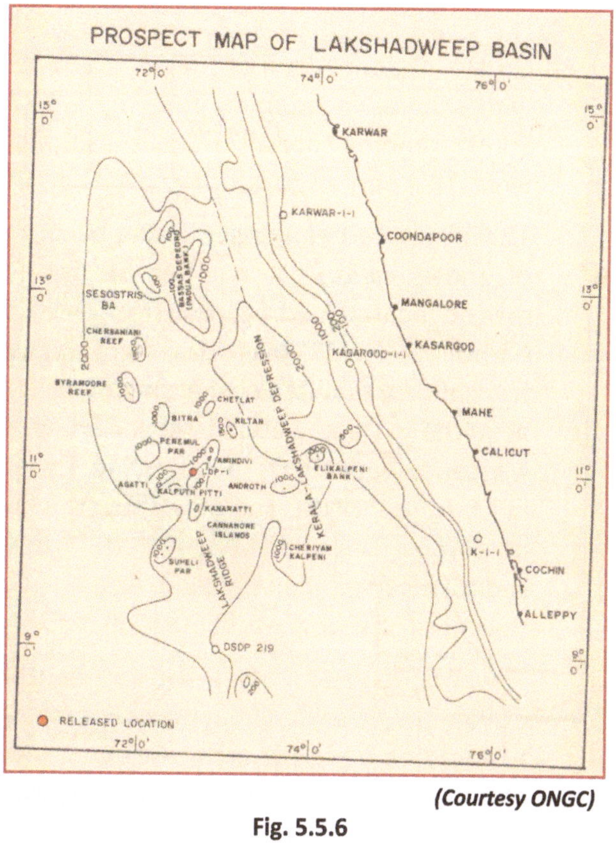

(Courtesy ONGC)

Fig. 5.5.6

(f) The maximum sedimentary thickness in the Lakshdeep islands is expected to be around 2500m. There is no shelf surrounding the islands in this area. The islands are generally aligned North - South. Each island has a very shallow lagoon towards its west (except Andtoth)

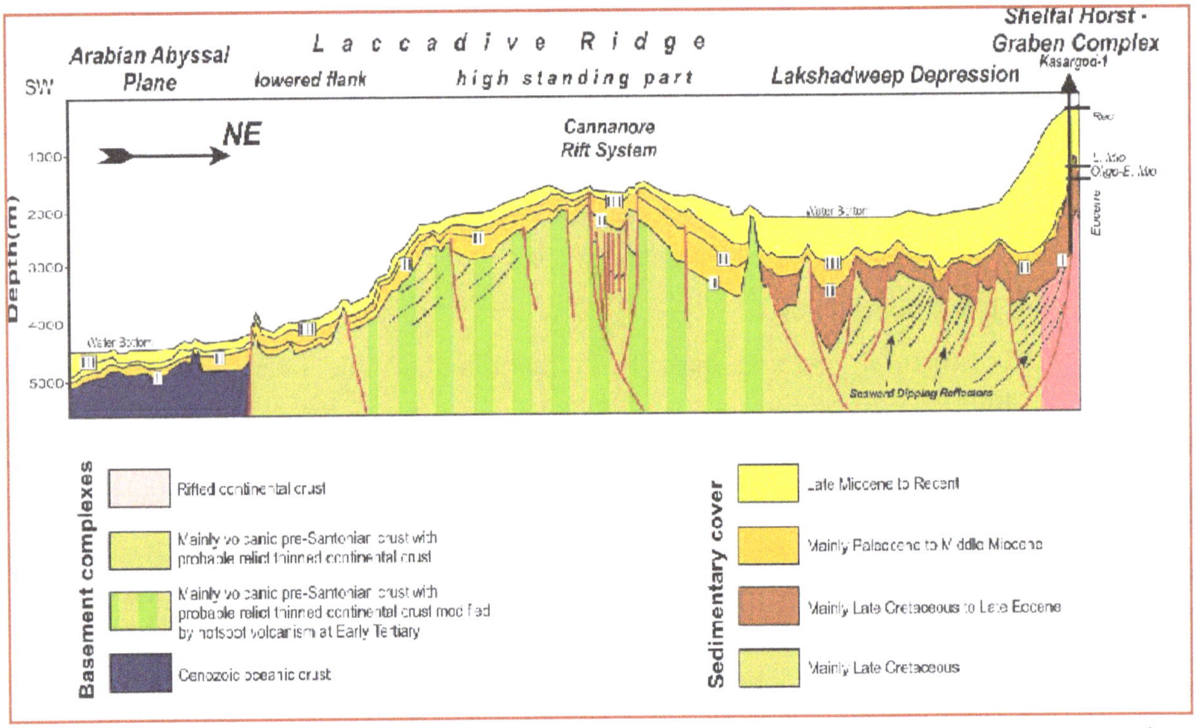

(Courtesy ONGC)

Fig. 5.5.7

On the regional profiles (RE 11 & 15), it was brought out clearly the existence of a "Fracture zone" with a possible orientation in the ENE-WSW direction and down thrown to the north. The North-South trending Laccadiv ridge is also offsetted by the fault, with the northern part offsetted towards west. This part also needs testing by drilling, after obtaining some more information.

12 wells have been drilled in the basin. Significant hydrocarbon shows have been observed in some of these wells. A number of positive absorbed gas anomalies have been identified. The prognosticated hydrocarbon resources are of the order of 660 MMT. Carbonate build-ups and Carbonate Banks over basement highs in the shelf margin, and Cretaceous wedge outs, and truncations against Eocene unconformity are the principal exploration targets. Recent seismic data acquired in deep waters revealed some attractive hydrocarbon play types like Carbonate build-ups in Miocene, wedge outs in Paleocene, Eocene channel sands in Eocene Oligocene, and Slump structures in Miocene. ONGC and other private players drilled to date, 17 exploratory wells in the basin. So far no commercial find is discovered. The surveys of Lady Glorita 1973 indicated that the thickness of sediments over shelf between Goa and Cochin is of the order of 3000 m. On the shelf edge, a sudden thickness of sediments 3500-4000 m is observed in a progradational sequence.

Towards NNW of Karwar, the sedimentary cover decreases with the basement level coming to 50 m at Panjim. Over Panjim, an east-west ridge like feature is noticed in the Basal sedimentary structure, referred as Vengrula arch. Regional profiles RE-11 and RE-13 indicated a clear fractured zone, the extension of which, on the land towards ENE, limits the Deccan trap activity. To the North of this Vengurla arch in the offshore of Ratnagiri, the thickness of sediments abruptly increases. Probably this fractured zone appears to have played tectonically a vital and dominant role, restricting the hydrocarbon migration and accumulation to the North of Vengurla arch. Probably, the migration could not reach to the south of Vengurla arch into the Kerala-Konkan basin.

Another important feature is that, in the south of the basin, the Paleo shelf lies to the east of Miocene hinge which appears to have a bearing on the Neogene tectonics and hydrocarbon prospect. The peculiarity of Kerala basin is that, there is no present day river large enough to transport the sediments to the basin from the nearby landmass.

Indus cone sedimentation is restricted to the Northern part of the Arabian Sea to the west of Laccadive Ridge and could not influence sedimentation to the east of Laccadive Ridge. The sediments brought by Narmada and Tapti rivers into the Gulf of Cambay filled the Dahanu and Surat depression and could not reach this basin due to the uplift of southern block of Vengrula arch/fault.

Data on Kerala-Konkan Offshore Basin

Basin Area	–	5, 80, 000 sq.kms.
Basin Type	–	Passive continental margin
Sediments ranging from	–	Cretaceous to Recent
2D Seismic data	–	1, 54, 974 LKM
3D Seismic data	–	713 sq.kms.
Exploratory wells drilled	–	17
Oil/gas discoveries	–	Nil
Hydrocarbon shows	–	Present

Exploration & Production Status

BASIN AREA : 580,000 Sq. Km.

SEISMIC DATA : 2D-154,974 LKM

: 3D-9730 Sq. Km.

Data acquired by	2D (LKM)	3D (Sq.km)
DG H	10,208	-
ONGC	115012	5490
Pvt./JV	29744	4239

Other Surveys Gravity Survey : 44607 Stations

Magnetic Surveys : 73410 Stations

Satellite Gravity Survey

Well Data Number O f Exploratory Wells : 17

Hydrocarbon Shows/discoveries : Gas shows in Miocene Carbonates and sandstones in 3 wells and in Cretaceous sand stones in 1 well

Basin Type : Passive continental margin basin

Hydrocarbon Resource Potential: **Onland** **: Not Applicable**

Offshore **: 280MMT**

Hydrocarbon Reserves : **Nil**

Production : **Nil**

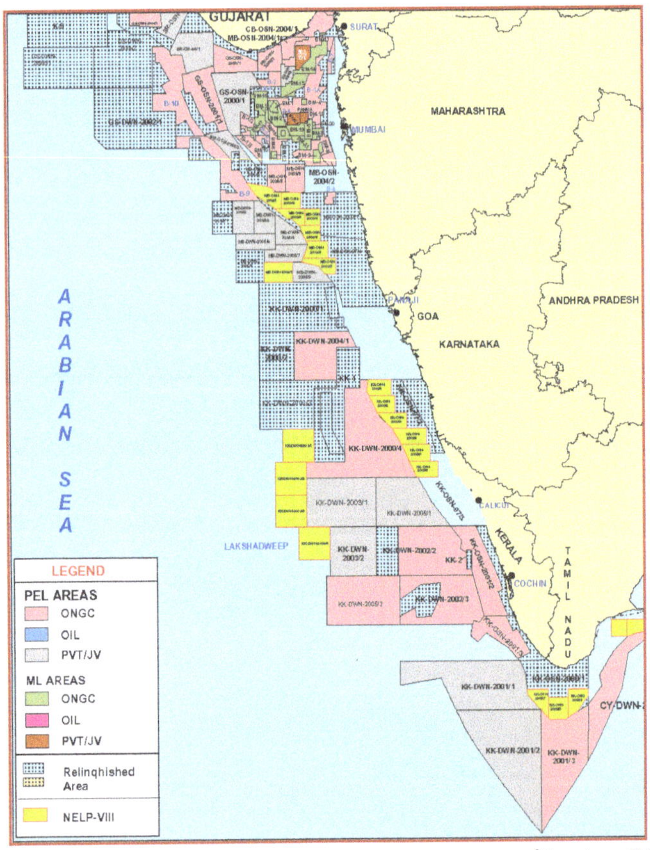

(Courtesy DGH)

Fig. 5.5.8

Blocks on Offer (NELP I to VIII)

The blocks offered and awarded under production sharing contract from NELP I to VIII are shown in the Table 5.5.9.

Table 5.5.9

NELP	Blocks Offered	Blocks Awarded	Awardees	Date of Signing/ Effective Date	Present area (sq.km)	Status
I		KK-OSN-97/2	RIL (100%-OP)	12-Apr-2000/		Relinquished
		KK-OSN-97/3	ONGC(100%-OP)			Relinquished
II		KK-DWN-2000/1	RIL (100%-OP)	17-Jul-2001		Relinquished
		KK-DWN-2000/2	ONGC (85%-OP), GAIL (15%)			Relinquished
		KK-DWN-2000/3	RIL (100%-OP)	17-Jul-2001		Relinquished
		KK-DWN-2000/4	ONGC (100%-OP)	17-Jul-2001		Relinquished
		KK-OSN-2000/1	ONGC (100%-OP)			Relinquished
III	KK-DWN-2001/1	KK-DWN-2001/1	RIL (100%-OP)	04-Feb-2003/ 03-Apr-2003	27315	
	KK-DWN-2001/2	KK-DWN-2001/2	RIL (100%-OP)	04-Feb-2003/ 03-Apr-2003	31515	
	KK-DWN-2001/3	KK-DWN-2001/3	ONGC (100%-OP)	04-Feb-2003/ 12-Mar-2003	21775	
	KK-OSN-2001/1					
	KK-OSN-2001/2	KK-OSN-2001/2	ONGC (100%-OP)	04-Feb-2003/ 12-Mar-2003	10588	
	KK-OSN-2001/3	KK-OSN-2001/3	ONGC (100%-OP)	04-Feb-2003/ 12-Mar-2003	6446	
IV	KK-DWN-2002/1	KK-DWN-2002/1	ONGC (70%-OP), OIL (20%), BPCL (10%)	06-Feb-2004/	20910	
	KK-DWN-2002/2	KK-DWN-2002/2	ONGC (80%-OP), HPCL (20%)	06-Feb-2004/ 17-Mar-2004	22810	
	KK-DWN-2002/3	KK-DWN-2002/3	ONGC (80%-OP), HPCL (20%)	06-Feb-2004/ 17-Mar-2004	20910	
V	KK-DWN-2003/1	KK-DWN-2003/1	RIL (100%-OP)	23-Sep-2005/ 23-Jan-2006	18245	
	KK-DWN-2003/2	KK-DWN-2003/2	RIL (100%-OP)	23-Sep-2005/ 05-Dec-2005	12285	
VI	KK-DWN-2004/1	KK-DWN-2004/1	ONGC (45%-OP), CIL (40%), TATA (15%)	02-Mar-2007/ 09-May-2007	12324	
	KK-DWN-2004/2	KK-DWN-2004/2				
	KK-DWN-2004/3	KK-DWN-2004/3				

Table 5.5.9 *Contd...*

NELP	Blocks Offered	Blocks Awarded	Awardees	Date of Signing/ Effective Date	Present area (sq.km)	Status
VII	KK-DWN-2005/1	KK-DWN-2005/1	BHP Billiton Petroleum International Pty.Ltd (26%-OP) GVK Oil and Gas Ltd. (74%)	22-Dec-2008/	14675	
	KK-DWN-2005/2	KG-DWN-2005/2	ONGC (90%-OP), GSPC (10%)	22-Dec-2008/	19234	
VIII	KK-DWN-2009/1					
	KK-DWN-2009/2					
	KK-DWN-2009/3					
VIII	KK-DWN-2009/4					
	KK-OSN-2009/1					
	KK-OSN-2009/2					
	KK-OSN-2009/3					
	KK-OSN-2009/4					
	KK-OSN-2009/5					
	KK-OSN-2009/6					
	KK-OSN-2009/7					
	KK-OSN-2009/8					
	KK-OSN-2009/9					

(Courtecy DDH)

CHAPTER - 6

THE EAST-COAST OFFSHORE SEDIMENTARY BASINS

The East Coast offshore from Cape Comorin in South to West Bengal in North has been divided into the following basins (Fig. 6.1)

1. Cauvery Basin – Mannar Sub-basin
   Ramnad Palk Bay Sub-basin
   Cauvery Basin
2. Palar Basin
3. Krishna-Godavari Basin
4. Mahanadi and North East Coast (NEC) Basins
5. Bengal Basin
6. Andaman and Nicobar Basin

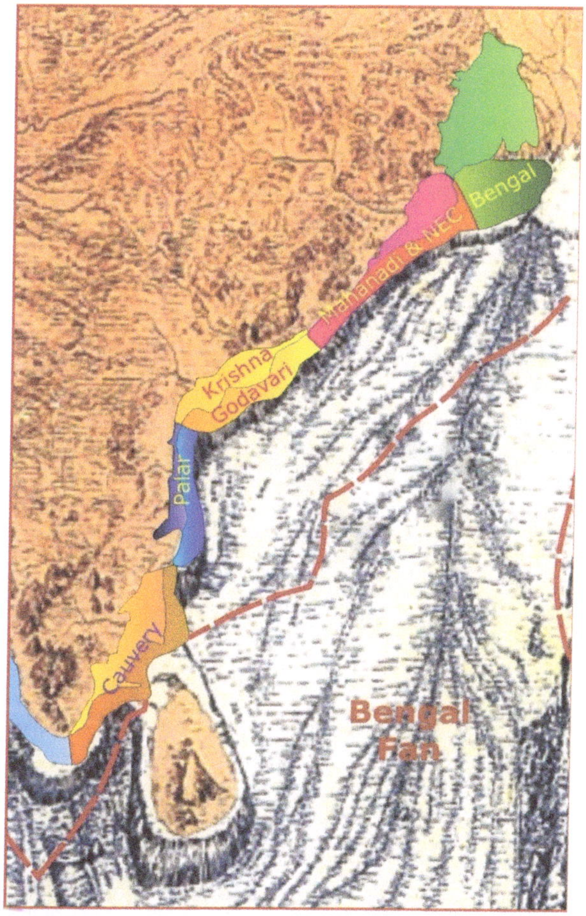

Fig. 6.1

6.1 Cauvery Basin

Introduction

The Cauvery basin which includes the Mannar Sub-basin and Ramnad-Palk Bay sub basins covers an area of 55000 Sq.Kms, of which 25000 Sq.Kms is onland and 30,000 Sq.Kms is in offshore up to 200 m isobaths, and in addition, there is about 40000 Sq.Kms of deep water offshore areas. The prognosticated resources of hydrocarbons are placed at 700 MMT (430 MMT for onland and 270 MMT for offshore). 5000 to 6000 m of sediments ranging in age from Late Jurassic to Recent are expected in the basin. Number of oil and gas fields have been discovered both onshore and shallow waters of the offshore. The most important discoveries are

Onland	Offshore
Narimanam	PY-1
Nannilam	PY-3
Adiyakamangalam	PH-9
Kovil Kalappal	
Bhuvanagiri	
Thiruvarur	
Tulsapatnam	
Kutralam	

Perungulam-Periapattanam

Kamalapuram

Pallivaramangalam

The Cauvery basin occupies most of the coastal plains of Tamil Nadu and Pondicherry and extends into the offshore of Bay of Bengal. The Southern part of the basin extends across Palk Bay and Gulf of Mannar into the Western Coastal area of Sri Lanka. (Fig. 6.1.1) The Gulf of Mannar on the southern part of East Coast is well connected to the Allepy platform (Kerala-Konkan basin) on the West Coast through the open sea of Indian Ocean. The geological setting of the Mannar Gulf is almost similar with the Allepy platform on the West Coast. Open marine conditions prevail in this part of the basin. Cauvery basin is the Pericratonic passive margin basin on the East Coast of India. The major tectonic units of the Cauvery basin are parallel to the NE-SW Eastern ghat trend and the basin is characterised with horst and graben morphology.

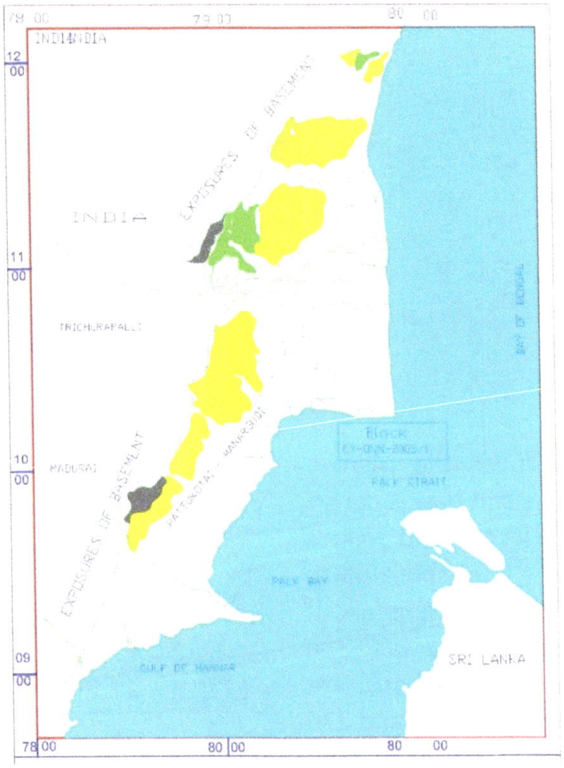

(Courtesy DGH)

Fig. 6.1.1

The Cauvery basin evolved during the break-up of Gondwana land and underwent different phases of deformation in its geological history. These can be referred as

1. Late Jurassic / Early Cretaceous Syn Rift extension
2. Late Cretaceous Rift followed by Early drift stage
3. Late Cretaceous Paleogene Drift stage and
4. Neogene passive margin subsidence.

The basin shows alternating NE-SW trending narrow, enechelon basement ridges and wide troughs.

Stratigraphy

Sediments over 6000 m thick are present in the basin ranging in age from Permian to Recent. However, the sedimentary rocks ranging in age from Late Jurassic to Recent are exposed along the Western margin of the basin fringing the Pre-Cambrian metamorphics. These out crops were mapped by ONGC.

The detailed stratigraphy of the basin is shown in Fig. 6.1.2. Seismo stratigraphic sequence and facies analysis guided by well data is the most suitable method to analyze the stratigraphic details, depositional system tracks and litho facies distribution (Figs. 6.1.3, 6.1.4, 6.1.5 and 6.1.6).

The sedimentary section of the Cauvery basin is divisible into seven stratigraphic units, each bounded by Hiatus, and they are:

1. Late Jurassic to Early Cretaceous Unit
2. Late Cretaceous (Pre companion Unit)
3. Late Cretaceous (Post Companion Unit)
4. Paleocene-Eocene Unit
5. Oligocene-Early Miocene Unit
6. Early Miocene Unit
7. Post Early Moicene Unit

The oldest exposed sediments are coarse gritty and pebbly sandstones on basin margins and are known as Siva Ganga beds of Upper Jurassic–Cretaceous age. The Lower Cretaceous rocks consisting of Reefoidal lime stone overlying black shale are exposed in thin patches. The Upper Cretaceous outcrops are divided into three formations namely, Utattur, Tiruchinapally and Ariyalur formations. The Tertiary rocks in the outcrop area consist of Ocherous sandstone, limestone of Paleocene age and Cuddalore sandstones of Miocene-Pliocene age. In the subsurface stratigraphy, five distinct major unconformities could be mapped.

1. Albian unconformity between Andamadam formation and Sattapadi shale (Late Cretaceous age) – Uttatur group.
2. Turonian unconformity between Bhuvanagiri formation and Kudavasal shale (Late Cretaceous age)
3. Maastrichtian unconformity between Portonovo shale (Late Cretaceous) and Kamalapuram formation (Paleocene-Eocene)
4. An unconformity between Tirupundi formation (Eocene) and Niravi sandstone (Oligocene)
5. The unconformity between Madanam limestone (Lower Miocene)and Cuddalore formation (Middle Miocene)

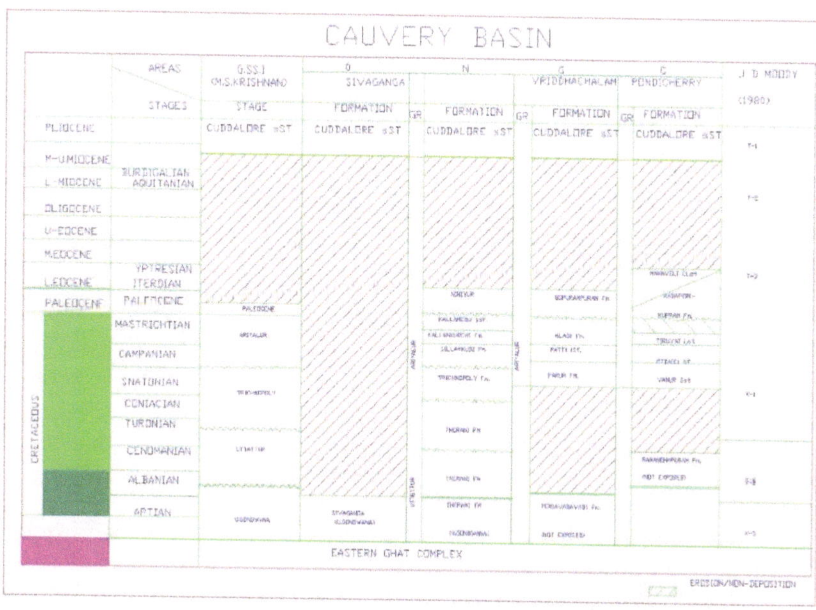

(Courtesy ONGC)

Fig. 6.1.2

Fig. 6.1.3 Time Structure Map of Basement

Fig. 6.1.4 Isopach Map of Upper Cretaceous

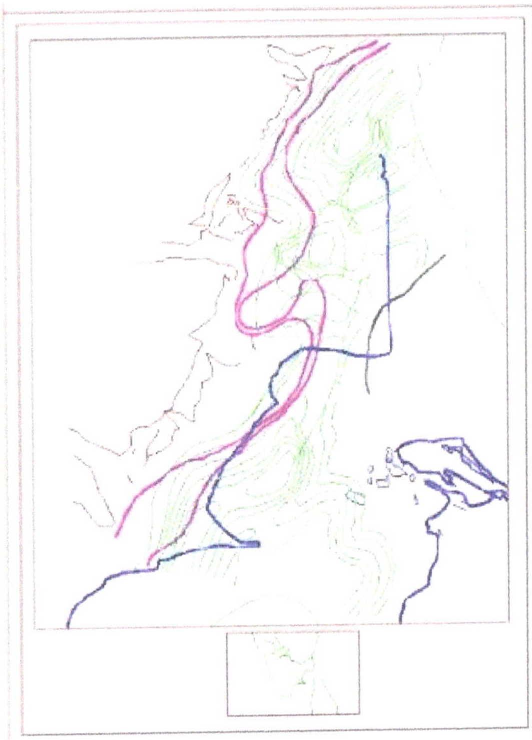

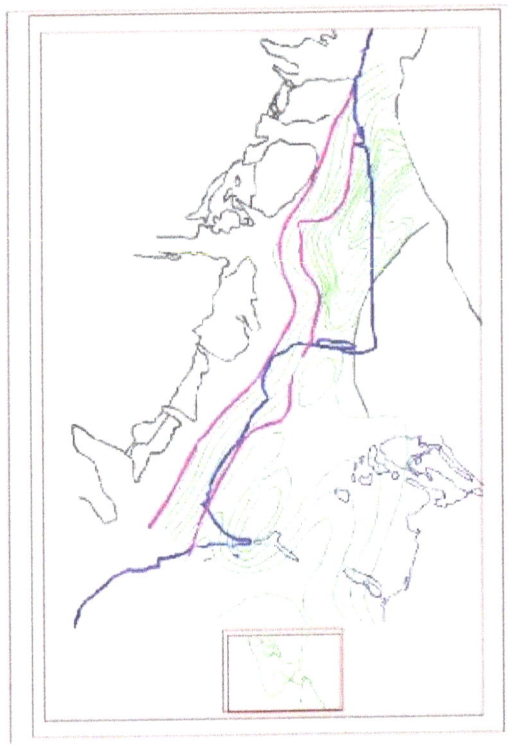

(Courtesy ONGC)

Fig. 6.1.5 Isopach Map of Paleocene -Eocene

Fig. 6.1.6 Isopach Map of Post Eocene

The Cretaceous sediments occur in all depressions and are thin over basement highs / ridges.

Tectonic Set-Up

Geo-scientific data based on Bouguer gravity, Residual gravity and Magnetic anomalies (Figs. 6.1.7, 6.1.8 & 6.1.9) suggest that the Pattukotai-Mannargudi ridge, extending into offshore beyond Karaikal and merging with the northerly prolongation of the Sri Lanka massif, divides the Cauvery basin, essentially into two parts.

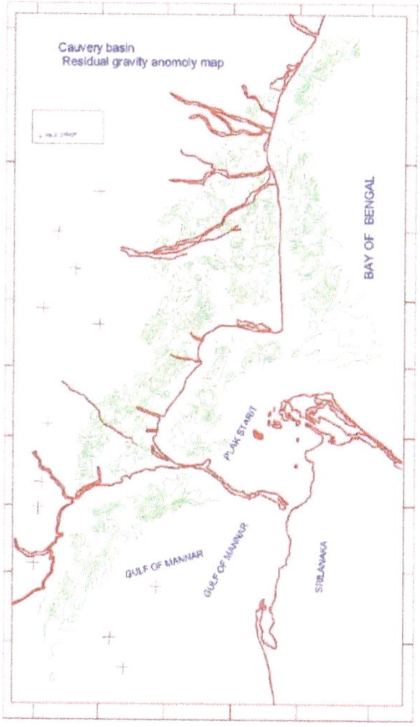

Fig. 6.1.7

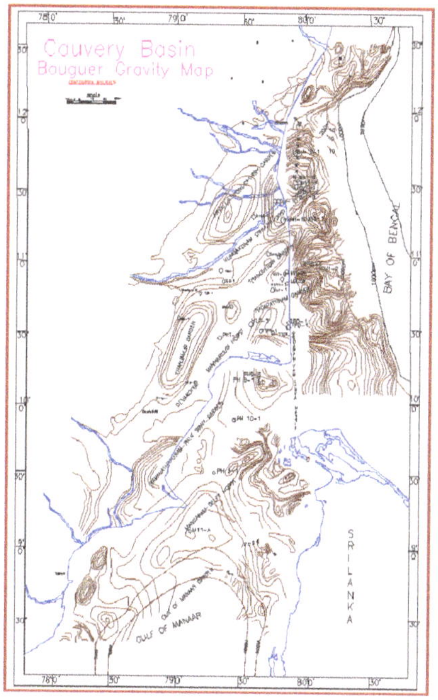

Fig. 6.1.8

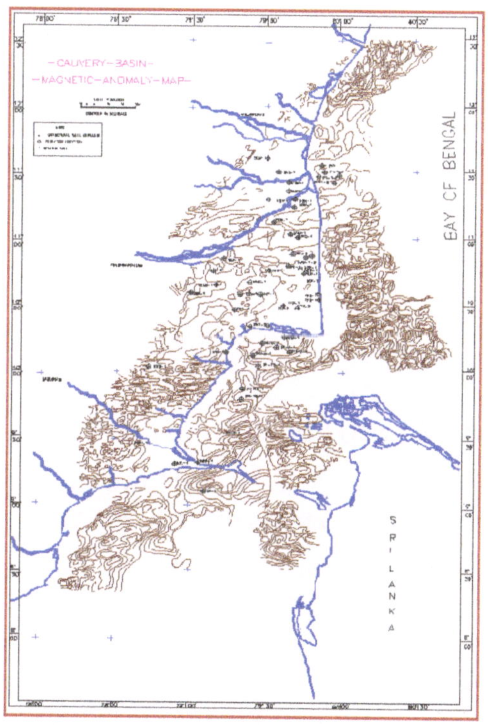

Fig. 6.1.9

(Courtesy DGH/ONGC)

1. Ariyalur-Pondicherry-Tanjore-Tranquebar to the North and
2. Nagapattinam-Palk bay-Ramnad-Mannar sub basin to the South.

The Pondicherry-Tranquebar basin opens up Northeast into the offshore where as the Nagapattinam depression is terminated towards east by merging of Karaikal high and Sri Lankan massif. The Mannar basin extends and widens towards south into Indian Ocean and merges with Allepy platform of West Coast. The Palk Bay-Mannar basin has been described as a 'pull –apart' basin formed by the rifting apart of the Indian and Sri Lanka Cratons with the Karaikal high as the fulcrum during the Cretaceous time and gradual sinking of the basin floor. Basalt flows in the deeper parts of the basin during rifting was evidenced from the Mannar well. However, the Cauvery basin which is crossed by horst and graben tectonics can be divided into a number of sub-basins / grabens. These are from North to South as (Fig. 6.1.10)

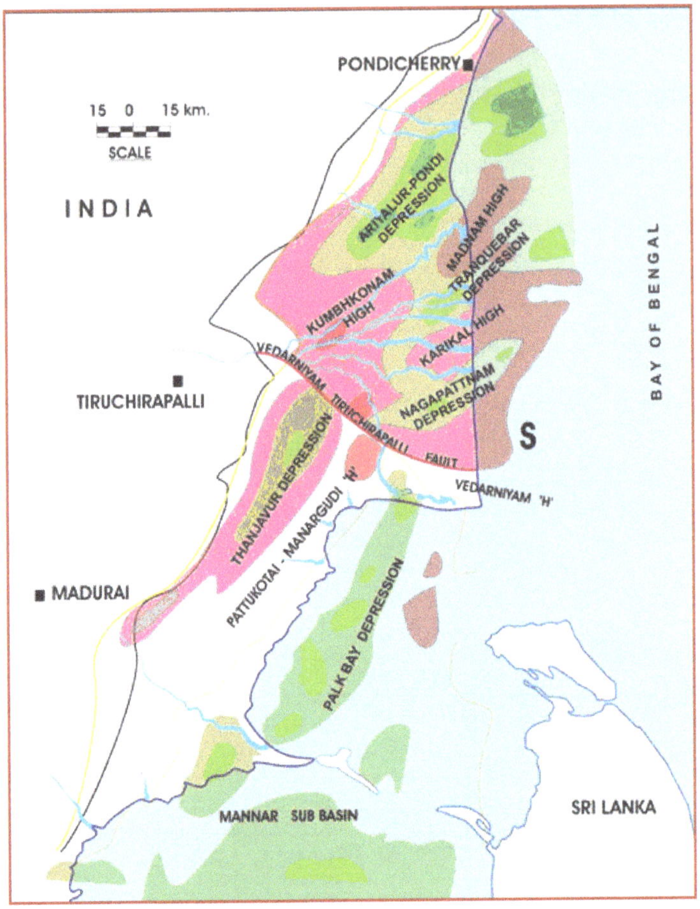

(Courtesy ONGC)

Fig. 6.1.10

- Ariyalur-Pondicherry Graben
- Kumbakonam Shiyali Ridge
- Tranquebar Depression
- Karaikal Ridge
- Tanjore-Nagapattinam Depression
- Vedaranyam-Sri Lanka High
- Palk Bay-Ramnad
- Mannargudi Devakottai Ridge
- Mandapam Delft
- Gulf of Mannar

Geochemistry and Hydro Geology

Geochemical studies were carried out by ONGC in detail over a number of samples collected from different wells and attributed to the various sub-basins (Table 6.1.1).

Organic Carbon content varies widely and rapidly, though the average content is more than 1% which is fair enough for a source rock to generate hydrocarbons. The type of organic matter is humic in Palk Bay-Tanjore and Gulf of Mannar grabens (Cretaceous section), and mixed in Nagapattinam graben and Karaikal horst (Cretaceous-Eocene).

In Tranquebar graben the organic content is marine (Cretaceous), whereas In Ariyalur-Pondicherry graben, it is marine to mixed. In Madanam, it is Humic.

Vitrinite reflectance values show that Cretaceous sediments are matured except in Gandharvakottai, which is quite near to the basin margin. The source rock potential of the Cretaceous sediments is generally fair, except in PH-9-1 and PY-1-3, where it is poor.

The vitrinite reflectance values for the Tertiary sediments rangefrom 0.41 in Madanam, 0.45 in Mayavaram to 0.64-0.92 in Karaikal. Organic matter is marine in Mayavaram, Pondicherry and Chidambaram but Humic in Madanam. The source rock potential is fair to good in most of the areas.

In general, in the Cretaceous sediments, about 1% organic matter of humic type is indicated to be matured enough to generate primarily gas. However, few organic matter samples suggest presence of wet gas and / or oil prone. The source potential indicates both oil and gas for the Eastern part of Palk Bay area. The carbonate facies of Cretaceous and Tertiary sediments off Vedaranyam well offer better source potential for generation of hydrocarbons. The Tertiary sediments which show maturity in Karaikal wells are likely to be potential source rock.

Thus, it can be broadly summed up that the source rocks of Cretaceous and Early Paleogene age are both oil as well as gas prone, as is evidenced in the wells of Pondicherry, Karaikal, Narimanam and Palk Bay.

Figs. 6.1.11, 6.1.12 & 6.1.13 show the organic carbon distribution in the Cretaceous and the Tertiary sediments and iso-maturation contours for the Cretaceous section respectively. The salient feature of these figures is that, the area wise identified objectives / hydrocarbon plays fall within the favourable areas, indicated by geochemical parameters.

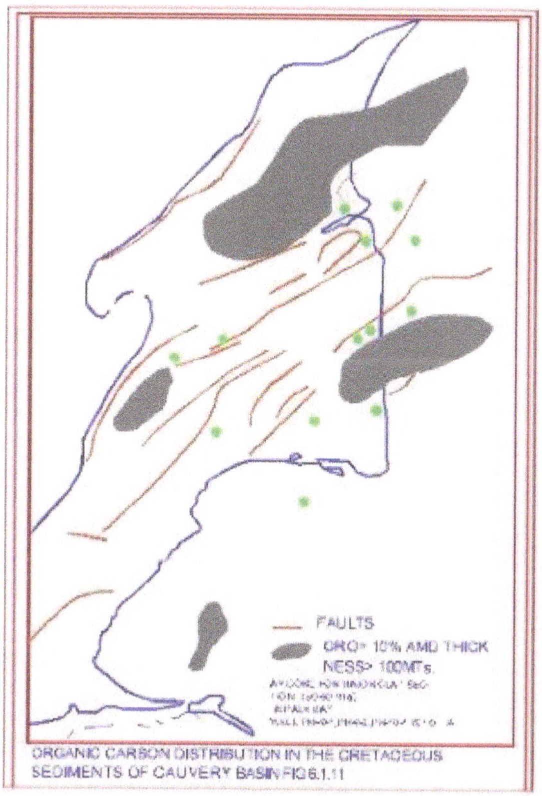

Fig. 6.1.11

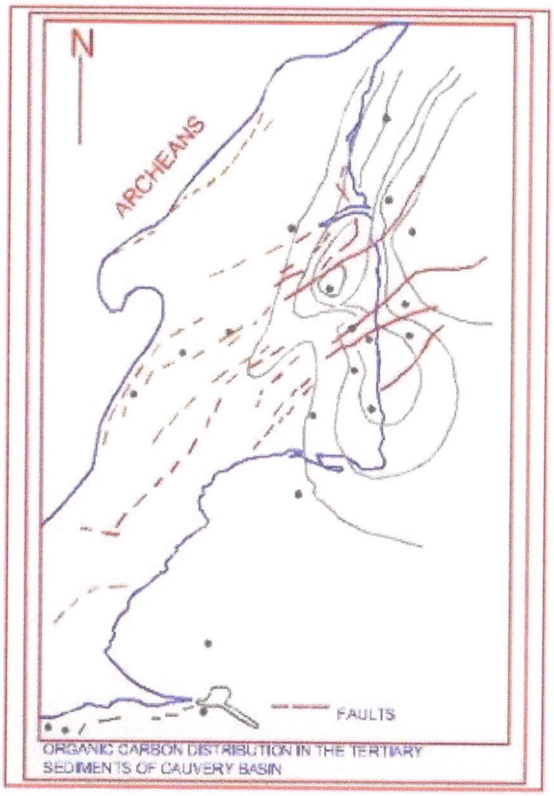

(Courtesy ONGC)

Fig. 6.1.12

(Courtesy ONGC)

Fig. 6.1.13

Hydro Geology

Hydro geology is an additional geological tool for exploration of oil and gas in a basin. Based on the general hydro geological indices, litho-facies data, and study of organic matters, two broad hydro geological systems are established for the basin.

1. The Mesozoic Hydrogeological system (Fig. 6.1.14).

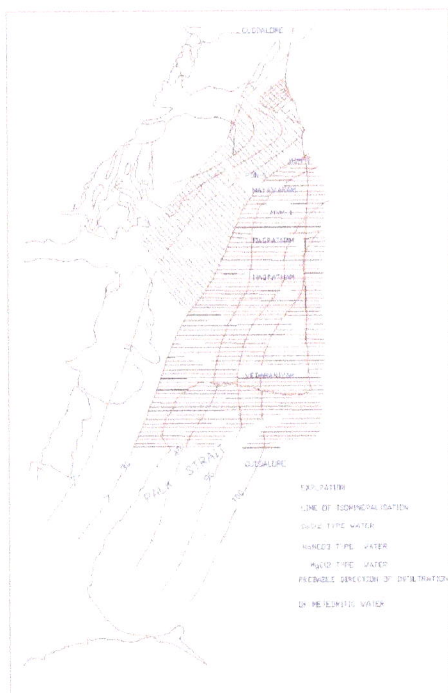

(Courtesy ONGC)

Fig. 6.1.14 **Fig. 6.1.15**

2. The Tertiary Hydrogeological system (Fig. 6.1.15).

In both the systems, the hydro geological conditions improve considerably to the Southeast of Nagapattinam depression and also to the East of Tranquebar depression i.e. towards offshore. The mineralization of waters touched values of around 100 gms /litre near these places, depicting $Cacl_2$ type waters with less So_4 content and high degree of metamorphism of waters.

Data obtained from various drilled wells of the basin indicate that the subsurface waters with degree of mineralization varying from 1.8 to 26.0gms/litre are of Sodium Bicarbonate type, and the So_4 content varying between 2.5 to 12.5 MEQ. These types of waters occur from the Ariyalur-Pondicherry depression to the northern part of the Nagapattinam depression. It is worth mentioning that even at depths of 2956-2982 m in Chidambaram well the formation waters are of Sodium Bicarbonate type (5-15 gms/litre) with a So_4 content of 4.7 meq.

Considerable improvement in hydro geological conditions for entrapment in the Mesozoic sequence to the East and Southeast of Nagapattinam depression is observed, with the degree of mineralization of waters ranging from 71 to 100.6 gms/lit, corresponding to Calcium Chloride type and sufficiently metamorphosed, where Na/Cl = 0.86 and Cl-Na/Mg = 1.9, and So_4 content is 0.3-0.8 meq.

In the Tertiary system also the mineralization of waters range from 17.5 – 99 gms/lit i.e. Sodium Bicarbonate to Calcium Chloride type with a trend in increase of mineralization of waters and also their metamorphism (Na/Cl = 0.86, (Cl-Na)/mg=5.0) towards East and South East of Tranquebar and Nagapattinam depressions specially towards the offshore part of the basin.

The hydro geology studies suggest that

1. In both Mesozoic and Tertiary systems, the hydro geological conditions improve considerbly to the Southeast of Nagapattinam depression and also to the East of Karaikal ridge and Tranquebar i.e., towards offshore.
2. Ramnad Palk Strait and Nagapattinam depression appear to represent a single broad tectonic unit within the Cauvery basin thus dividing the basin into two sub-basins as discussed earlier.
3. The offshore Northern parts of the Jaffna - Vedaranyam High and the Eastern part of Ramnad - Palk Strait – Nagapattinam depression appear to be more prospective.
4. Hydro geological studies suggest for West and Northwest migration of hydrocarbons from the Easterly subsided parts of the large depressions having huge thickness of sediments. The infiltrated fresh waters move from west to east. There is an interaction between these two movements of water. Infiltration processes together with other factors substantially reduce the prospects in the western parts of the basin (especially in the Western parts of Ariyalur-Pondicherry and Tanjore depressions).
5. In RAMNAD – PALKBAY South sub basin, the hydrocarbon movement is towards East – North East, towards offshore.

Petroleum Systems and Hydrocarbon Habitat

All the elements for good petroleum system are available in the basin, namely the source, reservoir, migration and entrapment.

The geological evolution of the Cauvery basin presents favourable areas that comprise the requisite inputs for formation of hydrocarbon pools both during the rift and drift phases.

The Early rift phase sequence holds good hydrocarbon prospects due to favourable thermal history. The Western flanks of intrabasinal horsts (like the Western flank of Devokottai-Mannargudi horst, Mandapam-Delft horst, Kilpalvur-Jayamkondam-Srimushnam high trend) hold good prospects. Drape in rift fill sequence over local basement highs makes good structural prospects. As the basin experiences thermo tectonic subsidence, faulting against the uplap rift fills as well as truncated rift fill sequence by

Tertiary sediments can provide good hydrocarbon play. Besides these prospects, the fractured basement itself can accumulate hydrocarbons as is the case in Pondicherry high wells (PY).

Evaluation of the depositional systems during the early stage of drift phase (Late Cretaceous & Early Tertiary) indicates that the prospects are restricted to deeper water turbidite fans of the slope / basin transition which may include the sand filled canyons. This type of plays which can be referred as stratigraphic are expected in the deep waters beyond the shelf edge especially off Tranquebar depression. To the East of Tirupundi-Vedaranyam-Sri Lanka massif, the continuation of Cretaceous clastic sedimentation below the Bengal fan is doubtful. However, on the continental margin, the prospects like carbonate buildups, turbidites and distal marine fan deposits, wedge outs and on-laps play a prominent role both during Cretaceous and Tertiary periods.

The hydrocarbon prospects in the Tertiary sediments are to be expected in turbidite fans and delta front sands covered by transgressive shales. The prograding systems of delta / delta front / slope / turbidites were repeatedly eroded and again covered due to the fall and rise of Sea levels in different geological times.

Mobility of generated hydrocarbons becomes obvious from the hydrocarbon finds obtained so far. Oil / gas occurs in fractured basement Pondicherry High, Karaikal , in Late Cretaceous (Palk Bay) in Paleocene (Kovilkalappal), Early Eocene (Narimanam, Karaikal , Madanam) and in the Late Oligocene (Narimanam) sediments. The generated oil from the various depressions could migrate through the fair ways, such as faults / unconformities.

Time temperature indices calculated on the basis of subsidence history and geothermal gradients and corroborated by VRO data show that oil generation from Cretaceous sediments continued up to Eocene in Ariyalur-Pondicherry graben, Oligocene in Tranquebar graben and Miocene in Nagapattinam graben. A good understanding on the entrapment mechanism for oil / gas accumulation appears to be the main requisite in this basin especially when stratigraphic traps in combination with or without structural element are mainly expected.

Stratigraphic and combination type of traps have been recognized as important exploration targets. Combination type traps like clastic wedges against basin margin faults and stratigraphic traps like mounds of the sub-aquous fan systems are recognized in the basin as favourable targets. The hydrocarbon trap anomalies on the land part are essentially structural with some amount of stratigraphic play, where as the hydrocarbon plays in offshore shallow and as well as deep waters are mainly to be stratigraphic or at the most strati structural.

The oils of Cauvery basin are Paraffinic-Napthanic in composition. The API gravity is around $41\text{-}49^0$ with 5.4% wax and 0.5% asaphaltenes.

Petroleum System Elements

Source Rock

Detailed geochemical studies carried out indicate that each sub-basin is having source rocks to generate hydrocarbons and trapped in the nearby structures indicating a short distance migration. These basins have very good source rock with type-III Kerogene having very good source rock potential.

The post rift Cretaceous shales (Portonovo, Kudavasal, Sattapadi and Syn rift Andimadam formations) are the potential source rocks in different sub-basins. Maturation modeling indicated that Andimadam formation is having better source rock potential than the others with a Catagenetic stage of maturation with VRO of 0.5%.

Reservoir Rock

Oil and gas accumulations have been observed in the fractured basement Pondicherry High and in the sandstones of Cretaceous to Oligocene age. The main type of sandstone reservoirs are Bhuvanagiri formation in Pondicherry sub-basin, Nanilam formation in Tranquebar / Ramnad sub-basins, Niravi formation in Tanjore sub-basin. The good type of sandstone reservoirs are seen on the southern flank of Karaikal Ridge in fields like Narimanam, Adiyakamangalam, Kamalapuram & Tiruvarur.

Cap Rock

Each basin has its own cap rocks. Kudavasal shale in Ariyalur Pondicherry sub-basin, Komarakshi shale in Tranquebar and Ramnad sub-basins, and Shiyali clay stone in Nagapattinam form the cap rocks.

Entrapment

Oil and gas accumulations in the basin have been found both in structural as well as combination traps.

Migration and Accumulation

The onset of generation of hydrocarbons might have occurred at the end of rift phase itself (probably companion period). Peak generation could have occurred by Paleocene- Eocene. Post Cretaceous tectonics (drift stage) and the subsequent Neogene tectonics would have played in primary / secondary migration and entrapment in the structures as discussed earlier.

PY-1 Field

PY-1 field is situated 50 Kms south-southeast of Pondicherry town in Tamilnadu offshore, in a water depth of 50 m and it is a northeast extension of Kumbakonam, Madnam ridge. The structure is separated from Kumbakonam, Madnam ridge by two faults which develop a graben in between the two highs. The field was discovered by ONGC during 1980 and the field was offered for JV partners for further development in the small size field category. Presently, HOEC is the operator with 100% Participating Interest (PI).

(Courtesy ONGC)

Fig. 6.1.16

The Pondicherry structure NW-SE trending structural high at Eocene level and also dissected with a number of NW-SE oriented faults (Fig. 6.1.16).

The weathered basement composed of Horn Blende-Biotite Syenite is the main gas bearing reservoir. The Eocene limestone / sandstone is the subsidiary oil bearing reservoir.

The inplace gas reserves are estimated at 7.28 bcm of which 4.5 bcm is recoverable. The inplace oil is about 0.28 MMT and is not planned for development production immediately. The production has not yet started from the field but as per the profile given by the JV Company, 45 MMSCFD are to be produced for about 7.5 years. The composition of gas is mainly Methane (93%) and oil is having API gravity of 53.

PY-3 Field

Field is situated southeast of PY-1 structure and is about 18 Kms offshore from the coast. The field was discovered by ONGC in 1998 and the field was awarded to JV Company during 1991. The structure is located in the offshore part of Tranquebar depression (Fig. 6.1.17) and is very complex at the cretaceous level and is an elongated southwest oriented anticline in the offshore part. The reservoir comprises of moderate to low quality Cretaceous sandstone of Nannilam formation. The sands were deposited as debris flow deposits in a trough formed by basement high to the east and west of the field with lateral, significant variantion in the reservoir quality. The initial phase of the developmental activity commenced in 1996. The development comprises Sub Sea wells tied back to a floating production unit. Total 4 wells have been used to develop the field. Water injection and infill drilling is planned to maintain the reservoir pressures.

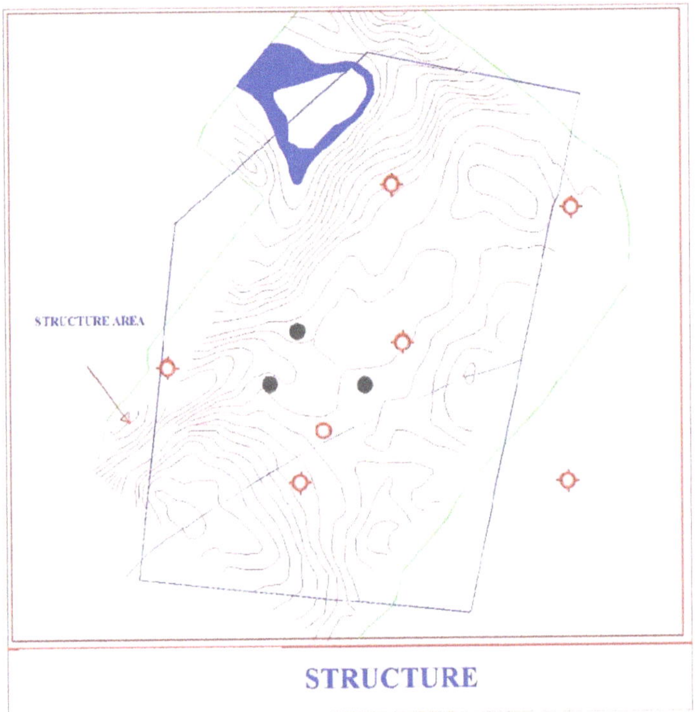

(Courtesy DGH)

Fig. 6.1.17

The PY-3 field came on stream during Nov 1997 with 4 sub sea wells. The initial production started at 1200 bbl/day of oil, which subsequently declined to 660 bopd. The JV company implemented water injection for maintaining reservoir pressures and maximize the recovery from the field and also to extend the field life.

The facility at PY-3 consisted of a floating production system (Tahara). There are 4 subsea wells tied back to Tahara. Tahara has a three stage crude oil separation system where actual liquid processing is possible. Water injection rate was approximately 6640 bwpd to maintain voidage replacement / compensation. The field produced associated gas of 3.7 MMSCFD along with the oil production. Crude oil from PY-3 was sold to CPCL at its refinery at Nagapattinam.

The reserves of PY-3 are as below

Proved 5.09

Proved + Probable 17.57

Proved + Probable + Possible 23.81

Exploration & Production Status

Basin Area **Onland** : **25,000 Sq. Km.**

Offshore : **30,000 Sq. Km.**

Seismic Data

2D - Onland : **40,999 LKM**

3D – Onland : **4030 Sq. Km**

2D – Offshore : **101,341 LKM**

3D – Offshore : **7425 Sq. Km.**

Other Surveys

Geological Surveys : **8500 Sq. Km.**

Gravity Magnetic Surveys

Onland : **14,094 Stations**

Offshore : **32,899 LKM**

Well Data

Exploratory wells Drilled Onland : **431**

Offshore : **76**

Hydrocarbon Discoveries / Field : **Onland : Oil Fields - 16, Gas Fields - 10**

Offshore: Oil Fields - 02, Gas Field - 03

Basin Type : **Pericratonic Rift Basin**

Reserves & Production Status

Hydrocarbon Resource Potential:

Onland : **430 MMT**

Offshore : **270 MMT**

Total : **700 MMT**

Hydrocarbon Reserves: Nil

Production: Nil

Blocks on Offer (NELP I to VIII)

The blocks offered and awarded under production sharing contract from NELP I to VIII are shown in the Fig 6.1.17 & Table. 6.1.T1.

Table 6.1.T.1

NELP	Blocks Offered	Blocks Awarded	Awardees	Date of Signing/ Effective Date	Present Area(sq.km)	Status
I		CY-OSN-97/1	HOEC (80%-OP), MIL (20%)	08-Jan-2001/ 16-Mar-2001	Relinquished	
II		CY-OSN-2000/1	ONGC (100%-OP)	17-Jul-2001/ 16-Aug-2001	Relinquished	
		CY-OSN-2000/2	ONGC (100%-OP)	17-Jul-2001/ 16-Aug-2001	Relinquished	

Table 6.1.T.1 ***Contd...***

NELP	Blocks Offered	Blocks Awarded	Awardees	Date of Signing/ Effective Date	Present Area(sq.km)	Status
III	CY-DWN-2001/1	CY-DWN-2001/1	ONGC (80%-OP), OIL (20%)	04-Feb-2003/ 12-Mar-2003	12425	
	CY-DWN-2001/2	CY-DWN-2001/2	RIL (100%-OP)	04-Feb-2003/ 03-Apr-2003	14325	
	CY-PR-DWN-2001/3	CY-PR-DWN-2001/3	RIL (100%-OP)	04-Feb-2003/ 03-Apr-2003	8600	
	CY-PR-DWN-2001/4	CY-PR-DWN-2001/4	RIL (100%-OP)	04-Feb-2003/ 03-Apr-2003	10590	
IV						
V						
VI	CY-DWN-2004/1	CY-DWN-2004/1	ONGC (70%-OP), GSPC (10%), HPCL (10%), GAIL (10%)	02-Mar-2007/ 28-May-2007	10302	
	CY-DWN-2004/2	CY-DWN-2004/2	ONGC (70%-OP), GSPC (10%), HPCL (10%), GAIL (10%)	02-Mar-2007/ 28-May-2007	12059	
	CY-DWN-2004/3	CY-DWN-2004/3	ONGC (70%-OP), GSPC (10%), HPCL (10%), GAIL (10%)	02-Mar-2007/ 21-May-2007	12017	
	CY-DWN-2004/4	CY-DWN-2004/4	ONGC (70%-OP), GSPC (10%), HPCL (10%), GAIL (10%)	02-Mar-2007/ 21-May-2007	12025	
	CY-PR-DWN-2004/1	CY-PR-DWN-2004/1	ONGC (70%-OP), GSPC (10%), HPCL (10%), GAIL (10%)	02-Mar-2007/ 15-May-2007	13451	
	CY-PR-DWN-2004/2	CY-PR-DWN-2004/2	ONGC (70%-OP), GSPC (10%), HPCL (10%), GAIL (10%)	02-Mar-2007/ 23-May-2007	9994	
VII						
VIII	CY-OSN-2009/1	CY-OSN-2009/1	BEII(100%-OP)	30-Jun-2010/	1362	
	CY-OSN-2009/2	CY-OSN-2009/2	OIL (50%-OP), ONGC (50%)	30-Jun-2010/	1621	
	CY-OSN-2009/3					

(Courtesy DDH)

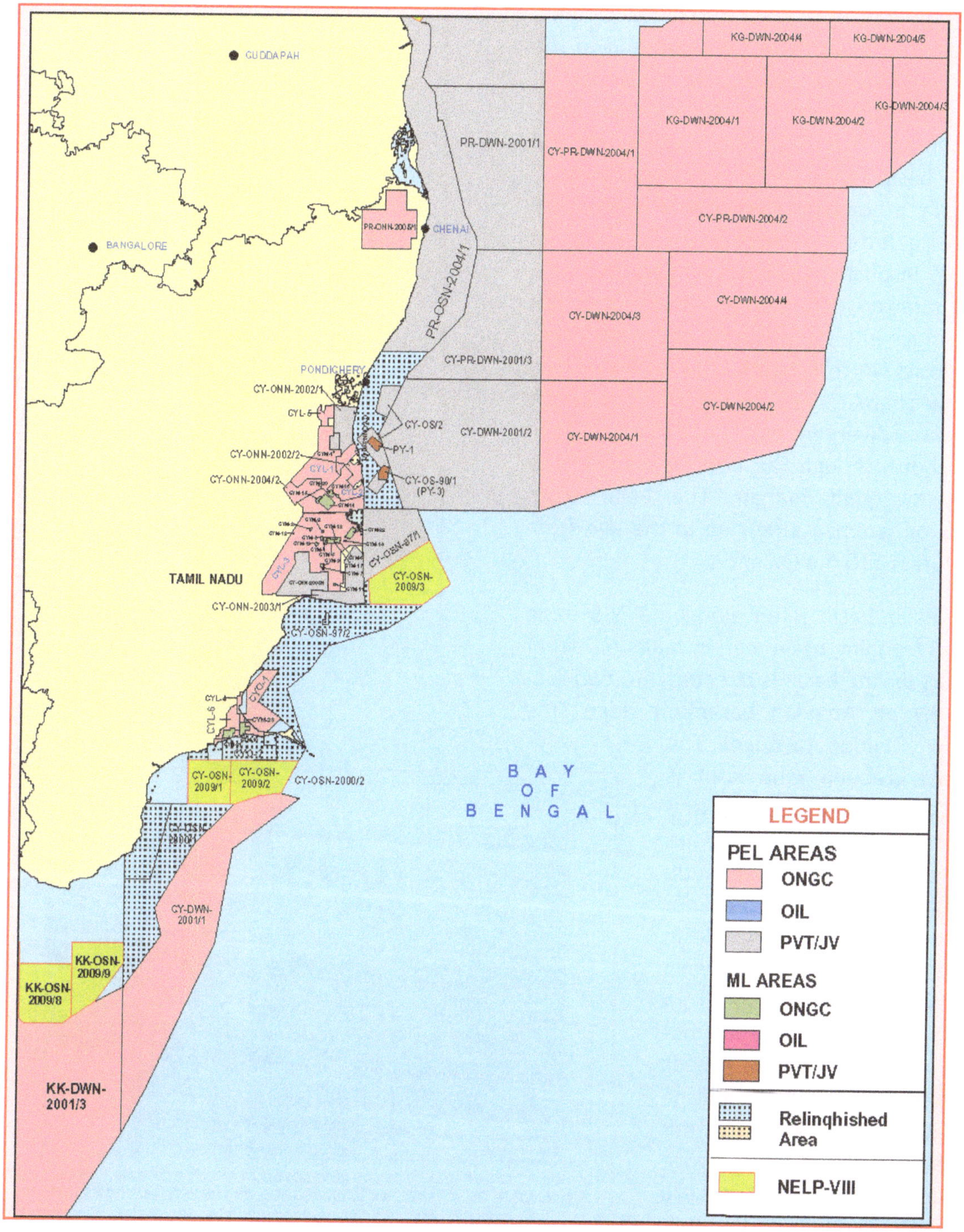

(Courtesy DGH)

Fig. 6.1.18

6.2 Palar Basin

Palar basin is situated on the East cost of India bounded to the North by KG-Basin and to the South by__Cauvery Basin. It is mainly a Pericratonic rift basin and sedimentalogically quite different from KG basin in the North and Cauvery basin in the south. The Palar basin is essentially a gondwana grabben / depression. Mainly situated on onshore part (Fig. 6.2.1) its influence in offshore is quite limited. Rather it can be inferred that the Palar basin does not extend much into the offshore, unlike the other basins to its North & South. The onshore basin is mainly characterized by the rift and synrift fills, and the divergent part is not clearly seen in the offshore, though the basin is classified as passive divergent margin. The Palar basin consists of sediments ranging in the age from upper Jurassic to the Recent.

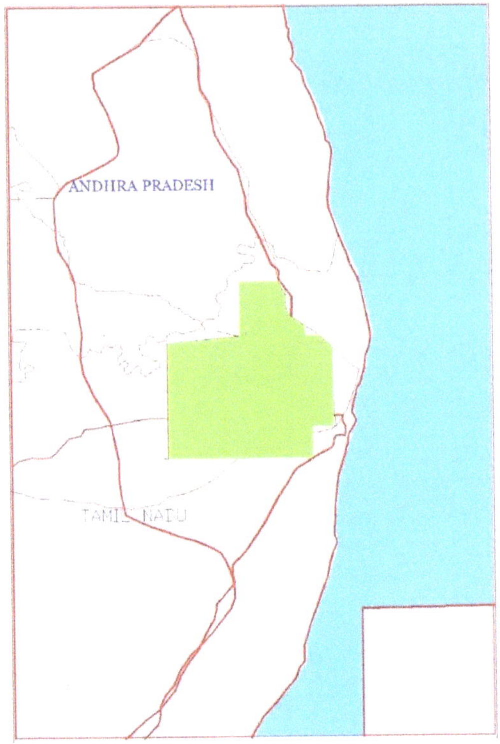

(Courtesy DGH)

Fig. 6.2.1

The Palar basin is an elongated N-S basin with N-S trending basin margin faults on either side. Further the basin is bisected into two sub-basins by an Archean basement ridge (The Archean - Naidupeta ridge). The Northern grabben /sub-basin is called Satyavedu depression and Southern as the Sripeumbadur. Compared to Satyavedu depression, Sriperumbadur depression is deeper. Palar Basin mainly consists of gondwana sediments ranging in age from Jurassic to upper cretaceous, but very thin Tertiary beds unconformly overlie the gondwana sediments. In Sriperumbadur depression the sediments range from Jurassic to Resent (Triassic?). The total sediments in the sub-basin is of the order of 3000 to 3500 mt. The basin mainly consist of over 2500 mt of gondwana sediments ranging in the age from Jurassic to upper cretaceous unconformably overlain by very thin Tertiary deposits (Fig. 6.2.2a). The gondwana sediments are mainly fluvial-Fangomerates. In the offshore part the limited part of the basin does not indicate any graben or graben tectonics, and the beds dip monoclinally towards East and Southeast. Earlier Chevoron-Amaco applied for permission to explore the offshore part of the basin. They carried out 2D seismic surveys and also drilled one or two exploratory wells without any success. The seismic data indicate that the shelf of Palar is very narrow, and suddenly goes into slope and abyssal plains. Subsequently Chevoron-

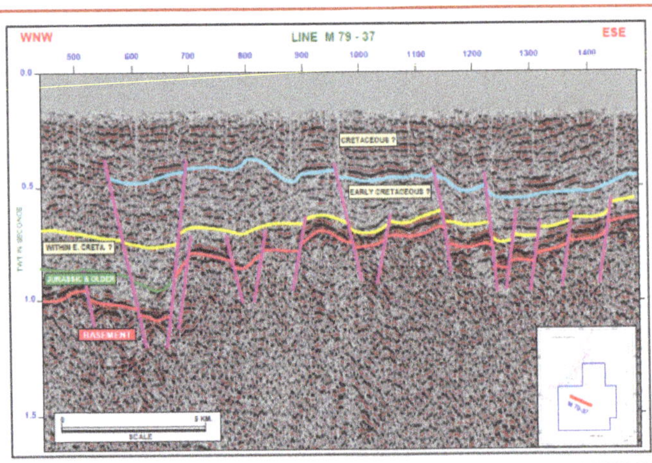

SECTION ACROSS FAULT CLOSURE DEVELOPED AT THE TOP OF EARLY CRETACEOUS. TRUNCATION OF JURASSIC & OLDER SECTION ON THE WNW PORTION IS EVIDEND

(Courtesy DGH)

Fig. 6.2.2

Amaco relinquished the block. However, during NELP-VI, Govt of INDIA offered one offshore block in the shallow waters, PR-ONS-2004/1(5) in the basin.

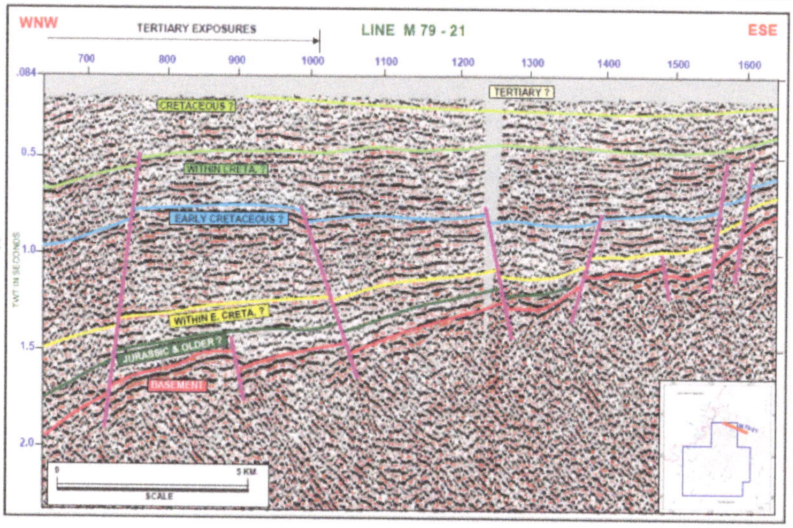

SEISMO-GEOLOGICAL SECTION ACROSS PALAR GRABEN. NOTE THE THIN TERTIARY COVE OVER MESOZOIC SEDIMENTS. PINCHOUT OF JURASSIC SECTION IS TOWARDS ESE.

(Courtesy DGH)

Fig 6.2.2a

The generalized stratigraphy is shown in the Fig. 6.2.3

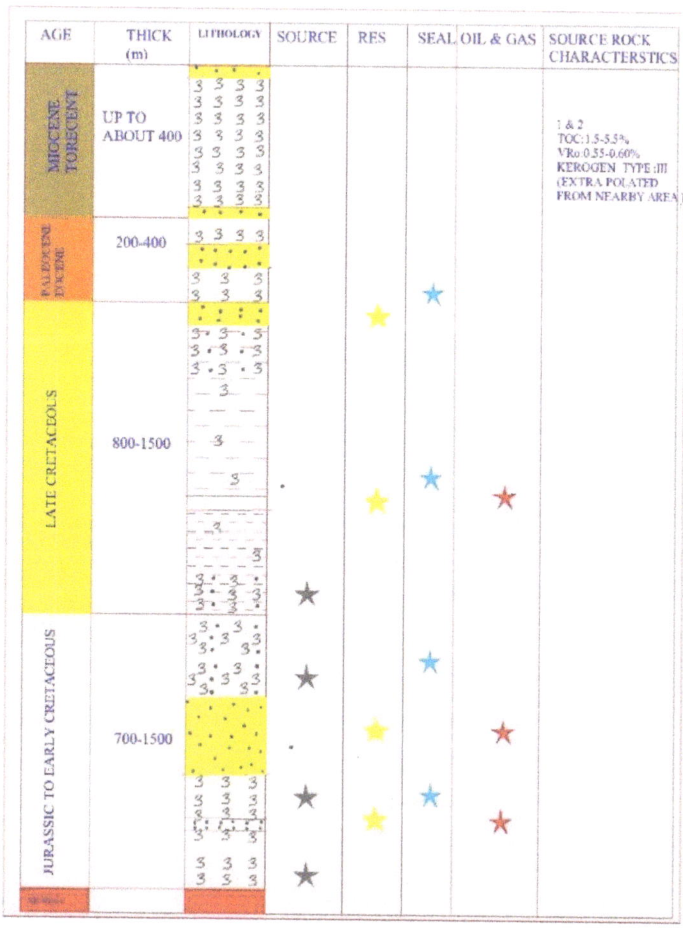

(Courtesy DGH)

Fig. 6.2.3

Exploration & Production Status

Basin Area : 36,000 Sq. Km.

Seismic Data Onland : 2D-2,214 LKM

Offshore : 2D-43,361 LKM

: 3D-4568 Sq.Km

Data acquired by	2D (LKM)		3D (Sq.km)	
	ONLAND	OFFSHORE	ONLAND	OFFSHORE
ONGC	2214	36244	-	-
Pvt./JV	-	7117	-	4568

Other Surveys Geological Survey : 2000 Sq.Km

Gravity Magnetic Surveys : 8000 Sq.Km

Satellite Gravity Survey

Well Data Number of Exploratory Wells : 7

Hydrocarbon Shows/ Discoveries: Oil and gas shows in Jurassic and Early Cretaceous Sandstones in 1 well

BASIN TYPE : Pericratonic Rift Basin

Reserves & Production Status:

1. **Hydrocarbon Resource Potential:**
2. **Hydrocarbon Reserves:**
3. **Production:**

Blocks on Offer (NELP I to VIII)

The blocks offered and awarded under production sharing contract from NELP I to VIII are shown in the table below and Fig. 6.2.4.

NELP	Blocks Offered	Blocks Awarded	Awardees	Date of Signing/ Effective Date	Present Area (sq.km)	Status
I						
II						
III	PR-DWN-2001/1	PR-DWN-2001/1	RIL (100%-OP)	04-Feb-2003/ 03-Apr-2003	8255	
IV	PR-OSN-2002/1					
V						
VI		PR-OSN-2004/1	CEIL(10%-OP), ONGC (35%), CAIRN INDIA (25), TATA (30)	02-Mar-2007/ 24-Apr-2007	9417	
VII						
VIII						

(Courtesy DDH)

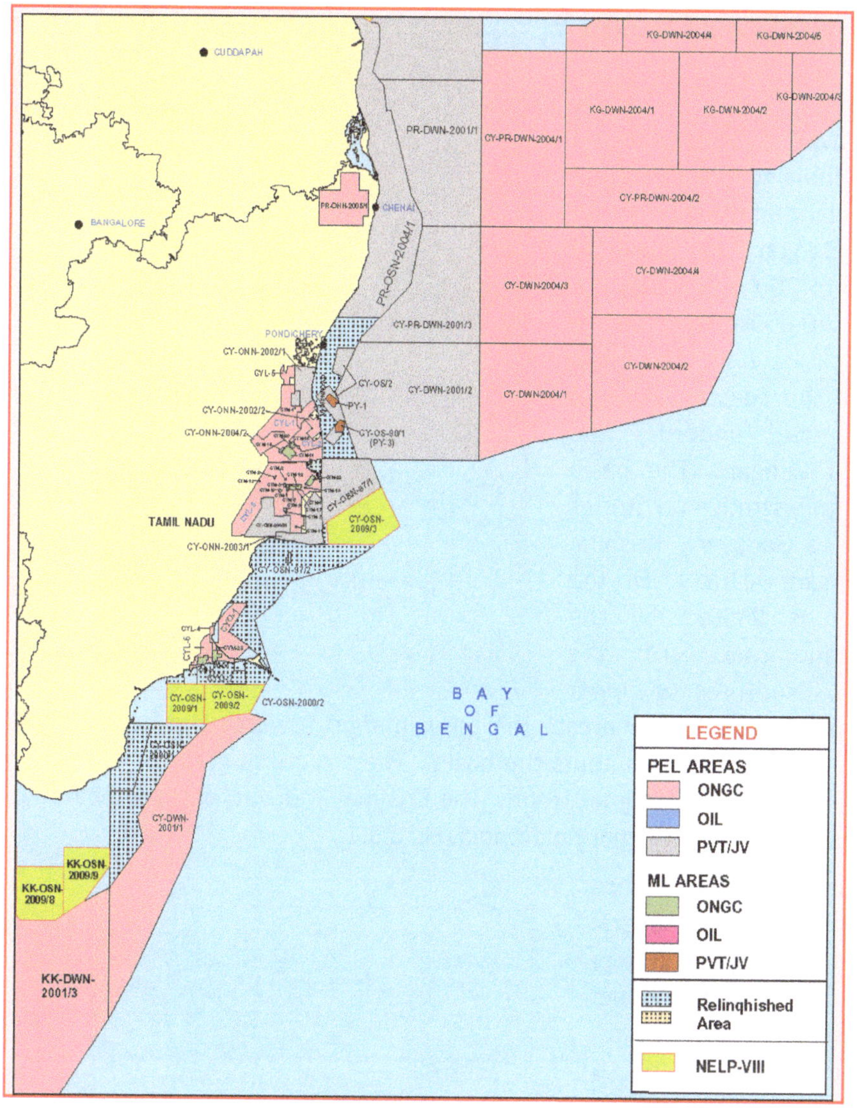

(Courtesy DGH)

Fig. 6.2.4

6.3 Krishna-Godavari Basin

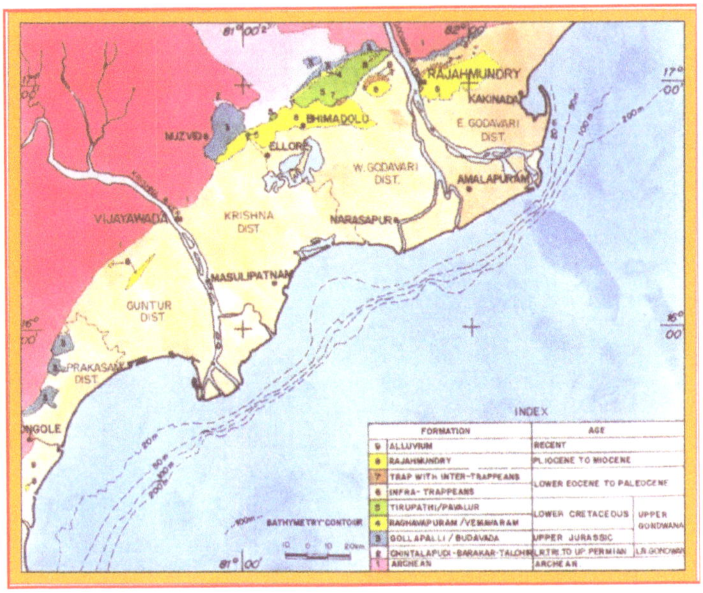

(Courtesy ONGC)

Fig 6.3.1

Introduction

The Krishna-Godavari basin located on the East Coast of India covers an area of 20,000 sq.kms onland and extends into the Bay of Bengal to envelop 21,000sq.kms up to 200 m isobath and then additional 20,000sq.kms up to 2000 m isobath and beyond. This basin occupies most of the coastal plains of Andhra Pradesh and underlies the waters of Bay of Bengal. The basin spans over the administrative territories of East and West Godavari, Krishna, Guntur and Prakasam districts. On the west, the basin is limited by the exposed Archean rocks of the peninsular shield. Exposures of Upper Cretaceous sedimentary rocks demarcate the basin margin towards the northwest where the NW-SE trending Pranitha Godavari graben abuts the basin. The structural grain of the basin is essentially NE-SW almost parallel to the Eastern ghat trend. The Krishna-Godavari basin is located in the central part of the Eastern passive continental margin of India. Fig 6.3.1.

The basin contains thick sequence of sediments with several depositional cycles, ranging in age from Late Carboniferous to Holocene. The major deltas with thick argillaceous facies (Godavari & Krishna rivers) prograded seaward, spanning from Late Createaceous through Tertiary to Recent and this regime had become the major hydrocarbon exploration target. 6-8 Kms of sediments, ranging from Upper Jurassic to Recent are expected in the basin.

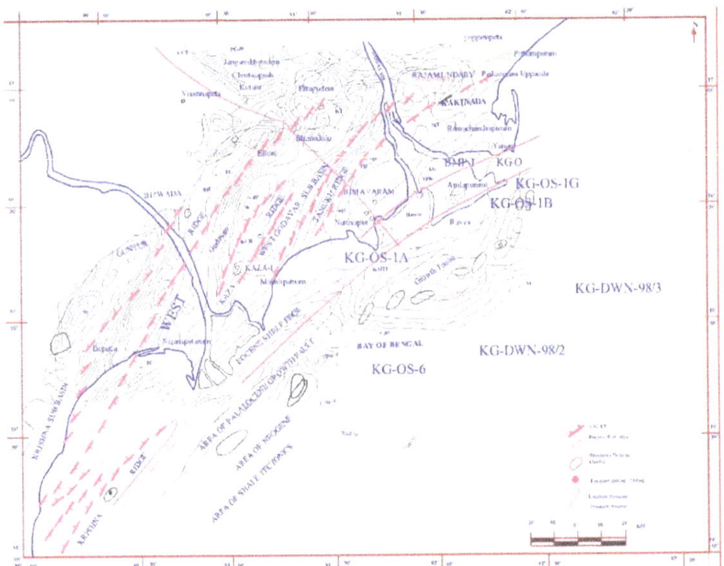

Fig 6.3.2 PROSPECT MAP OF KRISHNA AND GODAVARI BASIN

(Courtesy ONGC)

Fig 6.3.2

This proven petroliferous basin has potential reservoirs ranging in age from the Permian to the Pliocene. The exploratory efforts resulted in the discovery of a number of oil and gas bearing structures both onland and offshore (Fig. 6.3.2) (shallow as well as deep waters). Good source rock sequence is present ranging in age from Permo Carboniferous to Early Mio-Pliocene. Tilted fault blocks, growth faults, roll-over anticlines are the favourable structural traps where as turbidite fans, sub-marine canyon fans, wedge outs etc. play an important role as stratigraphic traps. The prognosticated resources of the basin are estimated to be around 1130 MMT, of which onland is around 575 MMT, offshore about 555 MMT. Resources in deep waters beyond 400 m water depth are likely to increase further by about 4000 MMT.

In Krishna-Godavari basin, on land more than 225 prospects have been probed, by drilling around 557 exploratory wells. Hydrocarbon accumulations had been proved in 75 of these prospects (22 oil and 53 gas). Notable discoveries are

Oil	Gas
1. Kaikalur	1. Razol
2. Vadali	2. Tatipaka
3. Mori	3. Narsapuram
4. Bantumilli	4. Pasarlapudi
5. Lingala	5. Mandapeta
6. Suryaraopet	6. Chintalapalli
7. Gopavaram	7. Adivipalem
8. Kesinapalli &	8. Elamanchali
9. Kesinapalli West	9. Enugupally
10. Kaza	10. Nandigama
	11. Endamuru
	12. Penumadam
	13. Ponnamanda
	14. Achanta
	15. Mulikipalle
	16. Magatapalli
	17. Gokarnapuram
	18. Kasavadaupalem
	19. Lakshmaneshwar
	20. Rangapuram
	21. Sirikattapalli

In offshore, so far more than 84 prospects have been probed by about 182 exploratory wells. Commercial Hydrocarbon accumulations have been proved in 22 of these prospects. In addition, about 90 discoveries have been made by private / Joint Venture companies so far in NELP blocks.

1. Reliance Industries Ltd discovered 15 Dhirubhai prospects in KG-DWN-98/3, KG-OSN-2001/2.
2. Cairn Energy India Ltd., discovered 3 prospects in block KG-DWN-98/2 within Mio-Pliocene.
3. ONGC discovered 3 hydrocarbon bearing prospects in the block KG-DWN-98/2 within Plio-Pliostocene sandsones of Godavari formation.
4. GSPC made 1 discovery in the block KG-OSN-2001/3 within Lower Cretaceous.

Geo Scientific Surveys

A. On Land

1. ONGC carried out detailed geological mapping in the area covering about 5000 sq.sms since 1959.
2. ONGC carried out Gravity Magnetic surveys in onland part covering an area of 19,200 sq.kms. Whereas M/s. Prakla Seismo and GSI acquired Gravity Magnetic data in the offshore area.

Seismic

1. ONGC in 1965 initiated the conventional single fold surveys and acquired about 2690 LKM of data by 1973.
2. CDP surveys commenced in 1973 and ONGC acquired about 34642 LKM with varying foldage (6 fold to 48 fold).
3. ONGC also acquired 2325 Sq.Kms of 3D seismic data (onland).

B. Offshore

In offshore area, the first surveys of regional nature started during 1964-65. These surveys were followed by multifold 2D/3D surveys in shallow to deep waters and also transitional zones. As on April 2005, more than 75,000 LKM of 2D and 26,500 S.Kms of 3D seismic surveys have been carried out. Additionally during 1972-74, 2028 kms of refraction seismic data was also acquired to study the basement configuration (Figs. 6.3.3 & 6.3.4)

Fig. 6.3.3

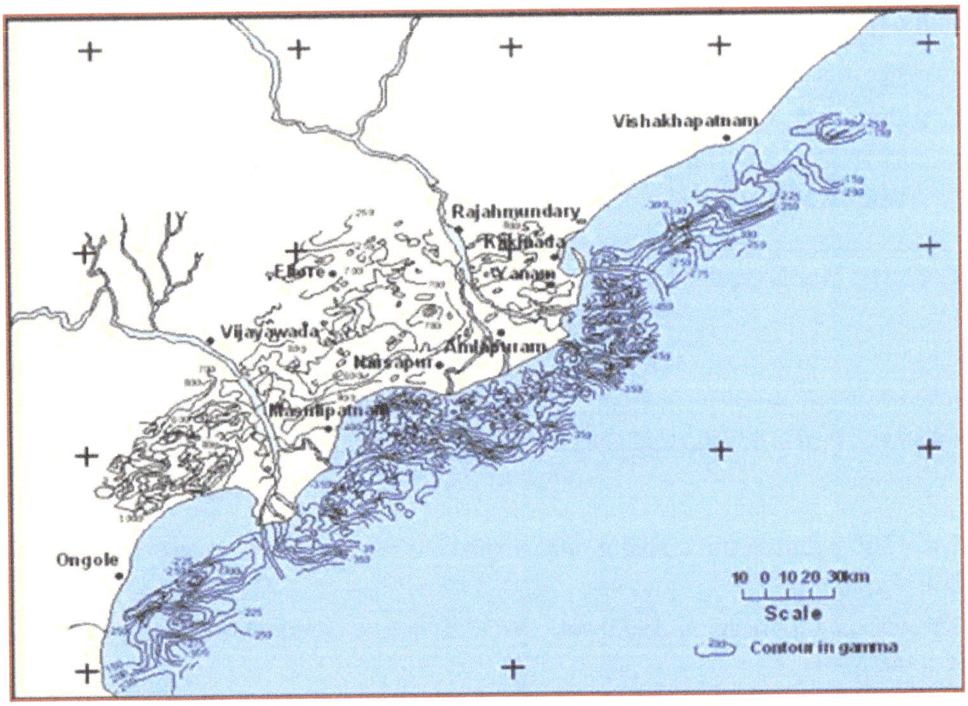

(Courtesy ONGC)

Fig. 6.3.4

In addition, DGH carried out the following surveys in the offshore

1. Satellite Gravity Surveys in Western, Eastern and Andaman offshores by Petro Scan in an area of 12,81,000 sq.km.
2. Speculative Surveys were carried out in the offshore of East Coast by DGH and Western Geophysical as J.v. Partners. Carried out 2D seismic and Gravity Magnetic Surveys of 7500 LKM.
3. DGH and Western Geophysical also acquired 2D Seismic / Gravity Magnetic Surveys of 3600 LKM in the Far East covering Andaman & Nichobar islands.

Tectonic History

K-G basin is a Peri-Cratonic continental passive margin. The basin came into existence following the rifting along Eastern continental margin of Indian craton in Early Mesozoic (Gondwana rift). The down to the basement faults which define the series of horsts and grabens cascading down towards the ocean are aligned NE-SW along the Pre-Cambrian Eastern Ghat trend. There are three major tectonic trends in the basin (Fig 6.3.5).

1. The Archean, Eastern Ghat trend which is NE-SW and is the oldest trend of the basin.
2. The next is Pranhitha Godavari NW-SE trend of Permo-Triassic which is younger to Archean trend.
3. The Neogene tectonic trend, essentially NE-SW almost parallel to Eastern Ghat trend and is controlled by delta regime and associated tectonism.

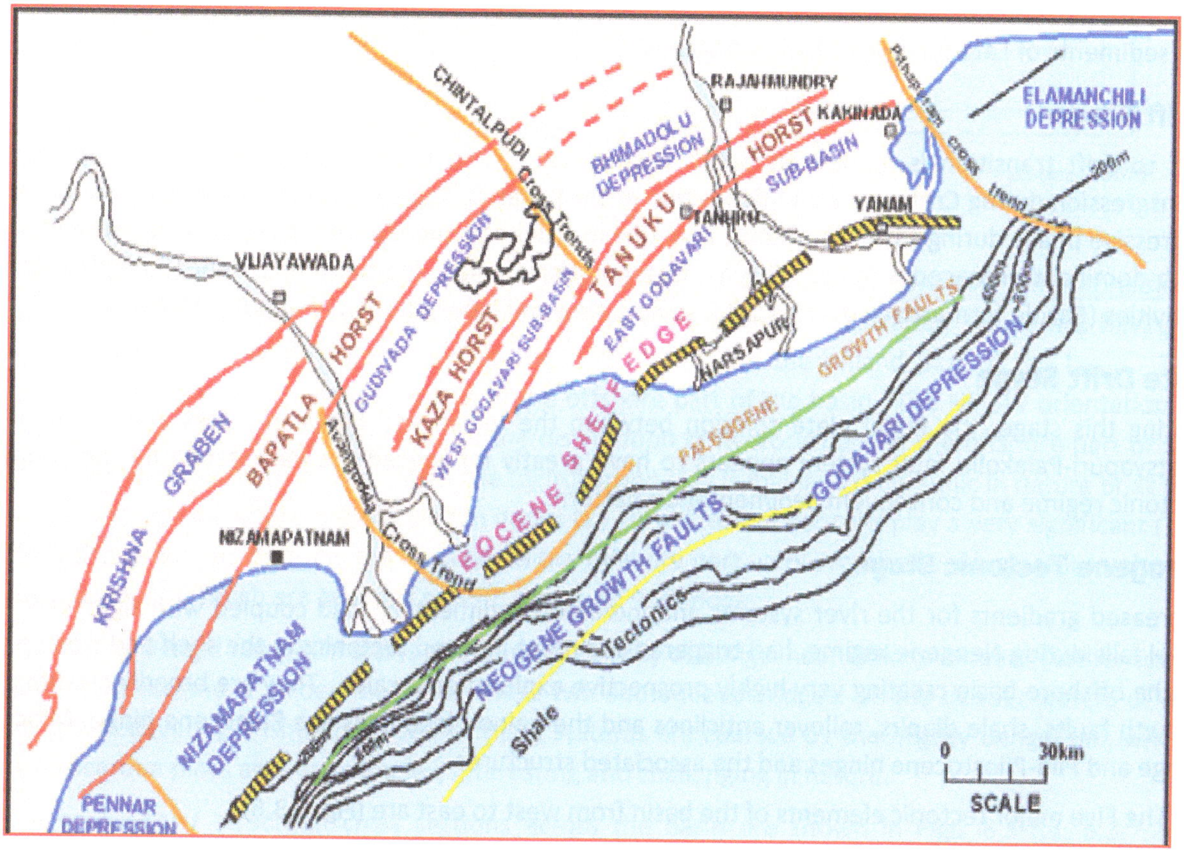

(Courtesy DGH)

Fig. 6.3.5

The Pranhitha Godavari NW-SE trend cuts across the Eastern Ghat trend near Badhrachalam and extends further East into the offshore basin covered by Neogene tectonic regime and fans as a funnel

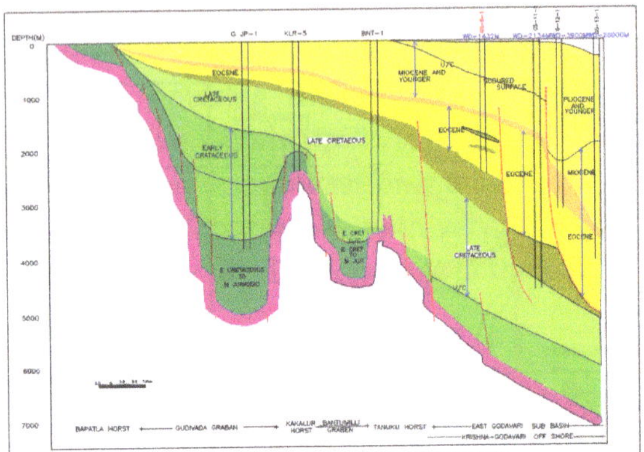

GEOLOGICAL CROSS SECTION, KRISHNA-GODAVARI BASIN

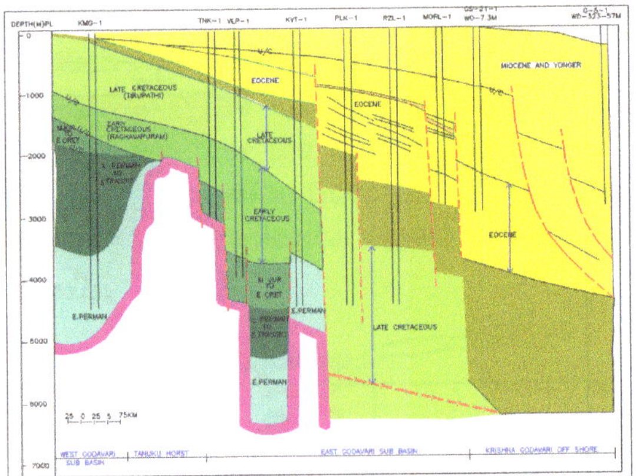

GEOLOGICAL CROSS SECTION, KRISHNA GODAVARI BASIN

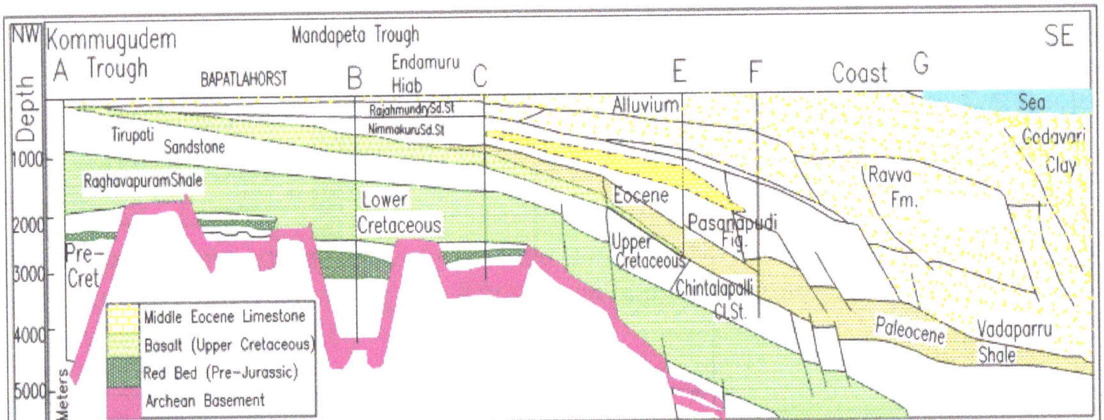

Structural cross section across the East Godavari subbasin showing thick Tertiary sediments toward the southeast of the MPFZ as seen in well G in the offshore part of the basin; toward the northwestern basin margin area at well A, thick per-Cretaceous sediments were encountered

(Courtesy ONGC)

Figs. 6.3.7.1, 6.3.7.2 & 6.3.7.3

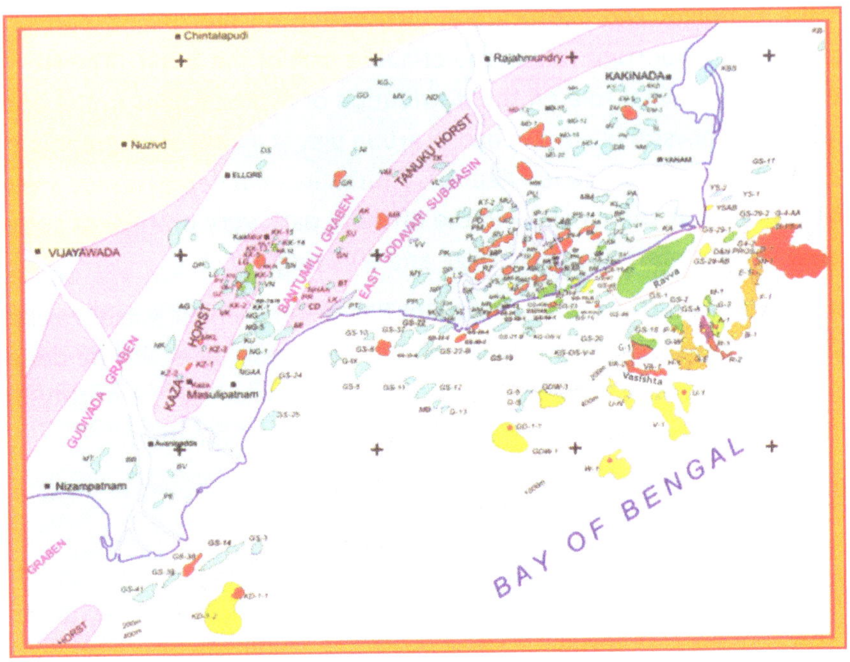

(Courtesy ONGC)

Fig. 6.3.8

Geo-Morphology and Sedimentary Environments in the K-G Delta

The major geo-morphological units identified in the basin are as follows (Fig. 6.3.9)

(a) Upland plains

(b) Coastal plains

(c) Recent flood plains

(d) Delta plains

The sedimentary environments in the Godavari delta can be classified as (Prof. A.S.R. Swamy, et al)

1. River channels and estuary
2. Flood plains
3. Natural levees
4. Man grove swamps
5. Tidal channels
6. Tidal flats
7. Lagoon
8. Kakinada Bay
9. Godavari Sand spit
10. Main land beach and dunes
11. Paleo sand ridges

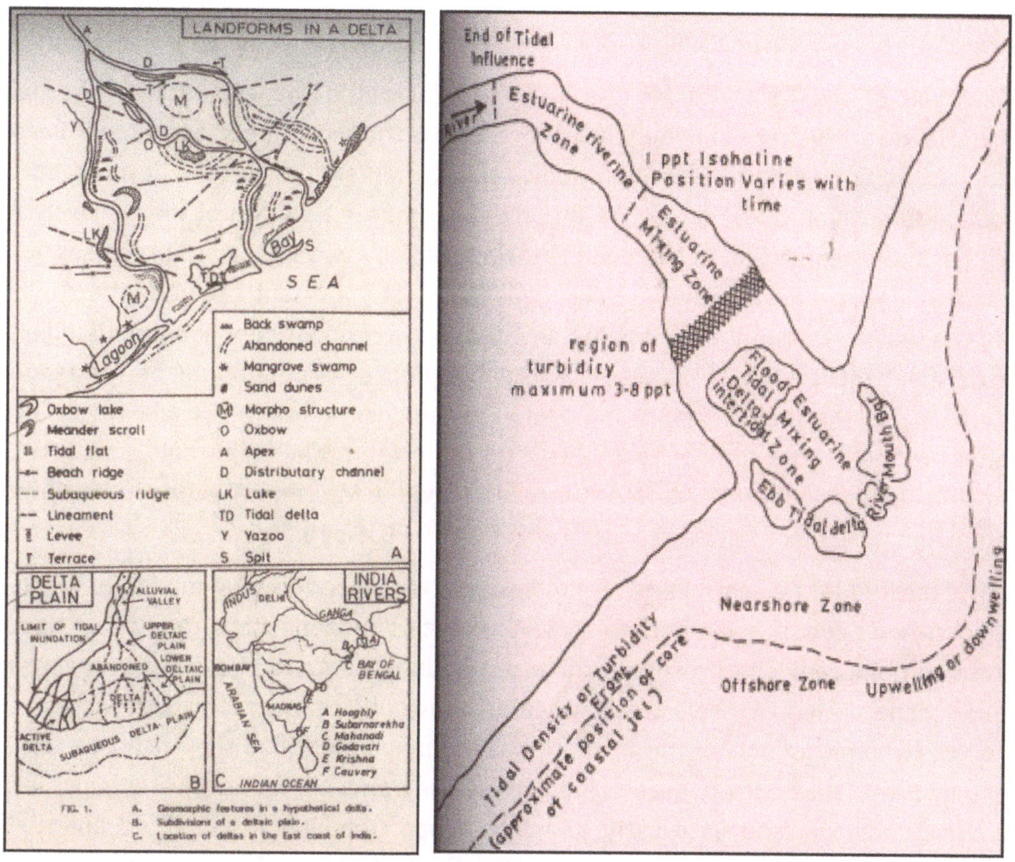

Fig. 6.3.9

The description and the dynamics of the various environments are not discussed and it is beyond the scope of this book. Godavari River is one of the major deltas in the East Coast, which shows many features commonly met within the deltaic plain. Numerous ancient channels and beach ridges with different orientations have been traced in this delta.

Krishna River

Krishna River flows NW-SE and the main river course is outside the NW-SE trending Pranhitha Godavari graben. As stated earlier, the Godavari river flows through the graben which is mainly controlling the delta building activity to the southeast in the Bay of Bengal. Unlike, the other deltas in the East Coast wherein the delta fronts have very limited bulge into the sea, Krishna River conspicuously protrudes into the Bay of Bengal. Whereas the apex of the delta, taking ancient channels also into consideration, is far away from the coast, the modern distributaries branch out from the main river nearer the sea. Krishna River can be categorized as 'lobate' wherein, the river action seems to have predominated over that off the waves and currents in the sea. Today however, many spits and bars are developing close to the delta front indicating the erosional and depositional action of marine agencies.

From the morphological features observed in some of the major deltas of the world that almost all of them contain similar features but their shape, size, locations and mutual relations differ, indicating the varying influence of the exogenic and endogenic forces in the growth and development of these deltas. World over, some of the Paleo deltas are the locales for hydrocarbon generation and accumulations. The study of the present deltas that show the way for proper appraisal of the potentialities of ancient deltas which are the major repositories of hydrocarbons. It will follow the dictum "***The present is the key to the past***" (Prof. R. Vaidyanathan, 2004).

Basin Morphology and Tectonics are the important factors in the evolution of the delta and its morphology. However, the rate of sediments discharged, is the most important factor. Reconstruction of the tectonic conditions requires consideration of all sedimentary units in the basin together with the size and geological position of the basin. Based on the geomorphic studies of the Godavari delta, it can be inferred that the progradation of the delta is taking place by two important processes viz. the shore line advance and lagoonal filling. Middle Miocene is the most important geological time in the Tertiary evolution of the Indian sub-continent. It is the time for continent-continent collision of Indian and Asian plates, thrust sheet movement in Himalayas and generation of large amount of Terragenous clastics sediments. It is also the time of uplift and eastward tilting of Indian peninsula which leads to high sediment input in the coastal zone of East Coast. All the present day major deltas on East Coast were established and expanded during Middle Miocene. Middle Miocene-Pliocene is a period of rapid sedimentation and building of thick sediment wedges (Prof. I.B. Singh.)

K-G basin is located in the central part of the Eastern passive continental margin of India. The basin contains thick sequences of sediments with several cycles of deposition ranging in age from Late Carboniferous to Holocene. This proven petroliferous basin has potential reservoirs ranging in age from Permian to Pliocene. Scientists from the National Institute of Oceanography, Mr. Murthy et.al (1995) studied the geo-dynamic aspects of the offshore K-G basin and identified the offshore extension of the NW-SE trending Pre-Cretaceous rift graben along the cross trends. The basin was a major intra-cratonic rift within the Gondwana land until Early Jurassic. When Gondwana land rifted apart, the Eastern margin of the Indian peninsula was positioned at Latitude 50° S and was oriented in an East-west direction (Chatterji and Hotton 1986).

Since the Cretaceous, the Indian plate had moved northward and the Eastern continental passive margin rotated 200° in a counter clockwise direction until it collided with Eurasia in the Late Eocene (Gordon Etal 1990, Srivastava & Chauhan 1987). The triple junction between the Indian peninsula, Australia and Antarctica was located at Machilipatnam Bay (Thompson 1976). The NW-SE trending Pranhita Godavari graben formed a failed arm of the triple junction (Burke & Dewey 1973). Based on Bouguer Gravity data, Murthy & Ramakrishna (1980) had identified three sub-basins separated by two basement horsts. From the southwest these are the Krishna, West Godavari and East Godavari sub-basins separated by the Bapatla & Tanuku horsts respectively.

The West Godavari sub-basin is further sub divided into the Gudivada and Bantumilli grabens by Kaza-Kaikalur horsts. The Kommugudem and Mandapeta troughs are situated on either side of the Tanuku horst. The Krishna sub-basin contains 1560 m of Cretaceous and older sediments above the Archean basement. Bapatla horst was uplifted during the Cretaceous. The Bapatla horst is discontinuous in the North Eastern part of the basin due to the presence of other lineaments obliquely. In the West Godavari Sub-basin, Cretaceous sediments are thin over the Kaza-Kaikalore horst compared to the very thick section on either side of the horst in the Bantumillli & Gudivada grabens. The sedimentary cover over the Tanuku horst is about 2500 m in the South Eastern flank of the horst which increases to 3500 m in the southwest indicating that the Tanuku horst plunges southwesterly into Machilipatnam Bay. In contrast, the Lower Cretaceous strata dip monoclinally towards Southeast over the Tanuku horst, suggesting the tectonic uplift of the horst since Cretaceous. In the East Godavari sub-basin, the sedimentary fill ranges from 3000 m over the pre existing basement horsts, to more than 5000 m in the deep basinal area in the southeast. Over 2000 m of argillaceous sediments of Cretaceous in the southeast might have been good enough to enhance tectonic subsidence and down faulting of sequences, forming enechelon faults in the East Godavari sub-basin during Late Cretaceous to Early Paleocene. This fault zone is termed as Matsyapuri-Palakollu fault zone. Tectonic readjustments had resulted in terraces and tilted fault blocks in the North Eastern area of the basin.

Discontinuation of the Bapatla horst in the Eastern part of the basin is associated with the presence of a northwestern orientation of gravity contours with strong negative gravity anomalies, suggesting possible continuation of the NW-SE trending Pranhita Godavari graben beneath the NE-SW trending Tertiary K-G basin. The NW-SE trending cross faults of earlier rift valley are termed as the Chintalapudi and Pithapuram cross trends in the NE-SW trending K-G basin. Based on the offshore magnetic data, these cross trends can be identified up to the deep water area of the K-G basin. The Ocean-Continent Boundary (OCB) was marked accordingly (Murthy et.al 1995), defining the southeastern termination of the NW-SE trending Pranhita Godavari graben.

Both the Kommugudem and Mandapeta troughs are deep Pre-Callovian down warps. In these troughs, the Upper Carboniferous to Lower Triassic sediments were deposited in varied Paleo environments ranging from glacial to marginal marine conditions. Mandapeda trough experienced Late Jurassic – Cretaceous down warping but the same is not seen in Kommugudem trough.

Sequence Stratigraphy

The general stratigraphy & Litho stratigraphy is shown in Fig. 6.3.10. & 6.3.11

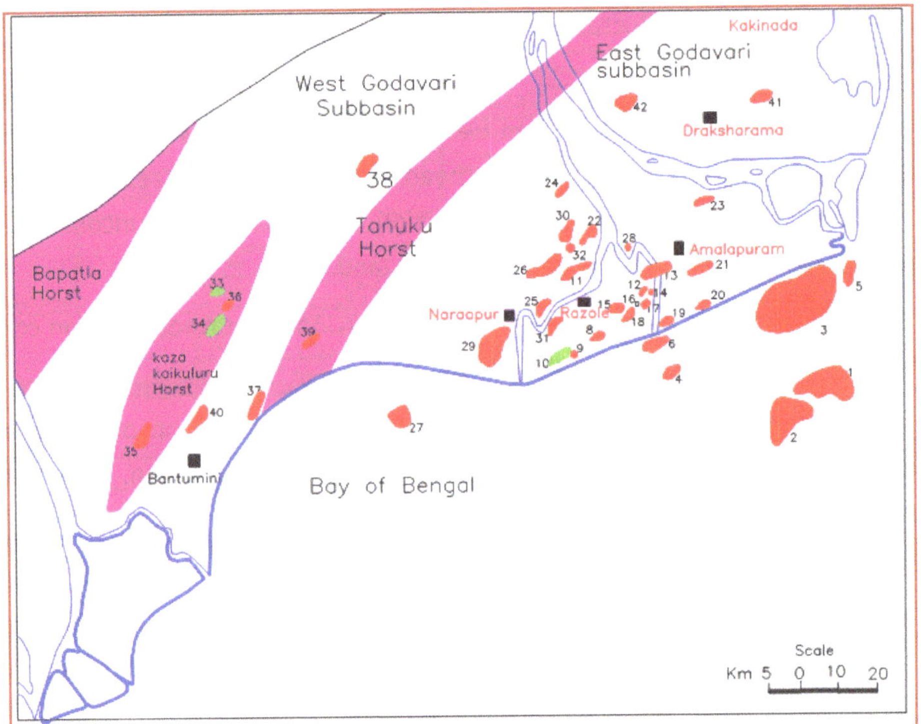

Hydrocarbon bearing prospects of the Krishna–Godavari basin.

(Courtesy G.N.RAO)

Fig 6.3.12

Kommugudem – Mandapeta – Red Bed System

This is the "Pre Cretaceous system" confined to the NW - SE trending rift valley extending beneath the East Godavari sub-basin. The hydrocarbon strike in Mandapeta well signifies a hydrocarbon potential of this system for the intensification of the exploration efforts. The hydrocarbon potential estimated is around 330 MMT of which only 20 MMT could be established and upgraded. This system is associated with many erosional unconformities which are having direct bearing on the hydrocarbon accumulations.

Raghavapuram – Gollapalli – Razole System

This is classified as "Cretaceous petroleum system". In the West Godavari sub-basin thin sands and lime stones act as reservoirs. Oil and gas pools are discovered in Kaza-Kaikalur, Bantumilli, and Narsapur-1 & 2. In Razole-1 the fractured volcanic rock with intertrappean beds produce gas in commercial quantities (Razole formation). Lenses of sand in Chintalapalli clay stone In East Godavari sub-basin also produce hydrocarbons. The Petroleum potential was estimated to be 230 MMT of which only 50 MMT have been established creating a vast scope for further exploration.

Palakollu – Pasarlapudi – Bhimanapalli System

Oland, this is the most prolific system in the K-G basin located further SE of Tanuku ridge. It is classified as "Paleogene petroleum system" which is mainly concentrated in East Godavari sub-basin. Hydrocarbon finds vary from Paleocene, Eocene and Lower part of Oligocene. Gas with condensate is found in Bhimanapalli-1 in Paleocene sandstone reservoir. Commercial gas and condensate production is observed from Thatipaka – Pasarlapudi structures in the basin. Oil is found in Mori, Kesinapalli, West Kesinapalli structures in the Eocene reservoirs. The estimated hydrocarbon resource from this petroleum system is around 300 MMT. About 100 MMT have been upgraded into the geological reserves, thereby indicating vast scope for further exploration.

Ravva-Godavari Petroleum System

Ravva-Godavari system is referred as "Neogene Petroleum System" and incidentally is the most important petroleum system in the offshore. Hydrocarbon finds are mainly obtained from Miocene to Pliocene and Pleistocene. The Mio-Pliocene sedimentary thickness is over 3500-4000 m in this part of the basin. The Ravva, G-1, G-2, G-3 & G-4, Padmavathi, Kanaka Durga and Annapurna structures and also Dhirubhai oil and gas finds in deep waters are the commercial finds in this Neogene petroleum system. However, in the offshore part even the Cretaceous sediments play an important role (Cretaceous Petroleum System). GSPCL found commercial gas in block No. KG-OSN-2001/3 and RIL found oil in Cretaceous section in MA structure of KG-DWN-98/2 in Cretaceous sandstone reservoirs. The hydrocarbon resources were estimated to be around 200 MMT of which 70 MMT of reserves have been established. If one sees critically into the prospectivity of the basin, the resource estimation of 200 MMT appears to be too low and warrants re-estimation, wherein the resource potential could even be 400 MMT, based on the recent finds of Reliance and ONGC in the ultra deep waters with giant commercial finds like UD-1A etc.

The Neogene tectonics is mainly controlled by prograding delta system of Godavari during various phases of Tertiary delta building activities. During the Neogene tectonic system, the Paleocene hinge, Miocene hinge and Plio-pliostocene hinges are clearly discernable. The growth faults, shale diaper and rollover anticlines are mapped in the shelf to slope systems and beyond. Strati structurally controlled submarine canyon fans and turbidite fans etc. are mapped with prolific Hydrocarbon potential in the abyssal plains. The various hinge zones ranging in age from Paleocene to Plio-Pleistocene play a dominant role in the hydrocarbon migration and entrapment in the offshore part of the basin.

The prospect mapped for this type of hydrocarbon plays is indicated in the Fig. 6.3.13 (HPC, HPP, HPE & HPM blocks) indicating the Cretaceous, Paleocene, Eocene, Miocene and Younger hydrocarbon zonations in the basin (Sheshavataram, 1983).

HPM Block

This block contains the youngest system of growth faults in which the Miocene, Plio-Pleistocene, prograding delta sequence sediments rolled over and formed anticlinal structures. These structures mainly lie in the deeper waters of Bay of Bengal i.e. beyond 400 m water depths. Oil and gas bearing structures like G-1, G-2, G-4, G-12, and K-2 are in this system. The Ravva, Annapurna, Kanakadurga, Padmavati structures lying over shallow shelf to slope system, to the east of Amalapuram coast are the type examples for Miocene prospect. Even the recent hydrocarbon strikes of Reliance/Dhirubhai Ambhani & ONGC fall in this system.

HPE Block – Eocene Prospect

The next hydrocarbon target is Eocene hing. The Eocene growth fault runs south of Amalapuram, through Kadali and Mori, almost nearer to the coast. Structures like Amalapuram, Kadali, Mori and Kesinapalli form against the Eocene hinge, which are proven to be hydrocarbon bearing and are the type examples for this system/block. Mori, Kesinapalli, Kesinapalli West fields are put on production for oil and gas. In this Eocene prospect zone, in addition to the clastic reservoirs, the shelfal lime stone is another potential reservoir.

HPP Block

The Matsyapuri fault acts as a Paleocene hinge and a number of structures lie in this block on land (between Matsyapuri fault to the west and Kadali-Mori fault to the East). Structures like Narsapuram, Razole, Matsyapuri, Moghalthuru, Palakollu, Kotipalli, Ramachandrapuram, Thatipaka, Pasarlapudi, Bhimanapalli etc. lie in this block. The hydrocarbon prospects in this block are mainly in Paleocene and the Late Mesozoic sections. The Paleocene / Mesozoic targets are at deeper levels (below 3200 m) and these prospects are associated with high temperature and pressure conditions which exist mainly in the gas window.Hence, the target of exploration is mainly for the gas and condensate in this block. While prospecting in this block, it is worthwhile to explore the fractured trap (Basalt and inter-trappean sections) in addition to the targeted clastic Paleocene rollovers. This block may cover the major part of East Godavari depression.

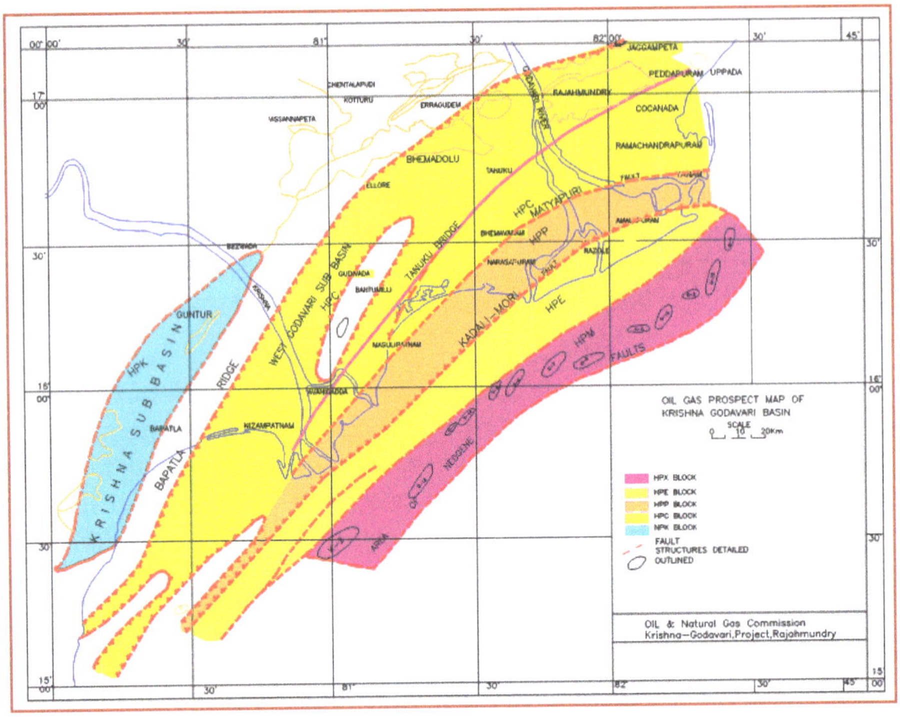

(Courtesy Seshavataram, 1980)

Fig. 6.3.13

HPC Block – Cretaceous Hydrocarbon Prospects

The entire West Godavari Depression is covered under this block i.e. Gudivada, Bantumilli sub-basins and the associated Kaza-Kaikalur high. The Tertiary section is quite thin and does not hold any prospect in this part of the basin. The main interest lies in Mesozoic section. A number of structures like Kaza, Kaikalur, Mokollu, Nidumolu, Vadali, and Bantumilli are detailed in this block. All these structures proved to be either oil or gas bearing in the Mesozoic section. Mandapeta prospect proved to be hydrocarbon bearing in the Gondwana section. Structures like drape over basement highs, stratigraphic wedge outs against existing highs, play a dominant role in this part of the basin.

HPK Block – Hydrocarbon Prospects of Krishna Depression

This block covers the entire Krishna depression to the west of Bapatla horst, and does not hold good hydrocarbon prospects either in Tertiary or Mesozoic sections. Because the basin is quite close and well connected to the outcrop area, where the infiltration regime and meteoric waters would be very active. In such a case, Even if hydrocarbons are generated in the basin, they are likely to be oxidized by the infiltrated waters.

Petroleum System

All the elements of a good petroleum system are available in the K-G basin like good source rocks, reservoir rocks, migration and good mechanism for trapping the hydrocarbons.

Source Rocks

Source rocks are available in the basin almost from Permian to the Recent.

Source rocks of about 200-250m are identified in Permian carboniferous, coal shale sediments. The source rock characters range from poor to very good and it is in oil generating window during Permian and in Metagenetic stage as at present.

Lower Cretaceous shales are oil prone which are over 800m thick in the basin The time temperature indices (TTI) generated in the basin suggest that the source rocks in this area are suitable for generating liquid hydrocarbons. The Cretaceous argillaceous rocks in the deeper parts of the basin are well matured to generate gaseous hydrocarbons and are rated to be very good source rocks.

Paleocene shales of East Godavari sub-basin contain dominantly Type-III Kerogene, with 2-32% organic matter and have fair to good source potential. The maturation is in the Early to peak stage for oil generation.

Lower Eocene shales located southeast of Amalapuram contain very rich organic matter concentrations with TOC varying from 3 to4%. The quality of organic matter is mainly Type-III and minor Type-II and they have the tendency to generate gas and subordinate oil and could have generated hydrocarbons since Early Miocene (Neeraja etal 1997). In Ravva area in the shallow offshore Oligocene-Miocene source rocks are identified, having bed thicknesses of over 400 m. They have very high organic matter concentration, TOC varying from 1.5 to 1.86%. In the shallower parts of the basin, the sediments are still immature but in the deeper parts they are matured enough to generate hydrocarbons which migrate to the structures nearby.

Reservoir Rocks

In K-G basin, reservoir rocks range in age from Early Permian to Mio-Pliocene. The oldest reservoirs are the Pre rift Permian sandstones that overlie the Permo-carboniferous source beds. Thin sandstone beds / lenses in Lower and Upper Cretaceous source shale beds proved to be good reservoir rocks capable of holding both oil and gas at high temperatures and pressures. Lower Eocene sandstones and also the shelfal limestones that overlie Paleocene source beds proved to be good reservoirs. These reservoir rocks are seen in Mori, Kadali and Kesinapalli structures near Eocene hinge. Mio-Pliocene deltaic sand beds are good reservoirs for both oil and gas as is the case in Ravva structure. The hydrocarbons might have migrated from deeper source beds. Sub-marine Canyon and turbidite fan deposits are potential reservoirs in the deeper waters of the basin. The hydrocarbon plays in these fan deposits are mainly stratigraphic in nature.

Hydrocarbon Entrapments

Though structural traps do exist in the K-G basin, most of them are only small to medium in size. In contrast all up-dip pinch outs, unconformity surfaces, permeability barriers and wedge outs play an important role in the entrapment of hydrocarbons.

Upper Jurassic, Lower Cretaceous sandstones are truncated against the pre-existing basement highs. The Raghavapuram shale which is overlying the Lower Cretaceous transgressive shale may be acting as the regional seal. The lime stones and sandstone beds within the Raghavapuram shale are also good reservoir rocks. Stratigraphic pays include turbidite fans located southeast of Paleocene hinge and also in the deeper waters of the basin.

Fractured basalts and also the intratrappean beds are good reservoirs and proved to contain commertial hydrocarbons (Razole). Thick and unfractured basalts covering the fractured basalts and intratrappeans act as local seals. Sandstone beds interspersed with clay beds, where the clay beds may act as local seals for entrapment. Growth faults with associated rollover anticlines having sandstone bodies act as potential reservoirs for hydrocarbon entrapment. The Mio-Pliocene unconformities also play a major role for hydrocarbon entrapment. There are a number of hydrocarbon plays in the basin. Some hydrocarbon plays are exclusive for the stratigraphic section and for its hydrocarbon play.

Annapurna was drilled to a depth of 2330 m from sea level and it encountered gas zone below 1870 m. The well flowed gas @ 2.3 MMSCMD through 1" choke. The commercial estimate and reserves was around 1 TCF. Annapurna prospect was located in a water depth of 750 m.

'Kanakadurga prospect P' was drilled in a water depth of 493 m and drilled to a depth of 2585 m. It encountered two significant hydrocarbon bearing zones. An interval tested at (2268-2285m), flowed oil @ 3650 bopd through $^5/_8$" choke. The initial estimate of recoverable reserves of oil was around 11.5 MMT, the associated gas from this zone and the reserves from the second hydrocarbon bearing zone are yet to be assessed (Fig.6.3.18).

Padmavathi (M) prospect drilled in a water depth of 500 m, has two intervals which proved to be oil and gas bearing. The deeper zone at 2092-2081m flowed 4500 bopd of oil and 2.7 MMCFD of gas through 5/8" choke. The recoverable reserves estimate was in the order of 5.5 to 11.0 MMT. Here also, the associated gas and the productivity of the second zone are yet to be assessed. In 2003, Cairn had farmed out 90% of its participating interests to ONGC. ONGC became the operator for the block at present. ONGC's efforts in the block led to a major discoveries.

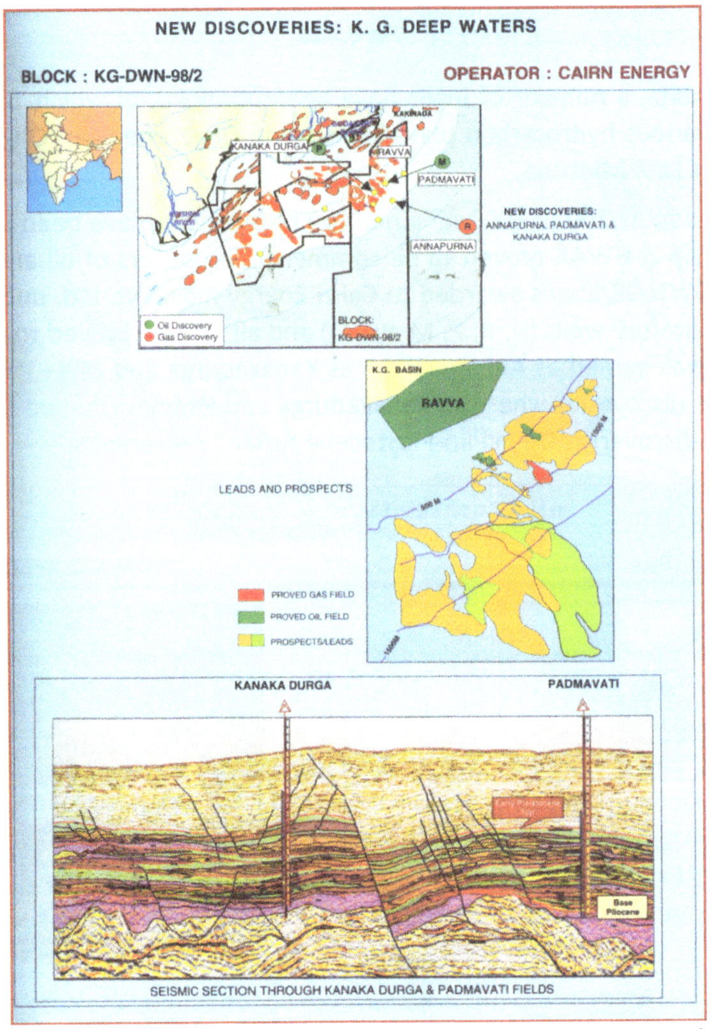

(Courtesy DGH)

Fig. 6.3.18

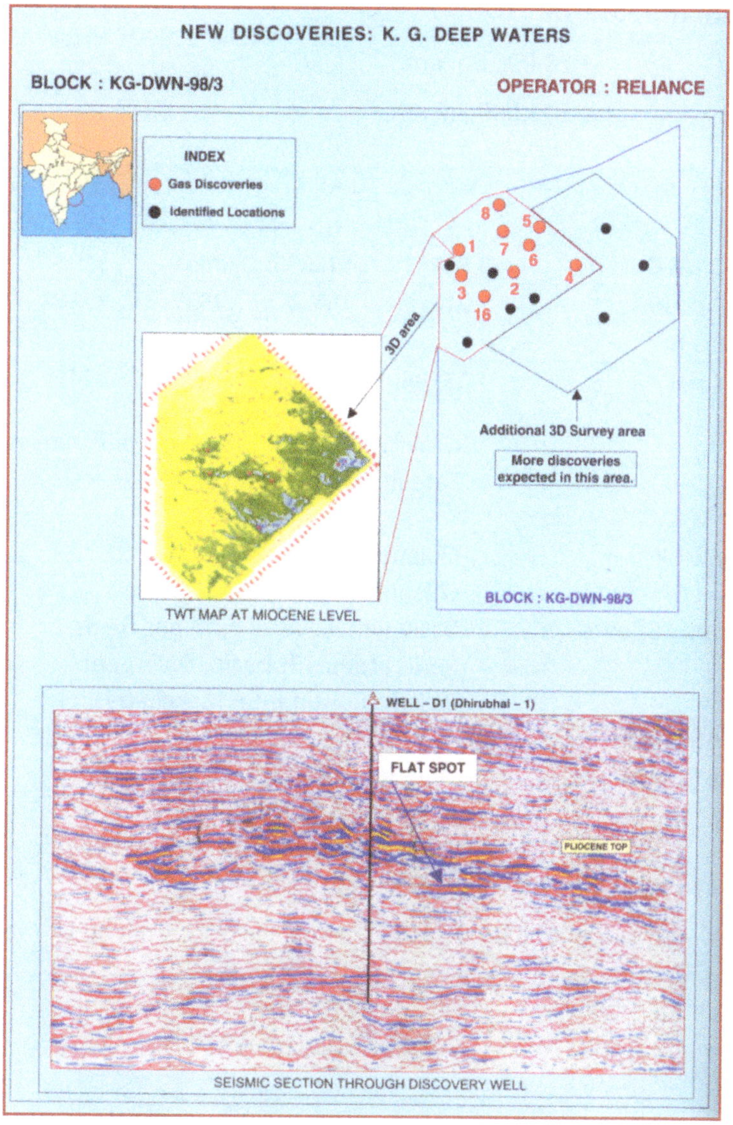

(Courtesy DGH)

Fig. 6.3.19

GS-23 Field

The field is situated 55 Kms south-southwest of Kakinada Town in 10-15 m water depth, off Amalapuram coast, and was discovered by ONGC in March 1994. The structure is the rollover anticline formed due to growth fault. The reservoirs are mainly Upper Oligocene to Miocene. The well penetrated 4 reservoir sands, in which three are oil and another one tested for gas. The field is having inplace oil reserves of 1.47 MMT of which 0.22 MMT is recoverable. The field is also having inplace gas reserves of 650 MMm3 of which 397 MMm^3 is recoverable. In the same lateral extent, another well GS-38 was drilled by ONGC and this well is having gas reservoirs (IGIP of 196 MMm^3 of which 118 MMm^3 gas is recoverable). Another well GS-8 drilled in the same tectonic set-up had IGIP of 203 MMm^3 gas with a recoverable component of 20 MMm^3 gas.

The exploration and development of deep water blocks of Krishna-Godavari offshore, beyond 400 mts water depth are discussed in Chapter 7.

Exploration & Production Status

Basin Area **Onland** : 28,000 Sq. Km.

Offshore : 145,000 Sq. Km.

Seismic Data

Data acquired by	2D (LKM)		3D (Sq. Km.)	
	On land	Offshore	On land	Offshore
O NG C	34481	611130	3051	5198
Pvt./JV	715	16813	453	17267
DG H	-	11748	-	-
Total	35196	89691	3504	32465

Other Surveys : Gravity Magnetic Surveys along with seismic surveys

: Satellite Gravity Survey

Well Data

Number of Exploratory w ells : Onland – 375

: Offshore – 198

Hydrocarbon Discoveries / Shows : Basin with several oil & gas fields

Basin Type : Pericratonic rift basin, Divergent Passive Continental Margin basin

Reserves & Production Status :

Hydrocarbon Resource Potential : Onland: 575 MMT

Offshore: 555 MMT

Total: 1130 MMT

Blocks on Offer (NELP I to VIII)

The Blocks offered and awarded under production sharing contract from NELP I to VIII are shown in the Fig 6.3.20 and Table.6.3.T1.

(Courtesy DGH)

Fig. 6.3.20

Table 6.3.T.2:

NELP	Blocks Offered	Blocks Awarded	Awardees	Date of Signing/ Effective Date	Present area (sq.km)	Status
I		KG-DWN-98/1	RIL(100%-OP)	12-Apr-2000/ 07-Jun-2000	6700	
		KG-DWN-98/2	ONGC (65%-OP), CEIL (10%), PIBBV (15%), HOEIBV (10%)	12-Apr-2000/ 12-Apr-2000	7294.60	
		KG-DWN-98/3	RIL(90%-OP), NIKO (10%)	12-Apr-2000/ 07-Jun-2000	7645	
		KG-DWN-98/4	ONGC (85%-OP), OIL (15%)	12-Apr-2000/ 19-May-2000	4970	
		KG-DWN-98/5	ONGC (100%-OP)	12-Apr-2000/ 19-May-2000	4490	
		KG-OSN-97/1	OIL(100%-OP)		Relinquished	
		KG-OSN-97/2	RIL(100%-OP)	12-Apr-2000	Relinquished	
		KG-OSN-97/3	RIL(100%-OP)	12-Apr-2000	Relinquished	
		KG-OSN-97/4	RIL(100%-OP)	12-Apr-2000	Relinquished	
II						
III	KG-DWN-2001/1	KG-DWN-2001/1	RIL (90%-OP), HEPI (10%)	04-Feb-2003/ 03-Apr-2003	11605	
	KG-OSN-2001/1	KG-OSN-2001/1	RIL (100%-OP)	04-Feb-2003/ 03-Apr-2003	1100	
	KG-OSN-2001/2	KG-OSN-2001/2	RIL (100%-OP)	04-Feb-2003/ 03-Apr-2003	210	
	KG-OSN-2001/3	KG-OSN-2001/3	GSPC (80%-OP), GGR (10%), JOGPL (10%)	04-Feb-2003/ 12-Mar-2003	925	
IV	KG-DWN-2002/1	KG-DWN-2002/1	ONGC (70%-OP), OIL (20%), BPCL (10%)	06-Feb-2004/ 17-Mar-2004	10600	
V	KG-DWN-2003/1	KG-DWN-2003/1	RIL (90%-OP), HEPI (10%)	23-Sep-2005/ 05-Dec-2005	3288	
VI	KG-DWN-2004/1	KG-DWN-2004/1	ONGC (70%-OP) GSPC (10%), HPCL (10%), GAIL (10%)	02-Mar-2007/ 15-May-2007	11951	
	KG-DWN-2004/2	KG-DWN-2004/2	ONGC (60%-OP), GSPC (10%), HPCL (10%), BPCL (10%), GAIL (10%)	02-Mar-2007/ 07-May-2007	11851	
	KG-DWN-2004/3	KG-DWN-2004/3	ONGC (70%-OP) GSPC (10%), HPCL (10%), GAIL (10)	02-Mar-2007/ 08-May-2007	6205	
	KG-DWN-2004/4	KG-DWN-2004/4	RIL (100%-OP)	02-Mar-2007/ 21-May-2007	11904	
	KG-DWN-2004/5	KG-DWN-2004/5	ONGC (50-OP), GSPC (10%), HPCL (10%), GAIL (10%), OIL (10%), BPCL (10%)	02-Mar-2007/ 23-May-2007	11922	
	KG-DWN-2004/6	KG-DWN-2004/6	ONGC (60%-OP) GSPC (10%), HPCL (10%), GAIL (10%), OIL	02-Mar-2007/ 23-May-2007	10907	

Table.6.3.T.2: ***Contd...***

NELP	Blocks Offered	Blocks Awarded	Awardees	Date of Signing/ Effective Date	Present area (sq.km)	Status
	KG-DWN-2004/7	KG-DWN-2004/7	RIL (100%-OP)	02-Mar-2007/ 23-May-2007	11856	
	KG-OSN-2004/1	KG-OSN-2004/1	ONGC (55%-OP), BG (45%)	02-Mar-2007/ 25-May-2007	1151	
VII	KG-DWN-2005/1	KG-DWN-2005/1	ONGC (70%-OP), IOCL (20%), GSPC (10%)	22-Dec-2008/	1727	
	KG-DWN-2005/2	KG-DWN-2005/2	BP Exploration (Alpha) Ltd (30%-OP), RIL (70%)	22-Dec-2008/	1949	
	KG-DWN-2005/3					
	KG-OSN-2005/1	KG-OSN-2005/1	ONGC (60%-OP), GSPC (20%), HPCL- Mittal Energy Ltd (20%)	22-Dec-2008/	2810	
	KG-OSN-2005/2	KG-OSN-2005/2	ONGC (80%-OP), HPCL (20%)	22-Dec-2008/	1881	
	KG-OSN-2005/3					
VIII	KG - DWN-2009/1	KG - DWN-2009/1	BGEPIL (30%-OP), ONGC (45%), OIL(15)%, APGIC (10)%	30-Jun-2010/	1800	
	KG -OSN-2009/1	KG -OSN-2009/1	ONGC (80%-OP), APGIC (10%), NTPC (10%)	30-Jun-2010/	1472	
	KG -OSN-2009/2	KG -OSN-2009/2	ONGC (90%-OP), Andhra Pradesh Gas Infrastructure Corporation Pvt. Ltd. (10%)	30-Jun-2010/	1471	
	KG -OSN-2009/3	KG -OSN-2009/3	CAIRN ENERGY INDIA PTY. LTD(10%-OP)-Cairn India Limited(CIL-90)	30-Jun-2010/	1988	
	KG -OSN-2009/4	KG -OSN-2009/4	ONGC (50%-OP), OIL (30%), NTPC (10%), Andhra Pradesh Gas Infrastructure Corp. Pvt.Ltd. (APGIC-10%)	30-Jun-2010/	835	

(Courtesy DDH)

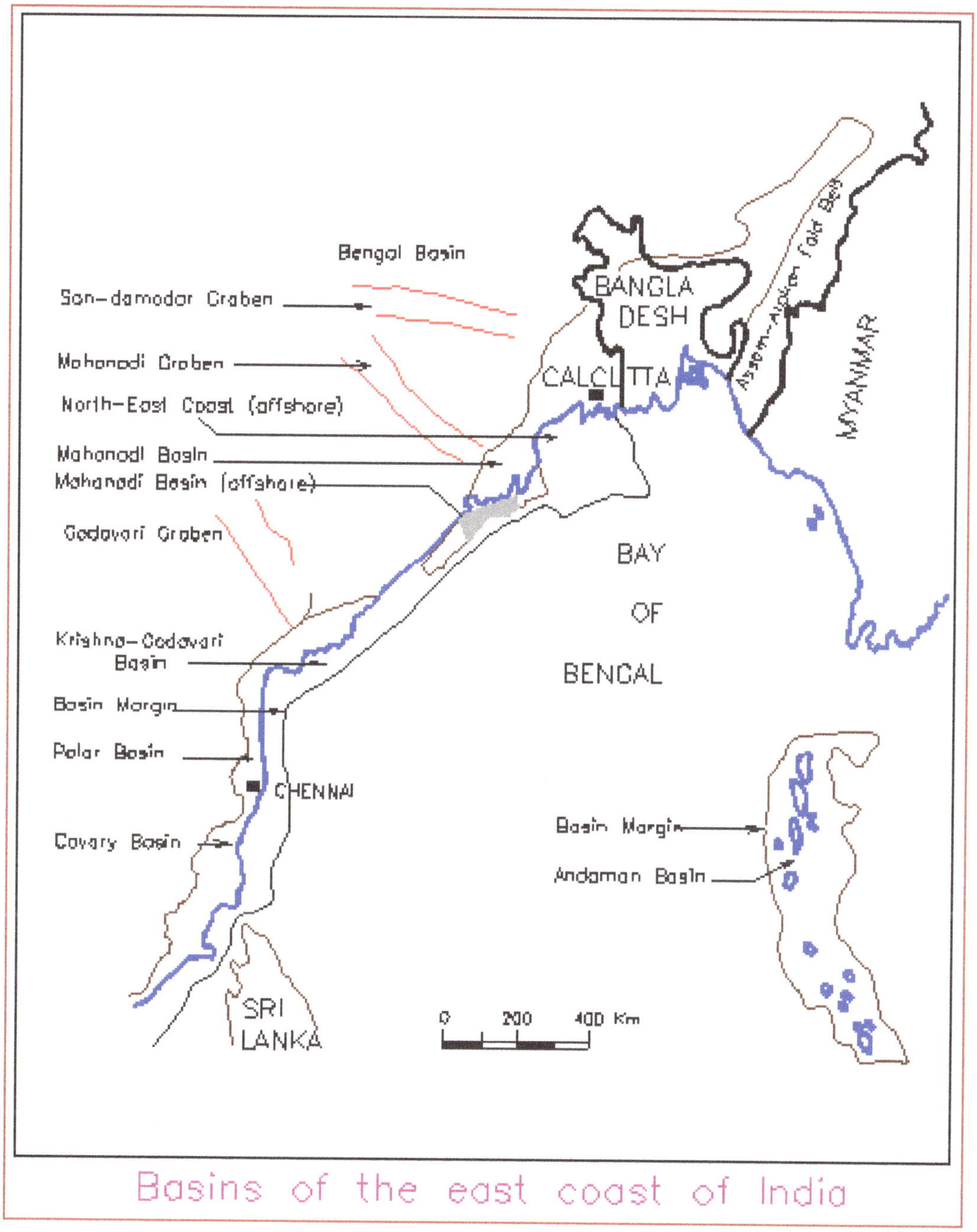

(Courtesy DGH)

Fig. 6.3.t.1

6.4 Mahanadi Offshore Basin

Mahanadi offshore basin is situated on the east cost of India, to the northeast of Krishna – Godavari basin and south of Bengal basin. It is peri-cratonic passive continental margin basin, formed due to rifting and breakup of Gondwana land and subsequent drifting of Indian Plate. The onshore Mahanadi basin has a geological extension towards east into offshore Bay of Bengal (Fig. 6.4.1). The basin covers a total area of 55000 sq.km of which 14000sq.km is in the shallow offshore area. The shelf- break occurs at around 150 mt isobath and the rest of the area is in the deep water part of the basin.

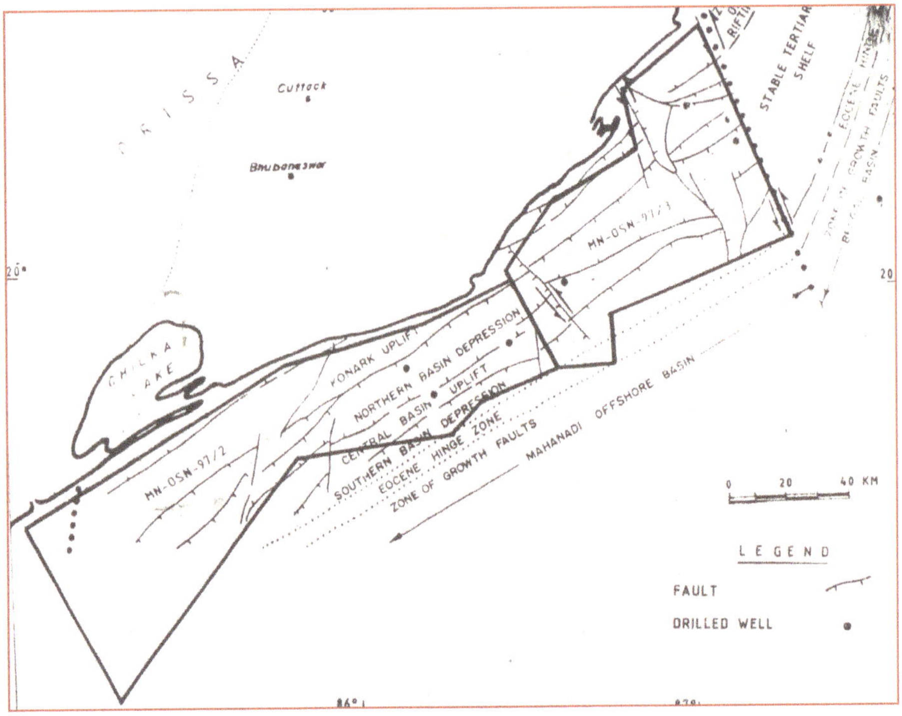

(Courtesy DGH)

Fig 6.4.1

The prognosticated resources of the basin are around 145 MMT, of which 100 MMT is estimated for offshore part of the basin up to 200 mt isobath. Deep water part of the basin holds huge and significant additional hydrocarbon potential.

The onland part of the Mahanadi basin is limited Northwest and West by Precambrian out crops belonging to the Indian crystalline shield. Towards NE it merges in the Northeast Coast (NEC) of Bengal basin. The SW limit is marked by basement high/ ridge known as 85^0 E ridge. To the West and Southeast of the ridge the basin deepens and is sealed, similar to KG-basin, till the end of Mesozoic Era. The shallow offshore part of the basin lies off the coast of Andhra Pradesh and Orissa. Mahanadi basin is immediately north Of KG basin abutting the Mahanadi on land graben of the East coast of India, and its passive margin came into existence with breakup of Gondwana land. Mahanadi basin is developing around one of the triple junctions on the East coast of India. Predominantly, continental sediments of Paleozoic and Mesozoic were deposited in the rift valleys around the triple junctions as synrift sediments. A transitional zone varying from continental crust to oceanic crust of varying width is present in various parts of the basin, immediately to the east of 85^0 E ridge. The age of the 85^0 E ridge which is a part of oceanic basement/trap may be of early cretaceous. During the late Oligocene/Miocene period, the Mahanadi-NEC area is very much influenced by Bengal deltaic sedimentation system, where the Mahanadi river delta coalesces with Bengal fan over 85^0 E ridge.

The geological section across Mahanadi Basin is shown in Fig. 6.4.2.

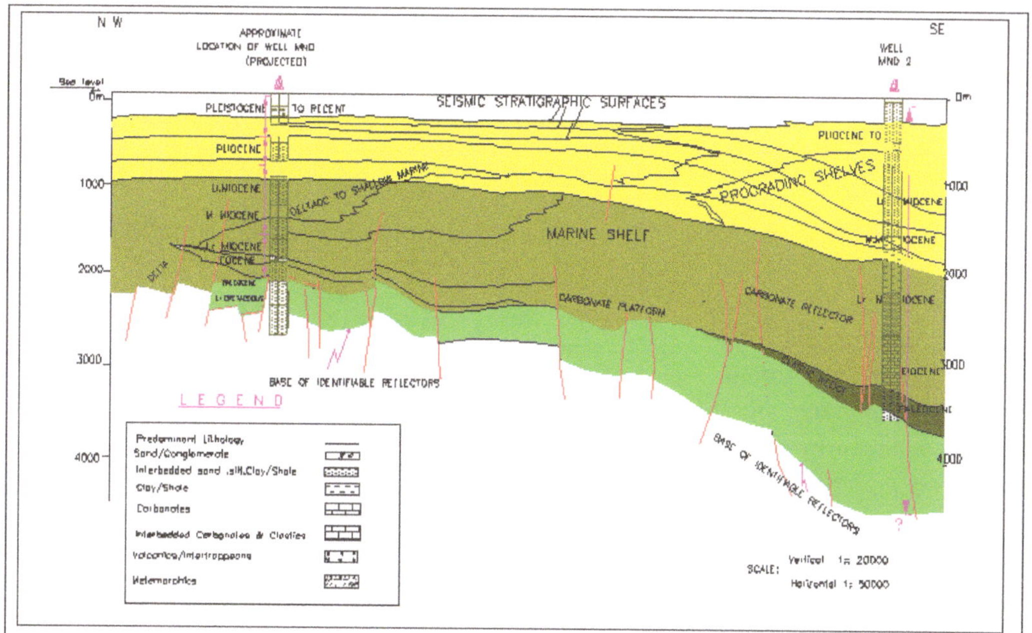

Seismo geological section across Mahanadi wells 1 and 2

(Courtesy DGH)

Fig. 6.4.2

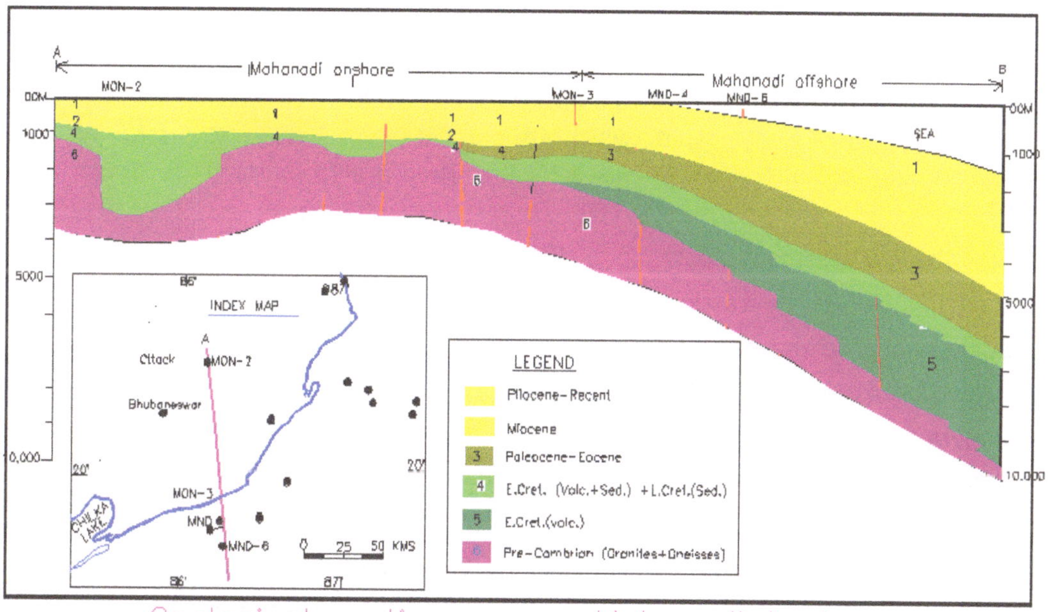

Geological section across Mahanadi basin

(Courtesy DGH)

Fig. 6.4.2.a

Basin Evolution and Stratigraphy

The evolution of east coast of India as a passive margin resulted in the formation of a number peri-cratonic basins namely Bengal, Mahanadi, KG, Pennar and Palar etc. The eastern continental margin of India represents a rifted passive margin evolved in response to continental rifting. India separated from Southern Super Continent Gondwana land during the late Jurassic to Early Cretaceous, though the rift initiation started much earlier. The NW-SE trending Mahanadi Graben is the failed arm of the rift associated with triple junction. Predominantly continental sediments of upper Paleozoic and Mesozoic age were deposited in Mahanadi basin. Rifting and synrift fill of the Mahanadi graben was continuing in Jurassic/ Early cretaceous.

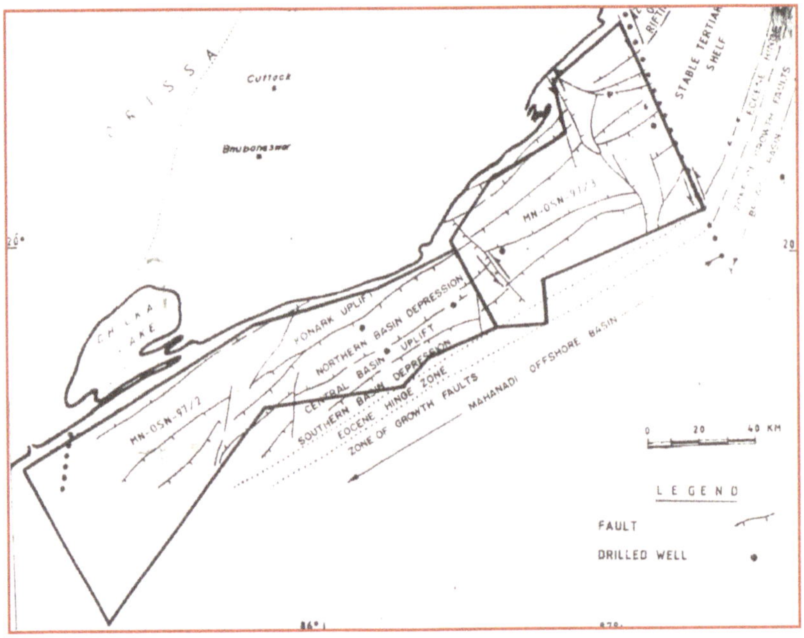

(Courtesy DGH)

Fig. 6.4.3

Rift Stage

Mahanadi Basin a peri-cratonic basin was initiated probably during Late Jurassic by rifting and subsidence of the Pre-Cambrian basement by a number of major faults with a dominant ENE-WSW trend and subordinate NNE-SSW and NNW-SSE trends (Fig. 6.4.4). The faults, which are parallel to the Eastern Ghats NE-SW trend divided basin into a number of linear depressions & Uplift. They are from North to South are as below (Fig. 6.4.5).

- Badrak shallow basement block
- Cuttack-Chand – Bali depression.
- Bhubaneswar – Kendra-Para up lift.
- Paradap Depression.
- Nimapara – Balikuda up lift.
- Puri depression
- Konark uplift.

The Konark uplift extends in to shallow offshore part of the basin. Further southeastward, the shallow offshore basin is also divided in to a number of Horst and Grabens up to Eocene hinge zone.

Further to NE, the Eocene hinge zone takes a sudden NE – SW swing and further northeastward, it assumes almost a north-south trend which is now referred as NEC area.

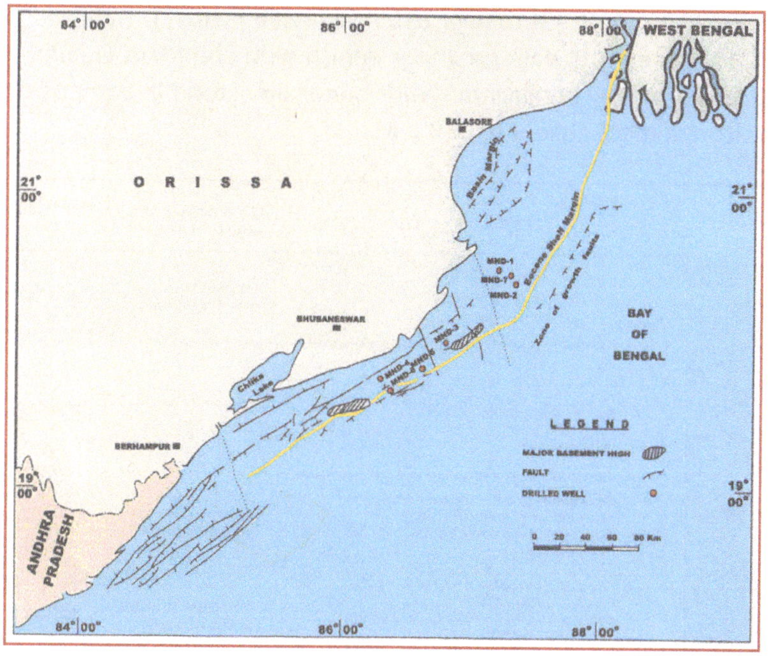

(Courtesy DGH)

Fig. 6.4.4

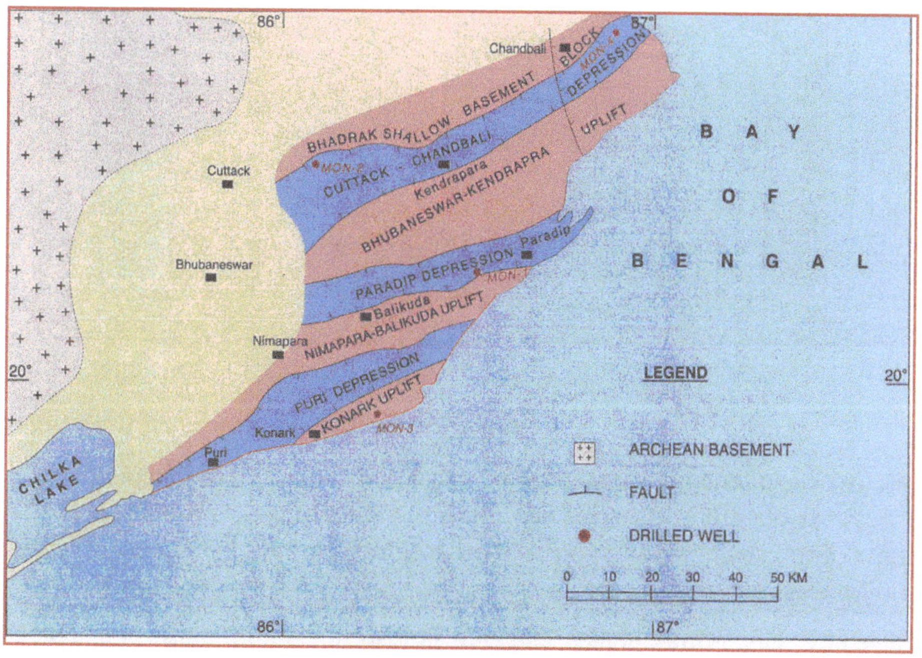

(Courtesy DGH)

Fig. 6.4.5

Generalised Litho Stratigraphy of Offshore Basin

Lithostratigraphic information obtained in different wells drilled in Mahanadi offshore and adjacent NEC area and the Generalised stratigraphy is shown in Fig. 6.4.6

Basement: The continental crust / basement in Mahanadi offshore basin is made up of Eastern Ghats, granulites and gneisses of pre Cambrian age. The Oceanic basement though it is not drilled in any of the wells, it is likely to be volcanic rock of early cretaceous age.

Early Crataceous: Predominantly volcanic sequences made up of basalts tuffs and intertrappeans unconformabily overlie the Precambrian basement. The thickness of volcanics varies abruptly from well to well, and ranges from 25 mt (MND-3) to over 850 mt (in well MND-7). The basalts varying from nearly glassy to medium grained in texture, dark grey to greenish with abundant Quartz, calcite and feldspars. The intertrappean beds are mainly argillaceous with minor limestone in composition and the thickness of inter trappens vary from almost absent to 150 mt.

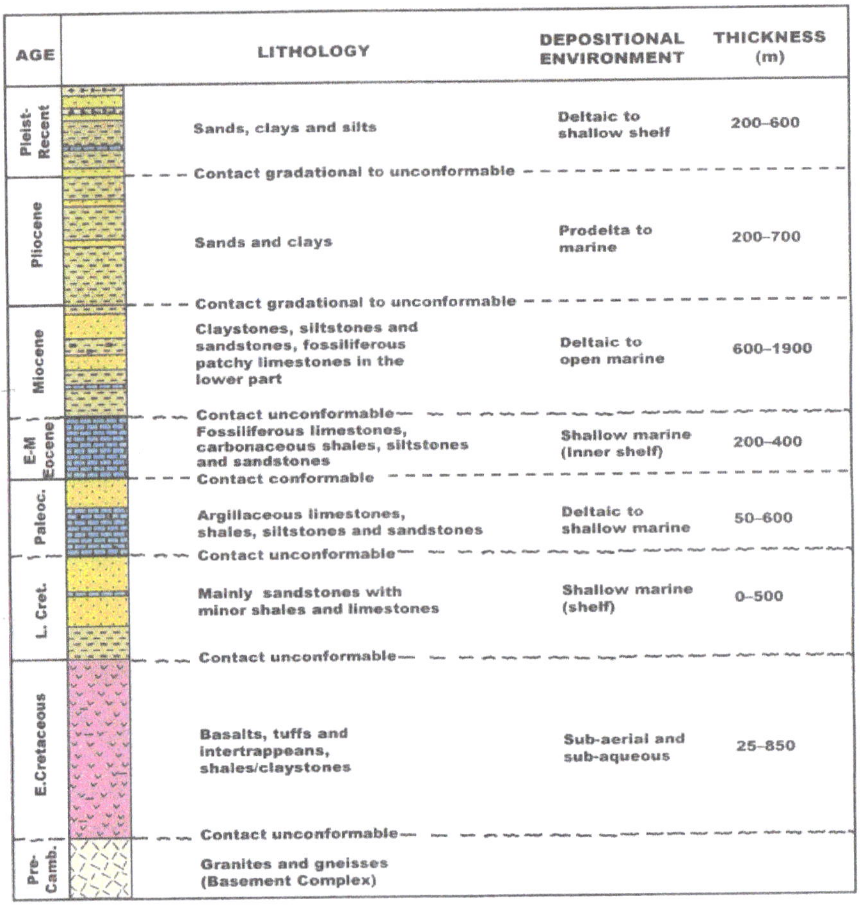

(Courtesy DGH)

Fig. 6.4.6

Late Crataceous: Thick sequence of sand stone and shales with minor limestone unconformably overlie the volcanoclastic rocks of early cretaceous age. The sediments range from shallow shelf in the offshore part. The thickness of this stratigraphic sequence varies widely in the offshore wells ranging from almost zero to 500 mt in MND-5.

Paleocene: Argillaceous limestones, shales, silts and sandstones are greenish in colour; unconformably overlie the cretaceous sediments with an erosional type unconfirmity. The thickness of Paleocene stratigraphic unit varies from 35 mt in MND-1 to over 560 mt in MND-3.

Eocene: Eocene rocks (fossiloferrous limestone with subordinate shale and fine grained sandstone), conformably overlie the Paleocene rocks, are mainly of shelfal facies, grade into shales beyond Eocene shelf edge.

Oligocene: On the top of Eocene there, is a wide spread erosional unconformity, both onshore and offshore part of Mahanadi basin. Oligocene sedimentation is not seen. However, in deep offshore, Oligocene sediments seem to wedgeout against Eocene shelf.

Miocene: Thick clastic sequence of Miocene rests unconformbly over the Eocene carbonates. In the southern part of the Mahanadi offshore in well MND-4 Argillaceous limestone (150mt) was developed in

Miocene. The thickness varies from 600 to 2000 mt towards the basin side in the offshore. Miocene clastic sediments were deposited broadly in deltaic to open marine conditions.

Pliocene: The Pliocene sequence is made up of coarse to medium grained unconsolidated sands, silts and clays and this unit rests unconformably over the Miocene clastics. The sediments were deposited as a deltaic sequence from shallow shelf to pro deltaic, with essentially marine conditions of deposition.

Pleistocene to Recent: This is the youngest stratigraphic unit made up of loose sands, silts and clays mainly of deltaic nature. And this stratigraphic unit conformably rests over Pliocene and often it is very difficult to differentiate these two units separately. For all practical purposes, this stratigraphic unit can be referred as plio- Pleistocene to recent deltaic unit. Since the shelf is narrow in this part of the basin, the larger part of Plio – Pleistocene and younger sediments by-pass the shelf and slope, and get deposited in the abyssal plane (submarine fans and turbidities fan deposits). These sediments should have major stratigraphic play for hydrocarbon generation, migration and entrapment.

2D seismic was obtained by various agencies (OIL, RIL and ONGC etc...) from 1978 to 2010. Over 25000 lkm of 2D data and 8650 sq.km of 3D data were obtained in the basin. Satellite gravity and also normal GM surveys were carried out in the basin. Over 25 exploratory wells were drilled mostly in the offshore part of the basin. Geochemical sniffer surveys were carried out for around 1500 lkm in the offshore part by M/S Inter Ocean of USA during 1988. The results of the sniffer surveys are given in table (6.4.T1). Geochemical anomalies had also been mapped in the basin. MND-1 and MND-2 were drilled during 1980. Only Hydrocarbon indications (Gas up to C4 and oil fluorescence) were observed while drilling, but no commercial discoveries on testing. Subsequently MND-4, 5 and 6 were drilled in MN-ONS-92/2 & MND 3&7 in MN-ONS-97/3 blocks. Again, hydrocarbon indications only were obtained during drilling. Formations pressures are hydrostatic and temperature gradient ranges from 2.5°C to 3.5°C per 100 mt. But just North of Mahanadi Basin in NEC block (NEC-ONS/92/2) RIL made a number of discoveries (like D9, 10, 11, 15,20,21,32 & 40) during the period 2005 to 2008 and the discoveries are at various stages of development.

Structure: Three major structural stages/ elements can be recognized in the geological evolution of Mahanadi Basin. They are

1. Mesozoic (Cretaceous)
2. Paleogene (Paleocene & Eocene)
3. Neogene (Miocene to Recent)

The Mahanadi onshore and shallow part of offshore basin upto shelf margin are traversed by a number of major faults running ENE – WSW, NNE –SSW and NNW – SSE trends of which ENE – WSW trend is prominent and almost parallel to the present day coast line. These faults have made the basin into a number of linear depressions and uplifts. NNW – SSE trending faults have significant lateral movements and strike slip components in the development of the basin. The Mahanadi Basin and NEC basins are separated by a major transverse fault, which has made a remarkable change fromNE-SW to N-S trend. In the Mesozoic and Early part of Paleogene, *'the horst – graben'* type tectonics prevail in the basin. At the top of Eocene, there is a wide spread unconformity in the basin. Several anticlines and fault controlled, inverted anticlinal structures have been mapped seismically.

Hydrocarbon Prospects

Source Rock

Significant hydrocarbon shows have been encountered in most of the exploratory wells drilled in Mahanadi Offshore Basin. Presence of adequate organic matter (TOC-1.5 to 2.5) was detected in most of the samples collected from the drilled wells. This is quite adequate for generation of hydrocarbons.

Organic matter within Paleogene sequence is indicative to be oil prone which reached adequate maturity; Ro is 0.55 @ 2800mt. and 0.65 @ 3300mt. depths indicating an Oil window in MND-2 well. Better source rocks may be available in the deeper parts of the basin.

Rock evaluation studies indicate migrated hydrocarbon in Miocene and older sediments of offshore wells. Geochemical studies indicated a number of positive sniffer anomalies in the deeper part of the basin. Good matured source rocks are also expected in deeper parts of the basin.

Reservoir Rocks

Good reservoir rocks are present throughout the stratigraphic section of shelfal part of the basin. Carbonate reservoirs with vuggy porosity and permeability and also sandstone reservoirs have been mapped in the Late Cretaceous, Paleogene and Early Miocene sequences. The carbonate rocks have good porosity, (>15%) and permeability and they become good reservoir rocks. Paleogene sandstones have good porosities and permeabilities to become good hydrocarbon reservoir rocks.

The intertrappean sequences with associated fractures and weathered volcanics are also potential reservoir rocks.

Reservoir sands with in channel/levee complexes of Pliocene and Pliostosene and Late Oligocene wedgeouts against Eocene Hinge in the offshore part of the basin are also the potential reservoir targets. Late Miocene-Pliocene submarine canyon systems and turbidite fan deposits in deep water basins are going to be excellent reservoirs.

Traps

Various types of structural, stratistructural and stratigraphic traps are present in the basin. These traps can be classified as below.

1. Basement controlled structures in the Mesozoic and Paleogene levels.
2. Structural closures against the shelf margin fault systems, affecting both Paleogene and Neogene sediments.
3. Carbonate mounds and shelf margin reef buildups in Paleocene and Eocene.
4. Wedgeouts in Mesozoic against the basement highs.
5. Eocene wedgeouts in the shelf margin.
6. Morphological erosional unconformities at the tops of Cretaceous and Eocene.
7. Channel cut and fills in the Mio – Pliocene section.

Mio-Pliocene submarine fans and also turbidite fans beyond shelf edge, Growth faults and associated Neogene tectonics like rollovers, shale diapers etc. are ideal candidates for hydrocarbon migration and entrapments.

Entrapment

Good entrapments of hydrocarbons can be expected in Structural and fault related traps in synrift cretaceous sediments having coarse grained sandstones (porosity 15-25%) with good permeability coupled with the availability of source facies within the same sequence. These potential traps may be sealed vertically by Paleocene/Eocene shales. In synrift basins, faulted anticlines, fault related structures and positive erosional features over the cretaceous unconformity play a dominant role in entrapping the hydrocarbons.

Wedge Outs & Pinchouts (Paleocene, Eocene & Miocene)

The sub aqueous 85° E ridge is a dominant structural feature with in the basin, which controls the entrapment of hydrocarbons in the basin since it acts as a barrier, onto which Paleocene, Eocene, Oligocene and Miocene sediments onlap, with good entrapment potential. The eastern tilt in the basin would facilitate hydrocarbon migration towards N-W, and the wedge out against 85^0 E ridge will be the potential stratigraphically controlled hydrocarbon play. Even the wedge outs against Eocene hinge zone could also have potential entrapment conditions.

Cap Rocks

Presence of clay stones and shales at different stratigraphic levels overlying the various types of reservoirs are considered to be good caps. The thick sequence of clay in Miocene section acts as a regional cap for hydrocarbon accumulation.

Submarine Canyons & Turbidite Fan Systems

Submarine canyon and turbidite fan systems are more characteristic for Mio-Pliocene and Pliocene sequences. This type of hydrocarbon play is expected in deeper waters of the basin as the sediment input brought into the basin by the rivers bypasses the shelf, slope system and gets deposited by way of sediment slump and fans out on the abyssal planes in high stand and rapid sedimentation conditions. Generally, such sedimentation is carried into the basin through major transverse lineaments. Erosional channels subsequently filled with sediments will have potential entrapment conditions. Thus Mahanadi basin, specially the shallow and deep water off shore area is endowed with good combinations of source, and reservoir facies with potential entrapment situations. Shelf to slope and the deep offshore part of the basin with characteristic sediment induced Neogene tectonics coupled with good reservoir and source facies and favourable entrapment conditions (Growth faults, rollover structures, drapover structures on shale diapirs) will be the excellent target for hydrocarbon exploration.

The presence of oil and gas had been recorded in a number of wells and several positive anomalies had been brought out by geochemical sniffer surveys in the offshore areas. The NEC basin is clubbed with Mahanadi basin for a broader outlook. The NEC basin is controlled by NW - SE trending Mahanadi graben in south and Dhamodar graben to the North. In the shallow offshore block NEC –ONS-97/2 RIL made a number of discoveries. They are referred as Dhirubhai Ambani discoveries (D9, 10, 11, 15, 20, 21, 32, and 40). A development plan for D9, 10, 11, 15, 20, and 21 was submitted to the Ministry which was approved and development drilling is planned.

Geocemical Sniffer Survey Results in Mahanadi Offshore Basin

Background Analysis

ALL DATA						
	Methane	**Ethane**	**Propane**	**I-Butane**	**N-Butane**	**THC**
Average	20.118	0.060	0.59	0.004	0.011	26.34
Sigma	13.798	0.043	0.057	0.007	0.030	14.78
AVG + Sig	33.911	0.103	0.116	0.011	0.042	41.13
Dta Pts	2336	2336	2336	1626	2131	2336
Maximum	151.450	1.347	1.920	0.214	1.031	161.94
Minimum	4.865	0.024	0.005	0.000	0.000	6.99
BACKGROUND						
Average	16.670	0.055	0.056	0.003	0.009	22.39
Sigma	6.044	0.015	0.015	0.002	0.006	7.31
AVG+SIG	22.715	0.071	0.071	0.005	0.015	29.70
Dta Pts	2132	2243	2306	1596	2160	2098

STRATIGRAPHIC BREAKUP OF WELLS DRILLED IN MAHANADI OFFSHORE AND ADJOINING AREAS

WELLS	MND-1	MND-2	MND-3	MND-4	MND-5	MND-6	MND-7	MON-1	MON-2	MON-3	MON-4	NEC-1	NEC-1A	NEC-3	NEC-4
BASE PLIOCENE	820	794?	991	408	1100	1052	927	441	540?	433	420	2146	2318	1064	4293
BASE MIOCENE	1835	2677	1755	1110	1727	1760	2463	1258	915	798	997	3201	3190	1542	4642+
BASE OLIGOCENE	--	--	--	--	--	--	--	--	--	--	--	3777	3620+	--	--
BASE EOCENE	2085	3055	1992	1407	2015	2029	2712	--	--	1131	--	4597	--	1588	--
BASE PALEOCENE	--	3488	2558	1894	2497	2248+	2961	--	--	1235	1243	4835	--	2001	--
BASE LATE CRTACEOUS	--	--	2801	2191	2984	--	--	--	--	--	--	5068	--	--	--
BASE EARLY CRTACEOUS	2435	3650+	2825	2238+	3810	--	3820	--	2993+	1914+	2222	5208+	--	2108	--
TD	2740 (BMT)	3650	3284 (BMT)	2238	4026 (BMT)	2248	3888 (BMT)	1393 (BMT)	2993	1914	2408 (BMT)	5208	3620	2302	4642

6.5 Bengal Basin

Introduction

The Bengal Basin, the present day largest delta of the world, formed by two mighty rivers Ganges and Brahmaputra flowing into Bay of Bengal. The delta system has also rivers like Damodar, Subarnarekha, Ajoy, Bhagiradhi, and Roopnarayan etc. The Bengal basin is located in the northeastern part of Indian peninsula (Fig. 6.5.1). The basin covers an area of over 1,70,000 sq.kms of which the onland part covers 32,000 sq.kms and offshore covers 33,700 sq.kms up to 200 m bathymetry . The remaining area is in the deep waters of the basin.

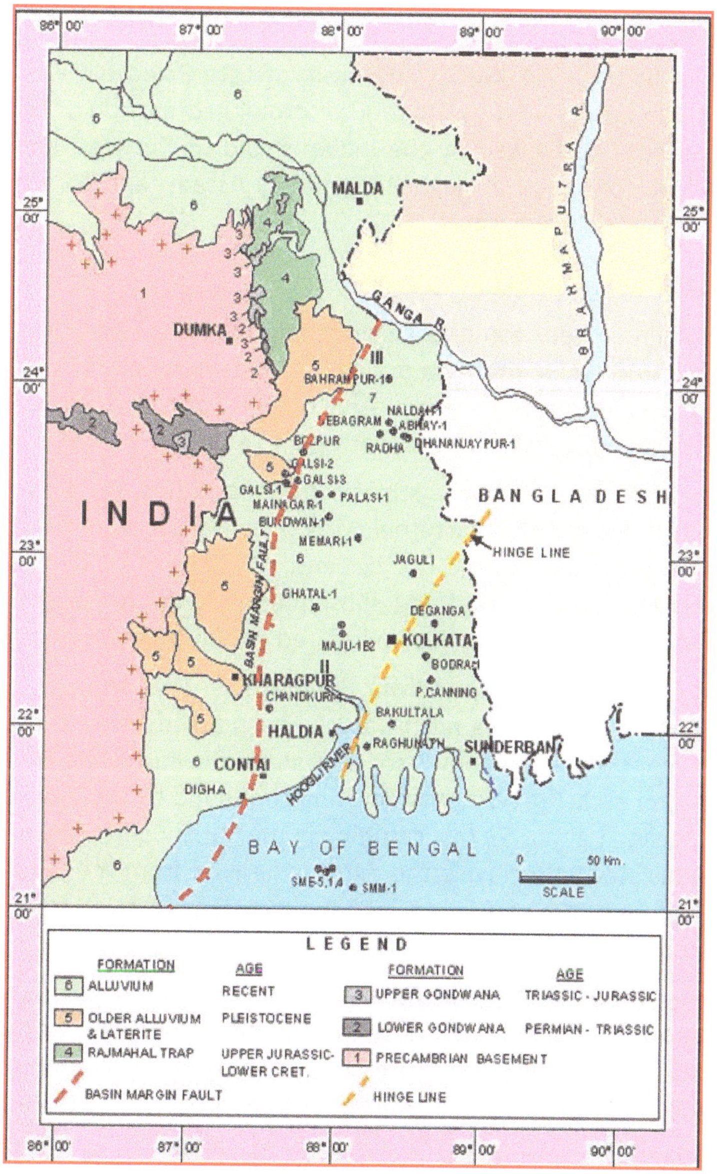

(Courtesy DGH)

Fig. 6.5.1

Hydrocarbon exploration in Bengal basin is going on for more than 50 years. Over this period, voluminous geoscientific data had been acquired and utilized for understanding the basin. Independent evaluation of the basin had been carried out by Multinational Companies like Total, CFP of France, Markevich Group of USSR and Machino Import of USSR etc., and rated the potential of the basin as very high for Hydrocarbon generation and accumulation.

The prognosticated hydrocarbon resources of the basin were estimated at 518 MMT of oil and OEG. Many of the 38 wells drilled so far in the basin by Indostanvac Co., ONGC and Carlsberg – Natamos in the basin have indicated many oil and gas shows. Liquid and Gaseous hydrocarbons were seen and tested in Ichapur and Golf green structures situated on the platform, near shelf margin area. The reservoir is Oligocene sandstone in Ichapur well and fractured Eocene limestone in Golf Green well.

Paleocene – Cretaceous and Oligocene sequences are the prime exploration targets in the platform areas whereas Miocene and Mio Pliocene reservoirs are the primary targets in the deep sea basins. Gondwana sequence is also an exploration target in the southern parts of shelf area. Porosity pods and fractured porosity within the Eocene limestone near hinge zone likely to serve as a good reservoir facies.

Tectonics and Stratigraphy

Bengal basin is a Poly-Cyclic basin i.e. Gondwana basin of Late Paleozoic – Early Mesozoic era, over ridden by the Tertiary basin, intervened by a period of prolonged erosion of Gondwana rocks, rift and volcanic activity. Bengal basin is flanked by the Indian shield to the west and Shillong massif in the northeast. To the east, it continues through Bangladesh, Assam, and to west Burma, forming the extensive Assam Arakan Geosynclinal basin.

Evolution of the Basin

Bengal basin owes its origin to very complex geo tectonic activities operating over a long interval of time. It developed as an intracratonic rift basin during Late Mesozoic to Mid Mesozoic times (Gondwana basin). Subsequently it changed into a modern passive margin basin during Mesozoic and Early Tertiary time, and late in the Late Paleogene time, got converted into a part of a convergent margin.

During carboniferous time, the Gondwana super continent underwent a partial fragmentation with the resulting formation of a system of intracratonic rifts. The Bengal basin thus receives the Terrestrial fluvio-Lacustrine Gondwana deposits in the rift grabens. The intensification of lifting resulted in the wide spread volcanic effusion, which filled and subsequently, leveled the entire relief of Gondwana paleo valley. The absolute age of volcanic effusion is dated to 155-170 MA.

The Indian plate got completely detached from Gondwana super continent during Early Cretaceous time at about 126 Ma and started drifting northwards. This marked the onset of deltaic sedimentation into the Bengal shelf in front of the easterly / south easterly flowing rivers like (Proto Damodar, Ajoy, Subarnarekha, Roopnarayan etc). The northward drifting continued till the Indian plate collided with the Eurasian plate during Eocene at about 53 Ma, resulting in the initial uplift of Himalayas. The Ganges and Proto Brahmapurtra deltas started receiving the detritus derived from the Eurasian plate in the north, whereas the western shelf part of the basin received the sediments derived from the plateau area in the west, by rivers like Damodar, Subarnarekha etc...

The Indian plate then went on drifting northeastward converging into the Burmese plate to the east. The oblique subduction of the Indian plate under the Burmese plate caused the Juxta position of the two delta systems. They subsequently merged gradually to form the complex modern delta system, wherein the mighty Bengal fan progressed towards south and even by passed the Ceylon continental mass.

Tectonic Frame Work

Based on the available geological, geophysical and subsurface data, the Bengal basin can be divided into four NNE – SSW trending structural zones. They are as under.

1. Basin margin fault zone
2. Shelf zone
3. Hinge zone (shelf / slope break) and
4. Deep basin

Basin Margin Fault Zone

The NE-SW trending basin margin fault zone demarcates the western boundary of the basin with the crystalline metamorphic basement complex. This fault zone is related genetically to the extensional tectonics during the Early Cretaceous time. The east-west trending block-faulting observed in Gondwana grabens, in the peninsular shield west of this zone may also be traced in the eastern part of the shelf at depths between 1500-3500 m.

Shelf Zone

The region between extreme western margins of the basin to the Eocene Hinge zone constitutes the western foreland shelf. It slopes very gently to the east and is affected by numerous down to basin faults, trending NE to NNE. Thickening of the sedimentary section with increasing marine influence with marine wedges in down dip areas are characteristic of this zone (Fig. 6.5.2).

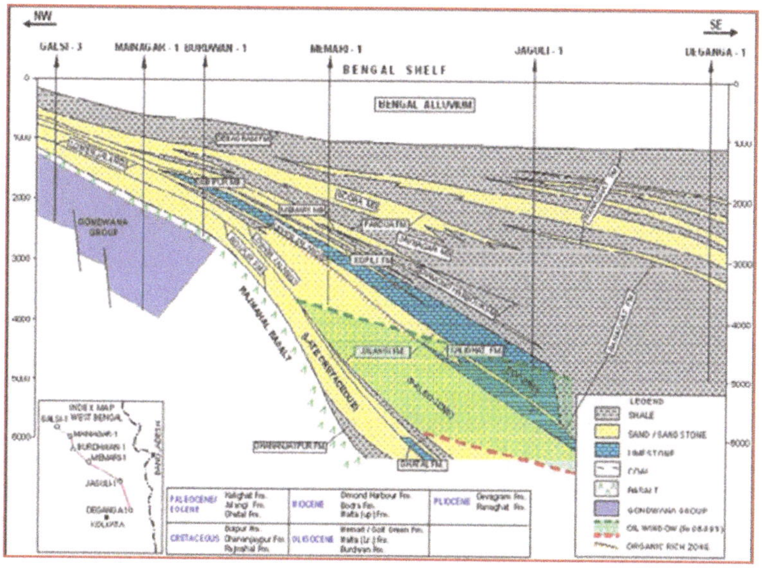

(Courtesy DGH)

Fig. 6.5.2

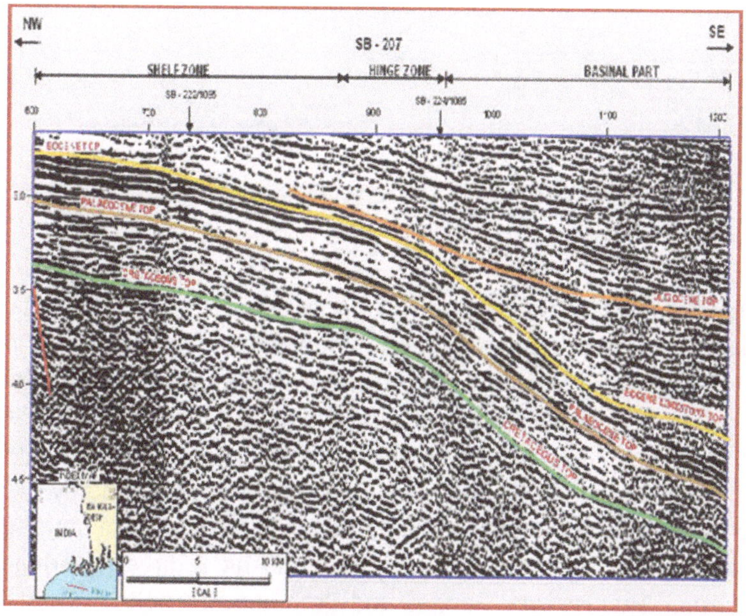

(Courtesy DGH)

Fig. 6.5.2.a

Hinge Zone

Commonly it is known as Kolkata – Mymensingh–Eocene hinge. This separates the western foreland shelf from the frontal flank (slope) of the subsiding trough to the east. This zone is characterized by gentle to moderate flexures, within Sylhet limestone of over 1000 m. (on seismic, it is referred as 'D' reflector).

This zone requires further discussion in detail, to understand the evolution of the Bengal basin. So far, the Calcutta Mymensingh high is referred as Eocene hinge. This part of the shelf and the slope beyond this hinge are characterized by the sediments brought from southeasterly flowing rivers like Damodar, Subarnarekah, Ajoy, and Roopnarayan etc. The hinge zone is characterized by the facies change of the shelfal limestone facies to shale facies in the basinal part. However, a close examination of the seismic data reveals that the D-Reflector can be traced very easily to the east of Kolkata Mymensingh high upto Barisal-Chandpur high situated in Bangladesh (Fig. 6.5.3) without losing its character. Possibility, the Eocene hinge zone extends upto "Barisal-Chandpur High".

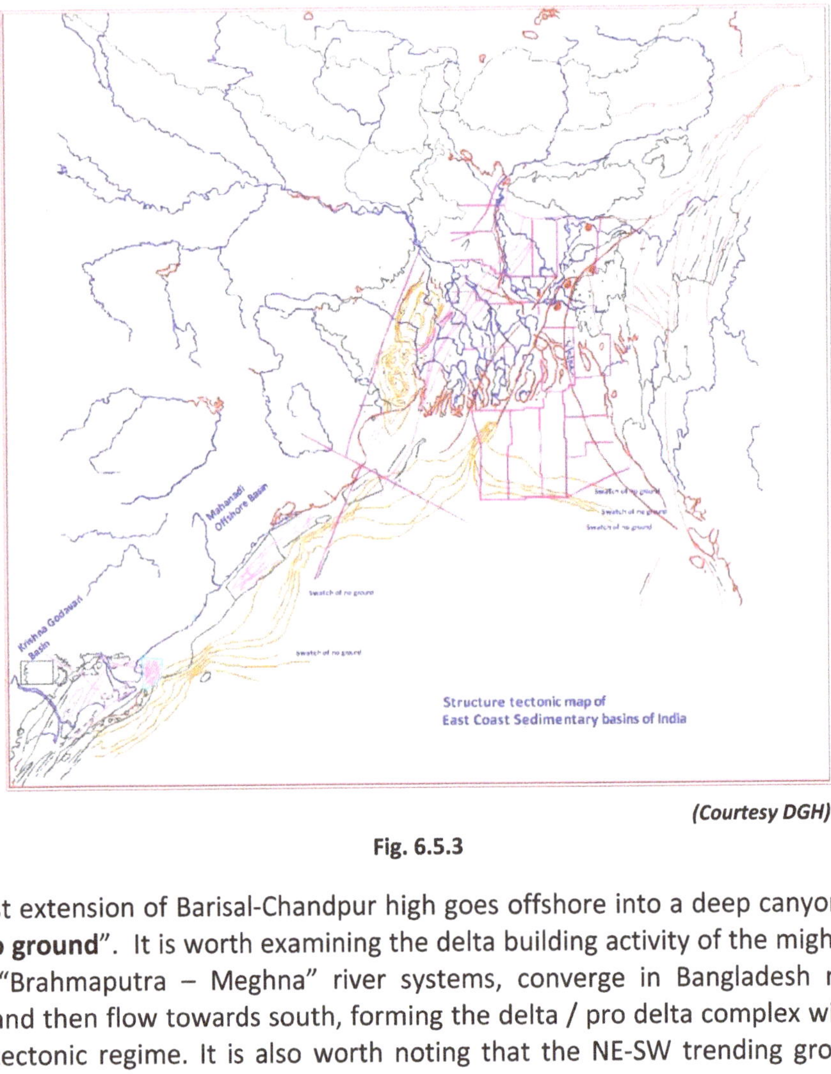

(Courtesy DGH)

Fig. 6.5.3

The southwest extension of Barisal-Chandpur high goes offshore into a deep canyon which is known as "**Swatch of no ground**". It is worth examining the delta building activity of the mighty rivers "Ganges – Padma" and "Brahmaputra – Meghna" river systems, converge in Bangladesh near the **Barisal-Chandpur High** and then flow towards south, forming the delta / pro delta complex with in the gigantic Bengal fan Neotectonic regime. It is also worth noting that the NE-SW trending growth fault regime which can be traced from K-G through Mahanadi to Damodar graben, suddenly turns towards east-west to the east of "**Swatch of no ground** (deep Canyon)", which is reffered as a major geomorphic anomaly. The best guess is that this linear tectonic lineament is having a direct bearing on the hydrocarbon prospectivity of the basin. The Tertiary Ganges and Brahmaputra delta system overrides the NW-SE trending Damodar valley graben delta. The delta building activity of the Damodar graben system loses its identity beyond Barisal-Chandpur high and its extension towards southeast. To the east of **'swatch of**

no ground' (deep Canyon), Neogene tectonics play a dominant role and the Neogene growth fault related, hinge lines of Eocene, Miocene and Pliocene age, run almost east-west, prograding and younging towards south. It is also necessary to understand that the major gas field Sangi of Bangladesh is situated over one of these growth faults of Bengal fan. In the light of these observations, the reinterpretation of Bengal Basin and its delta modeling has to be redrawn, keeping in view the hydrocarbon potential of the basin. It is necessary to note that all the wells drilled by ONGC and other companies are situated in the zone of Kolkata Mymensingh high and its southwest extension. Though hydrocarbon indications are seen, no major commercial find has been established in the basin yet. Probably due attention may have to be given to the Barisal-Chandpur High and the basin towards east, in the Indian Territory of "Swatch of No Ground". It is also worth noting that to the west of Arakan yoma fold belt, in the shallow offshore of Burma; a major gas field was discovered in the year 2005 by JVC DEBAU – ONGC - GAIL.

Deep Basin

The rapidly subsiding basement east of Eocene shelf slope break (Hinge zone to the east of Barisal-Chandpur high, horsts a prism of 10-15 Kms of sediments. The prominent D-reflector top of Sylhet limestone) loses its character in the basinal part because of its facies change to shale and occasional scoring away of limestone by episodically active submarine canyons. It is to be kept in mind that any hydrocarbon find of Eocene to Miocene will be beyond the depth of 5000 m in this part of the basin.

Stratigraphy

Except for a few Gondwana outcrops in the north western part of Indian shield, the whole of Bengal basin is covered with Alluvium. The stratigraphy is constructed with sub crop sampling from 38 wells drilled in the basin. The general stratigraphic column with depositional environments is shown in Fig. 6.5.4 & Fig. 6.5.4a.

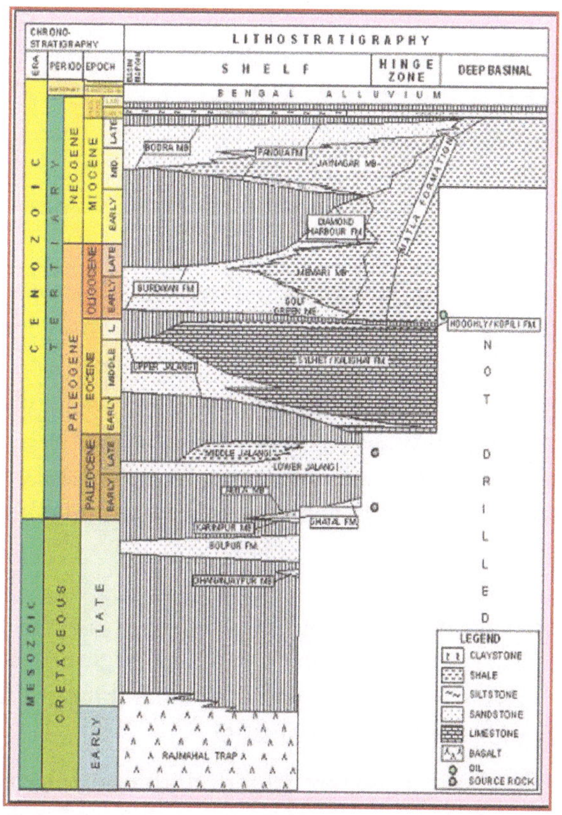

(Courtesy DGH)

Fig. 6.5.4

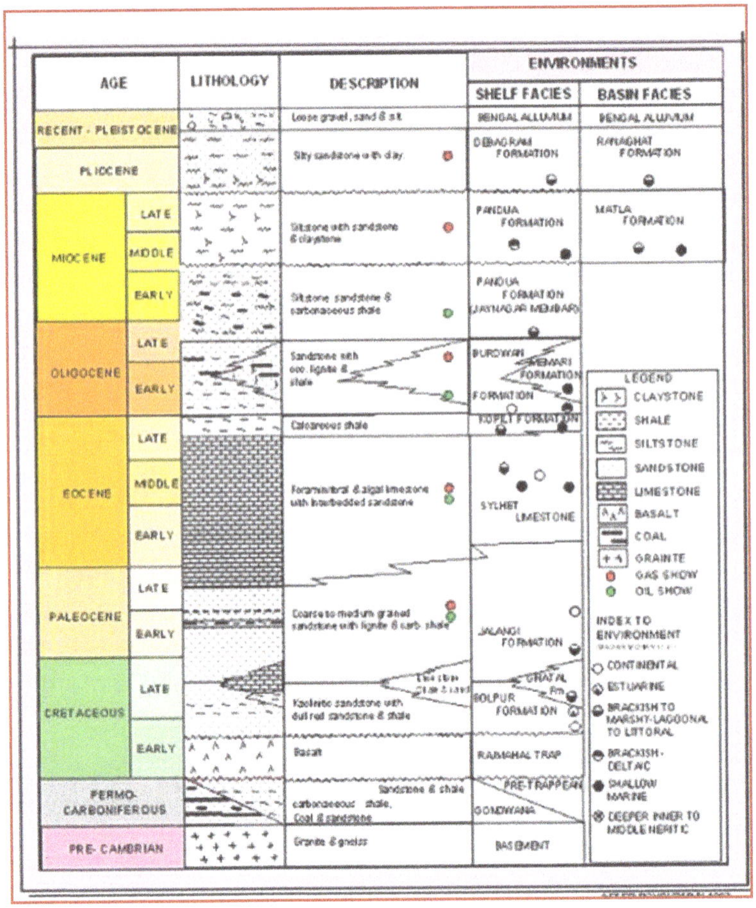

(Courtesy DGH)

Fig. 6.5.4.a

The oldest sedimentary unit is the Gondwana group which consists of mainly sandstone, shale, and coal interbeds/intervals in the lower section and monotonous thick coarse grained Arkosic sandstone in the upper section. The maximum thickness of this Gondwana is of the order of 2000 m. These Gondwana sediments are covered by Rajmahal Basaltic trap flow of Jurassic to Early Cretaceous age with clastic interbeds as intertrappeans.

Post trappean deposition in Bengal basin started in a continental environment. Gradual transgression by Late Cretaceous sea had given rise to facies change from continental (Bolpur) to marine sedimentation (Ghatal formation). During Early Paleocene, deposition was going on in a regressive phase. Late Paleocene is again marked by transgression and high stand phase of sea level changes while an extensive carbonate deposition continued through out Eocene time. The Early Oligocene time is represented by deposition in low stand phase. Neogene time is represented by rapid subsidence and deposition of admixture of clay, silt and sandstone in deltaic environment.

Geological History

Bengal basin is mainly a Tertiary basin overriding the Gondwana grabens lying in the western foreland part of plate edge, geo-syncline of Assam-Arakan region. Its origin is related with commencement of drift of Indian plate towards northeast, with Mesozoic sedimentation, the history of which is noticed below Rajmahal volcanics (Damodar and Purnea grabens).

In the Pre-rift stage during permo-carboniferous period, the huge extensional domain was affected by NW-SE Damodar trend and NE-SW Eastern Ghat trend resulting in block faulting. In the Pre-rift stage itself the basin experienced spasmodic outpour of Basalt to Andesitic volcanic flows of Rajmahal group.

In the Post-rift stage, the formation of Bengal basin started, during Middle and Upper Cretaceous times. The transgressive deepening of the basin is recorded during upper part of Upper Cretaceous. During Paleocene and Lower Eocene times, the major part of shelf area witnessed alternating continental clastic and Esturine environments. However, away from the Paleocene shelf edge, the delta out building was predominant with the development of organically rich Jalangi shale facies in the distal part (Fig 6.5.5).

Wide invasion of Middle to Upper Eocene seas in the shallow shelf recorded thick shelf facies of Nummulitic lime stone (Sylhet lime stone and Kopili shales). During Oligocene time, a basin wide regressive phase has been recorded especially as clastic wedges against Eocene shelf slope break.

In Late Tertiary time, the basin experienced a massive deltaic stage. The progradation of delta by Ganges / Brahmaputra river course, southward shelf progradation with superimposed interplay of turbidites and fan deposits. Numerous growth faults, down to basin faults, from time to time have affected the deposition and distribution system. Vast deposition and quick burial have given rise to high formation pressures especially in Mio-Pliocene sediments of the basinal area. There was a break in deposition in Late Miocene in the shelf area while an uninterrupted sedimentation continued the basinal part. Sedimentation in Pliocene is represented by Esturine environment in the shelf where as in the basinal area it is represented by prograding deltaic facies with enormous thickness of sediments, causing basin subsidence. Basin wide marine regression occurred during Middle to Upper Pliostocene.

Hydrocarbon Potential

Source Rock

Shales and limestones are the main source of the basin with good amount of organic carbons of Types-II and III with adequate maturation. TOC values vary from 0.4% to as high as 3.11% (Paleocene sequence). Vitrinite reflections (VRo) range from 1.4 to 1.9% in Gondwana coal shale sequence and 0.48 to 0.77% in the Upper Cretaceous / Paloegene shales. There is a general increase in source rock thickness basinward.

Reservoir Development

So far, the hydrocarbon shows and recent flows of liquid and gaseous hydrocarbons in the basin are mainly from sandstone reservoirs. Entrapments are due to block faulting, low amplitude faulting and draping in the shelf to slope part, whereas the entrapments in the basinal part are due to Neogene tectonics like growth faults, rollovers, drape over shale diapirs, turbidite fans / canyons / trubidite flow and slump deposits.

Habitat of Oil and Gas

The hydrocarbon indications in almost all the drilled wells and the presence of liquid and gaseous hydrocarbons in the recently drilled wells proved beyond doubt, the definite generation of hydrocarbons and their migration and entrapment in Bengal basin.

Systematic source rock studies, maturated stimulations of geo-chemical prospecting have established that the basin has witnessed atleast two major phases of oil and gas generation.

The presumed first generation from major lower Gondwana shale coal sequence is during Permian time (250 Ma). The second phase is from Paleocene Eocene shales with the migration during Late Mio-Pliocene time. The Lower Gondwana sediments in Galsi-Palasi have very good source rock potential with TOC ranging between 1.22 to 1.44% and VRo 1.4 to 1.9% with good hydrocarbon index.

Most extensive development of source rock facies has been observed in Paloegene sediments particularly Jalangi shales. The average TOC varies from 0.4 to 3.11% with admixture of Type-III and Type-II organic material. The VRo values range between 0.48 to 0.477%. Geochemical modeling has indicated that, oil window was formed about 8 MABP where as in the deep basinal part oil and gas windows formed 23 MABP and 10 MABP respectively. Source rocks of Sylhet limestones are also major contributors for the generation of hydrocarbons. Recently, Type-II organic matter has been found with thin oil window zone in offshore well SME-5. In the deep basinal part, the huge thickness of Mio-Pliocene shale deposit at depth, appear to be highly interesting for hydrocarbon generation and migration.

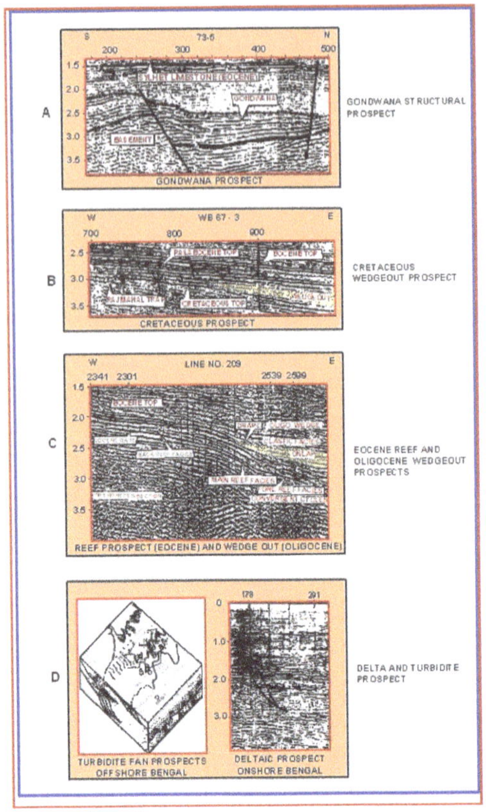

(Courtesy DGH)

Fig. 6.5.5

Offshore Wells (Indian Part)

SME-1: The well is located on Kolkata Mymensingh high zone (Fig. 6.5.6). The well was planned for exploration of reefal anomaly in Eocene limestone and drilled to a target depth of 5074 m. No indication of reefal anomalies was observed in the limestones. Both in Miocene and Eocene the well encountered high pressures. Dissolved gases up to C-4 and C-5 were observed during drilling of Miocene and Eocene sequences.

SME-5: This well is located on the Kolkata Mymensingh high and was drilled to 5830 m for reefal Eocene limestone with flat spot anomaly and also for Paleocene and Cretaceous. Dissolved gases up to C-2 were observed in Miocene section.

SME-4: This well is located in the basinal side East of Kolkata Mymensingh high and was drilled to 5575 m. Super formation pressures were encountered in Miocene clastic. Gases up to C-2 have been observed during drilling of Miocene section.

SMM-1 & SMM-2: These two wells (Fig. 6.5.7) are located in the basinal side, to the East of Kolkata Mymensingh high. The exploration target was to probe the Miocene deltaic sediments in growth fault related anticlines. Abnormal pressures were encountered in Miocene sequence. During drilling and testing, gases up to C-4 were observed. The Miocene section in SMM-2 is having fair to good organic matter with 20% Type-II organic matter, and with the TOC values varying from 0.53 to 2.78%. Early phase of maturation seems to have commenced around 3526 m in Miocene.

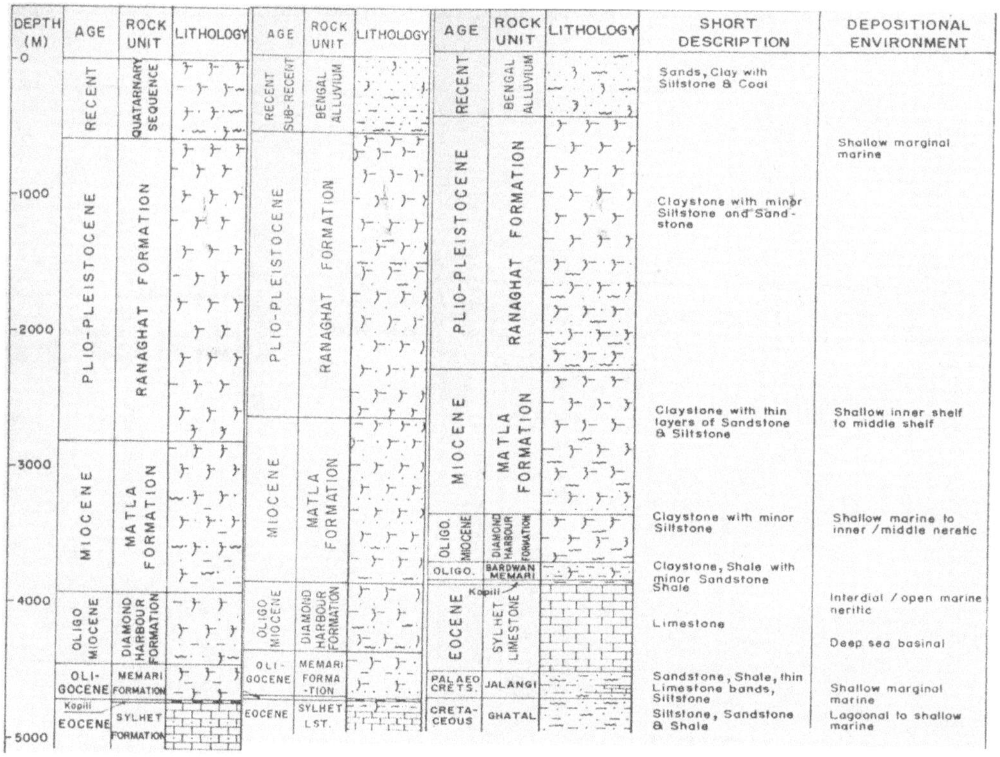

(Courtesy DGH)

Fig. 6.5.6

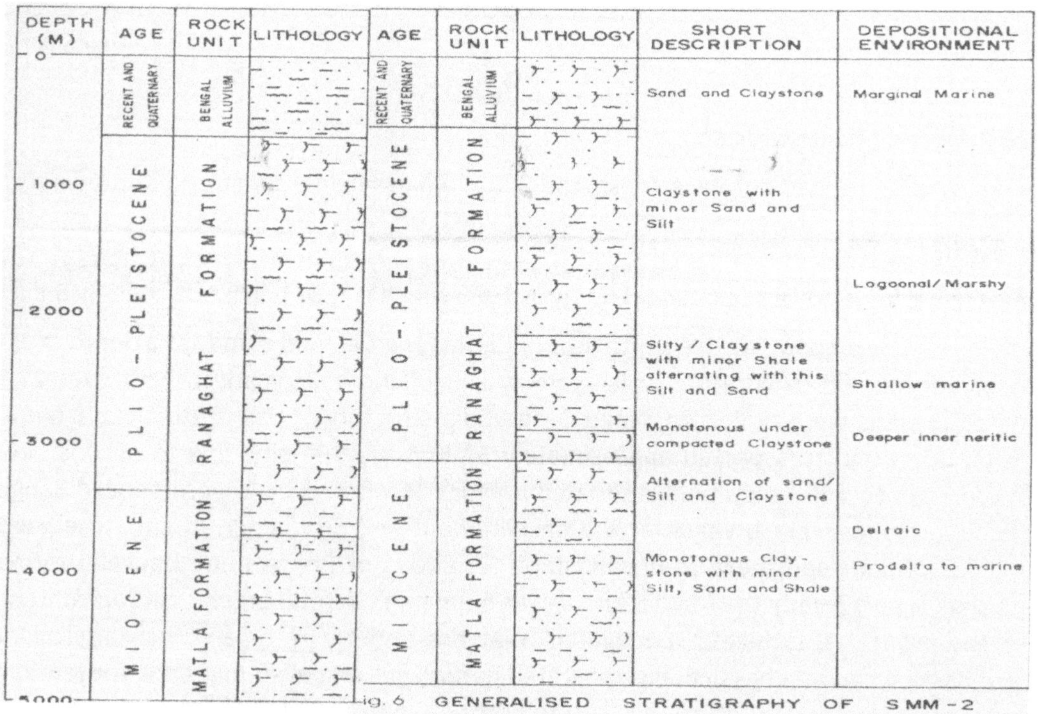

(Courtesy DGH)

Fig. 6.5.7

Miocene Prospects of Deep Basinal Area

There are a number of plays such as Neogene Deltaic prospects, Canyon / channel filled sediments and sub-marine fan / turbidites. A number of channels have been identified in this part of Miocene Pliocene section across various seismic lines. Various slope fan / turbidites mapped in the deeper parts of the basin are quite interesting.

(Courtesy DGH)

Fig. 6.5.8

Offshore Fields (Bangladesh)

In the shallow offshore parts of Bangaldesh, there are two gas fields- Sangu and Kutubdia (Fig. 6.5.8).

Sangu Field

Sangu gas field was discovered by Cairn Energy in 1996 which is now being operated by Shell Oil Co.

The field is located about 50 kms southwest of Chittagong City and stands at a depth of 10 m water in the Bay of Bengal. 58 exploratory wells were drilled offshore Bangladesh with over 23 discoveries. 17 exploratory wells were drilled on Sangu & Katubdia structures. Gas is produced from 4 wells from Sangu-A platform and transported approximately 50 Kms via 20" pipe line to an onshore processing plant at Chittagong. The Sangu field is situated in block-16 offered by Government of Bangladesh. The Sangu field is retained for 30 years for the development and production of gas. The gas finds in the various fields of Bangladesh are from sandstone reservoirs of Bhuban and Bhukabil formations. The Miocene shales, particularly the Middle and Lower Bhuban have total organic carbon content up to 0.9% and are gas prone. It is further postulated, that the underlying organic rich argillaceous rocks of Oligocene Barail formation has actually generated natural gas and after migrating for few kilometers up-dip, finally accumulated in the sandstone reservoirs of Bhuban and Bhukabil formations.

A total of 22 natural gas fields have been discovered in Bangladesh. The initial inplace gas in these 22 fields is about 26 TCF, of which about 16 TCF is considered recoverable. The gas occurs in Miocene and Pliocene sandstone reservoirs at depths of 1000 to 3500 m below the surface.

Table 6.3: Status of Drilled Wells in Bengal Basin

	Well	Company	Year	Drilled depth (m)	Prospect	Bottomed at	Age	Tested for	Hyderocarbon show	Observation
1	Bakuttala-1	ONGC	1976	3700	Pliocene fault closure	Ranaghate	Pliocene	Pliocene	Dissolved gas	Entrapment conditions absent
2	Bishnupur-1	ONGC	1982	3006	Mio-Pliocene Structural closure	Pandua	Miocene	Miocene	GYF at 2324 - 2994 m	Water bearing
3	Ranaghar-1	ISPP	1959	3965	Lower Pliocene nose	Pandua	Miocene	Miocene	Gas (1468-89, 2470)	No structural closure found
4	W.Ranaghat-1	ISPP	1960	2604	Basal Pliocene	Pandua	Miocene	Pliocene	Dissolved Gas 2526 – 44	Reservoir absent
5	W.Jaguli-1	ONGC	1988	3010	Bright spot in Miocene	Pandua	Miocene	Not Tested	Nil	Bright spot due to tight silt st.
6	Raghunath-1	ONGC	1989	5339	Growth fault in Miocene	Matla	Miocene	Miocene	Gas up to c6 4560m	Complications in testing
7	Bodra-1	ONGC	1966	4197.50	Structural Closure in Miocene	Matla	Miocene	Miocene	Gas (3805,3850,3881-4197) GYF(3925 -37, 3940 – 49)	Tested above 3600 due to high pressure
8	Bodra-2	ONGC	1987	4950	Growth fault in Miocene	Matla	Miocene	Miocene	Gas(3784 – 4849)	High Pressure Sand Produced water with dissolved gas
9	P.Canning-1	ISPP	1960	4042	Large nose in Pliocene	Matla	Miocene	Miocene	Gas (546 – 613,2375-78,2543-61)	Dissolved gas
10	SM-M-1	ONGC	1988	4951	Growth fault & rollover in Miocene deltaic sand	Matla	Miocene	Miocene	Gas (3715-34,4213-20,4568,4811-20)	Reservoir absent; dissolved gas

Table 6.3: ***Contd...***

	Well	Company	Year	Drilled depth (m)	Prospect	Bottomed at	Age	Tested for	Hyderocarbon show	Observation
11	SM-M-2	ONGC	1988	4975	Growth fault & rollover in Miocene deltaic sand	Matla	Miocene	Miocene	Gas (3296-3304,3473-80)	Reservoir absent; Dissolved gas
12	BB-B-1	C&N	1976	4073	Structural closure in Miocene	Matla	Miocene	Miocene	Feeble cut in SWC(2947,3433)	Produced dissolved gas up to c6
13	BB-A-IR	C7N	1976	4073	Structural Closure In Miocene	Matla	Miocene	Miocene	Gas (2984 – 3602)	Feeble free gas in cased hole DST
14	Deganga-1	ONGC	1987	5007	Fault closure in Miocene	Matla	Miocene	Miocene	Gas cut mud 3552,4082 4171)	Reservoir absent
15	SME-4	ONGC	1989	5575	Oligocene Wedge	Sylhet lst	Eocene	Oligocene	Gas in Miocene	Formation fluid data could not be obtained
16	SME-1	ONGC	1987	5079	Eocene Reef	Sylhet lst	Eocene	Eocene	Pale Cut in SWC (4773,4864) Gas 94792,4864)	High pressure sand layers produced dissolved gas
17	D.Harber – 1	ONGC	1981	5555	Eocene Reef	Sylhat Lst	Paleocene	Eocene	Gas(3484, 4773)	Dissolved gas after simulation
18	Domjur-1	ONGC	1988	5189	Cretaceous pitchout	Jalangi	Paleocene	Eocene	Dissolved gas in cased hole DST	Tight Lst
19	Jaguli-1	ONGC	1983	5637	Eocene carbonate buildup	Syihet Lst	Late Paleocene	Paleocene ,Eocene	Dull GYF (4461-64)Dissolved gas 5094-5579)	Dissolved gas in testing
20	Maju-1	ONGC	1984	4020	Seismic void in syihet Lst, similar anomaly below Lst	Jalangi	Paleocene	Eocene	+ve cut (3247-51,3319-6)Patchy GYF(3446-73,3832 - 36)also gas (4005 -09)	Dissolved gas in testing. Anomaly not proved

Table 6.3: ***Contd...***

	Well	Company	Year	Drilled depth (m)	Prospect	Bottomed at	Age	Tested for	Hyderocarbon show	Observation
21	Maju-2	ONGC	1885	4009	Seismic anomaly in sylhet Lst	Jalangi	Paleocene	Eocene	Gas In Oligocene-Eocene, Pale GYF & Cut (3344,3400,3476,3645,3783, 3795, 4002)	Anomaly proved to be alteration of clastic & Carbonates
22	Radha-1	ONGC	1981	3506	Reversal at sylhet lst & Ghatal lst level	Jalangi	Paleocene	Eocene	Fain cut (2912-160Gas (2955-53.3448)	Dissolved gas in testing
23	SME-5	ONGC	1989	5841	Reef and flat spot in sylhet Lst	Shale & SST	Late Cretaceous	Eocene	Nil	Dissolved gas in testing. Reef not found.
24	Memari-1	ISPP	1959	3807	Base of sylhet Lst & Oligocene thinning	Lalangi	Late cretaceous	Miocene	Gas (2015-2022)	Dissolved gas in testing
25	Budwan-1	ISPP	1958	2720	Fault closure at top of sylhet Lst	Basal/Andesite	Cretaceous	Eocene	Asphalt 1951	Entrapment condition absent
26	Bolpur-1	ISPP	1958	1524	Drope over basement high	Bolpur	Cretaceous	Eocene	Nil	Entrapment condition absent
27	Ghatal-1	ISPP	1960	3109	Gravity high with Oligocene thinning	Ghatal	Cretaceous	Cretaceous	Gas cut salt water at 2125	Dissolved gas
28	Abhay-1	ONGC	1982	2900	Fault closure in Eocene	Ghatal	Cretaceous	Paleocene, Eocene	GYF & Cut (2629,2647-53)	Dissolved gas in testing fresh water flusing
29	Dhananjaypur-1	ONGC	1985	5204	Mesozoic clastic wedge	Rajmahal trap	Early cretaceous	Cretaceous	Nil	Wedge not proved.

Table 6.3: ***Contd...***

	Well	Company	Year	Drilled depth (m)	Prospect	Bottomed at	Age	Tested for	Hyderocarbon show	Observation
30	Galsi-2	ONGC	1977	1238	Reversal over basement scarp	Rajamahal trap	Early Cretaceous	Cretaceous	Nil	Freshwater on testing
31	Naldaha-1	ONGC	1983	3506	Seismic anomaly at Eocene/ Paleocene level	Rajmahal trap	Early Cretaceous	Eocene, Oligocene	GYF & Cut 2240,2281-88,2389-90	Entrapment conditions absent
32	Jalangi-1	ISPP	1958	3576	Fault closure in sylhet Lst	Jalangi	Late Cretaceous	Miocene	Pale cut 2133 Asphalt at 2545. Gas at 3049	Late origin at structure
33	Debagram-1	ISPP	1959	3106	Fault closure above Eocene	Bolpur	Cretaceous	Eocene	Gas (1050,1428,2131) GYF at 1426	Freshwater in Lst
34	Chandkuri-1	ONGC	1989	3862	Paleocene – Cretaceous	Granite	Precambrian	Paleocene	GYF & Cut (2119-36)	Dissolved gas with fluorescent water of high salinity
35	Galsi-1	ISPP	1958	1296	Growth fault in post trappean clastics	Rajmahal trap	Early Cretaceous	Eocene	Gas (1057-68)	Freshwater on testing
36	Gailsi-3	ONGC	1989	4221	Pre trappean clastics	Dolerite	Precambrian	Gondwana	Nil	Entrapment conditions absent
37	Mainagar -1	ONGC	1989	4221	Fractured basement & Gondwana in horst block	Diorite	Pre Cambrian	Basement	Nil	Tight basement
38	Palasi-1	ONGC	1989	5640	Gondwana in deep graben	Dolerite	Pre Cambrian	Gondwana	Nil	Entrapment condition absent

Table 6.4:

Formation Pressure Temperature Data (Shelf & Hinge Area)

Pressure in kg/cm^2 temp in ^{o}c

Depth in parenthesis in meters

Formation		Palasi-1	Mainagar-1	Domjur-1	Abhay-1	Rada-1	Jaguli-1	SME-5
Plio-Pleistocene						78^0 C(2513)	63^0 C(2631)	
Miocene	Late				64^0 C(17020	70^0C(2012)		
	Mid							110^0 C(3748)
	Early							
Oligocene				118^0 C(2237)			121^0 C(4098)	
Eocene				396(3928) 153^0C(5139)		90^0C(2850) 100^0C(3100)	168^0C(5060)	484(4930) 140^0C4972)
Paleocene				498 (4744) 165^0C(5139)			172^0C(5396)	
Cretaceous						292(2623)		149^0C(5841)
Rajmahal trap			87.7^0C(2445)					
Gondwana		189^0C (5640)	133^0C(3515)					
Basement			147^0C (4221)					

Table 6.5:

FORMATION PRESSURE TEMPERATURE DATA (SHELF & HINGE AREA)

PRESSURE IN KG/CM2 TEMP IN ^{0}C

DEPTH IN PARENTHESIS IN METERS

Formation		SME-1	SM-M-2	SM-M-1	Deganga	Raghunath	Bodra-2	Bodra-1
Plio-pleistocene					96 $^{\circ}$C(3040)	3070 (3040) 76.6 $^{\circ}$C (2478)		346(2550) 513(3580) 95^{0}C
Miocene	Late		703(3480)	114^{0}C(3690)	105.5^{0}C 466(3599			
	Mid	630(3918) 109.4^{0}C(3961)		765(3733) 139^{0}C(4571)	118^{0}C 958(4430)	972(4050) 833(4150) 150^{0}C(4571)		
	Early		155^{0}C(4970)	152^{0}C(4946)	158.3^{0}C(5003)			
Oligocene		126.6^{0}C(4505)						
Eocene		1104(4880) 141^{0}C(4994)						
Paleocene								
Cretaceous								
Rajmahal trap								
Gondwana								
Basement								

Natural gas from Bangladesh is very pure with about 95 to 99% Methane and almost no Sulphur. The average compositions are as given below:

Methane	:	97.33%
Ethane	:	1.72%
Propane	:	0.35% &
Higher Hydrocarbons	:	0.19%

Gas in most of the fields is dry, but in few fields it is wet with considerable amount of condensate. The condensate reserves are estimated at 65 MMBBLS. The porosity and permeability of reservoirs are good to excellent and there are multiple sand reservoirs. In some wells, around 12 gas sands have been mapped. The source rocks of the gas are considered to be Oligocene shales belonging to the Barial group. In the Sangu field gas initial inplace reserves (proven + probable) is 1031 BCF. Initial reserves (Proven + Probable) are 848 BCF of which 58 BCF was produced from Mio-Pliocene sandstones.

Kutubdia Gas Field

This gas field was discovered in Bay of Bengal by Union Oil Company of USA in 1976 and is about 92 kms southwest of Chittagong port. Being small in size, no plan had been drawn for the development of this offshore field. In all probability, the Kutubdia gas field had to be produced along with Sangu field gas development. The initial gas inplace reserves (Proven + Probable) are placed at 780 bcf, of which recoverable reserves are of the order of 468 bcf from Miocene sandstones at a depth of 2650 m. No production so far from this reservoir.

Geology

- Bengal basin is a polycyclic basin in which Gondwana basin of Late-Paleozoic to Early-Mesozoic is overridden by Tertiary basin.
- The basin started as an intercratonic rift basin (Late-Paleozoic to Early Cretaceous) during Gondwana period and slowly changed to passive margin during Late-Mesozoic to Tertiary period
- The basin experienced complex geo-tectonic activity from Early-Mesozoic to Late-Tertiary period (Neogene) due to plate movements, rifting and drifting and subsequent Neogene tectonism (In the offshore part)
- The widespread volcanic activity observed during Jurassic to Cretaceous period (152-170 MABP-Rajmahal Traps)
- Indian plate got detached from Gondwana super continent during Early-Cretaceous (120 MABP) and started drifting towards north. During this period delta building started into Bengal basin shelf from western peninsula (Proto Damodar, Ajoy, subarnarekh rivers.)
- The northward drifting of the plate continued, till it collided with Eurasian plate during cretaceous (53 ma BP) resulting in initial uplift of Himalayas (Tethyan orogeny)
- Ganges and Proto Brahmaputra deltas are receiving detritus from Eurasian plates from North by Miocene onwards
- Subsequently, the plates started drifting towards North East and subducted under Burmese plate. The phenomena caused the juxtra position of the deltas pertaining to Ganges-Padma, Brahmaputra-

Magna, and Irawaddi Rivers over riding and concealing the prograded deltas of Proto-Damodar, Subarnarekha, Ajay and other rivers flowing towards east from Indian peninsula.

- Delta build up/Bengal fan development is mainly controlled and guided by two NE trending sub-oceanic ridges (85° E and 90° E Ridges) and the main Bengal fan complex building is to the East of Barisal, Chandpur high and its southward continuation into the "**swatch of no ground**" (Deep Canyon).
- The Bengal fan is mainly controlled subsequently by the submarine canyons known as **"swatch of no ground"**.
- To the east of Barisal-Chandpur high the Paleocene, Eocene, Miocene and Pliocene growth faults controlled by Neogene tectonics have EW trend. The NE-SW trend which was predominant to the west of **swatch of no ground,** subsequently takes EW trend due to the influence of delta building, by two North to South flowing mighty rivers, Ganges and Brahmaputra.
- The influence of Bengal fan development is seen beyond Srilanka Island.
- The NW-SE Dhamodar trend and NE-SW Eastern Ghat trend are two major tectonic trends which are responsible for major faulting, trigger rifting of the basin during Permo-Cretaceous times.
- In the offshore part, the Neogene tectonics play a dominant role where growth faults and shale diapirs and rollover anticlines, connected with Paleocene to Pliocene hinge lines responsible for further hydrocarbon entrapment.
- The basinal part accumulated more than 10 to 15 km thick pile of sediments
- The tilt of the basin towards SE and South is responsible for southward shelf progradation and delta build up in the Neogene time
- Inter play of turbidities and fan deposits are seen at various stratigraphic levels
- Channel cuts and subsequent channel fills are seen at various levels from Upper Cretaceous to Recent, where stratigraphically controlled hydrocarbon play is indicated
- The shales of Paleocene and Eocene, which are regarded as source rocks, have rich organic content of Type-II and Type-III, with adequate maturation. The total organic content is as high as 3.5% and the vitrinite reflectance (VR) ranges from 1.4 to 1.9%
- There are oscillations in the basin which are caused by marine transgression and regression at various phases in geological history
- Transgression is recorded in Late Cretaceous and regression in Early Paleocene. Another cycle of transgression is witnessed in the Late Paleocene during which high stand phase of marine transgression / sea level changes observed
- During Oligocene, deposition of low stand phase is clearly seen
- During Neogene period, rapid subsidence due to enormous rapid piling of deltaic progradational sediments and formation of Bengal fan is a very characteristic feature
- Another marine regression during Mid-Plio-Plistocene is recorded in the basin

Entrapment

- The following entrapment conditions are mapped in the basin. They are:

1. Block faulting
2. Low amplitude faulting and drapes over it
3. Porosity pods in Eocene limestone specially near Eocene hinge zone(Barisal-Chandpur high and on Eocene Shelf)
4. Fractured volcanic traps/Intertrappean in the hinge zone
5. Wedge outs in the Oligocene deltaic system, growth faults, associated rollover anticlines, shale diapirs and associated structures, turbidities, sub marine canyon deposits

Hydrocarbon Generation

- Two major phases of Oil generation are witnessed in the basin
    1. Gondawana Coal-Shale sequences of Permian age
    2. Paleocene-Eocene shales
- During Mio-Pliocene times, Oil and Gas windows are observed at 23Ma and 10Ma respectively in deeper basinal areas
- Turbidity fans in the shelf slope break are characteristic in the basin. Free gas is observed in BB-A-1R well. Pentane, Hexane and higher hydrocarbons are observed in BB-B-1 well.
- Oil and gas, in minor quantities have been produced from Ichapur and Golf Green wells.
- In offshore part, the Hydrocarbon Exploration targets are associated with Neogene growth fault regime and the exploration targets are beyond 5 km target depths.

Exploration Status

- The prognosticated reserves of the basin are 580 MMT Oil or OEG
- 38 wells had been drilled in the basin:
    1. 10 wells were drilled by Indo Stanvac Petroleum Company during 1950, in some wells oil and gas indications were observed.
    2. 2 wells were drilled by M/S Carlsberg Natomac during 1975-76 in the offshore.
    3. 26 wells were drilled by ONGC
        (a) On land - 21 Prospects
        (b) Offshore - 5 Prospects
- During 1960-70 Aero-Magnetic, Gravity, Magnetic, Analog and refraction seismic surveys were carried out
- During 1970, first CDP seismic survey (6 and 12 fold) was obtained in the basin
- During 1975-76, Carlsberg Natomac carried out 24 fold CDP seismic survey in the offshore part.
- During 1975-1990, CDP surveys (12, 24, 48 fold) were carried out by ONGC. 3D seismic surveys have been carried out on some important structures like Golf Green, Ichapur etc.
- 8806 sq.km of geological mapping and 5400 sq.km of structural and tectonic mapping were carried out

Seismic surveys

	ONGC	ISPP
On shore	18996 sq.km	7770 sq.km
Offshore	14807 sq.km	2500 sq.km

Gravity, Magnetic : 48878 Stations

Surface Geochemical Surveys : 3720 sq.km

6.6 Andaman and Nicobar Islands

The Andaman Nicobar basin located in the SE part of the Bay of Bengal occupies an area of 8270 Sq.km on land, 32000 Sq.km offshore up to 200 km isobaths and another 40,000 sq.km upto 3000mt water depth (Fig. 6.6.1). This basin forms a part of major island arc from Burma in the North (MARTABAN basin) and to Indonesia in the South. Andaman and Nicobar, the chain of Islands is 800 km long, arc shaped, largely covered by dense forests, forming low and gentle hillocks with second order topography. The islands are grouped as Andaman, the Archipelago, and the Nicobar group of islands system extending extending from Burma in the North (Martaban basin) and to Indonesia in the South. Andaman and Nicobar, the chain of Islands is 800 km long, arc shaped, largely covered by dense forests, forming low and gentle hillocks with second order topography. The islands are grouped as Andaman, the Archipelago, and the Nicobar group of islands Fig.6.6.1).The main Andaman Islands are further divided into North, Middle and south Andaman Islands and the Nicobar Islands as Little and Great Nicobars.

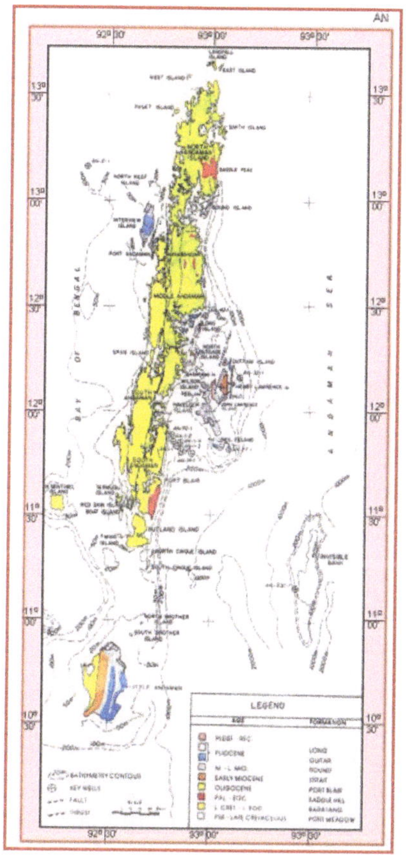

Fig. 6.6.1

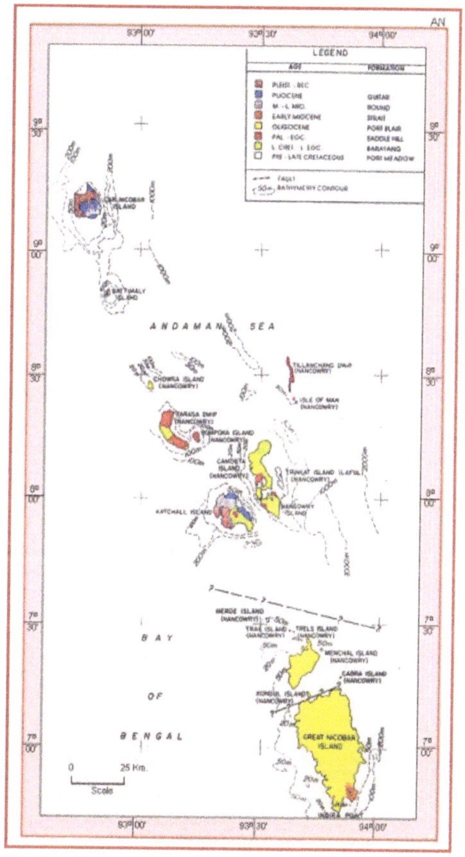

Fig. 6.6.1a

Exploration activities by ONGC in the Andaman – Nicobar basin date back to 1959, with the initiation of Geological Mapping. Most of the islands have been covered by reconnaissance, semi detailed and detailed mapping with good control of geological traversing. ONGC even conducted shallow drilling (up to 200m) for stratigraphic information. Wide grid reconnaissance seismic surveys were carried out by ONGC during 1977 to 1980. DGH under GOVT. of India's campaign, carried out speculative systemic surveys in JV partnership with M/s Western Geophysical, covering complete East cost and Andaman and Nicobar basin. This speculative survey has given immense and valuable information on the stratigraphy and structural frame work of the Andaman–Nicobar area and its Hydrocarbon prospectivity Figs. 6.6.4.a, 6.6.4.b, 6.6.4.c, 6.6.4.d, 6.6.4.e, 6.6.4.f & 6.6.4.g.

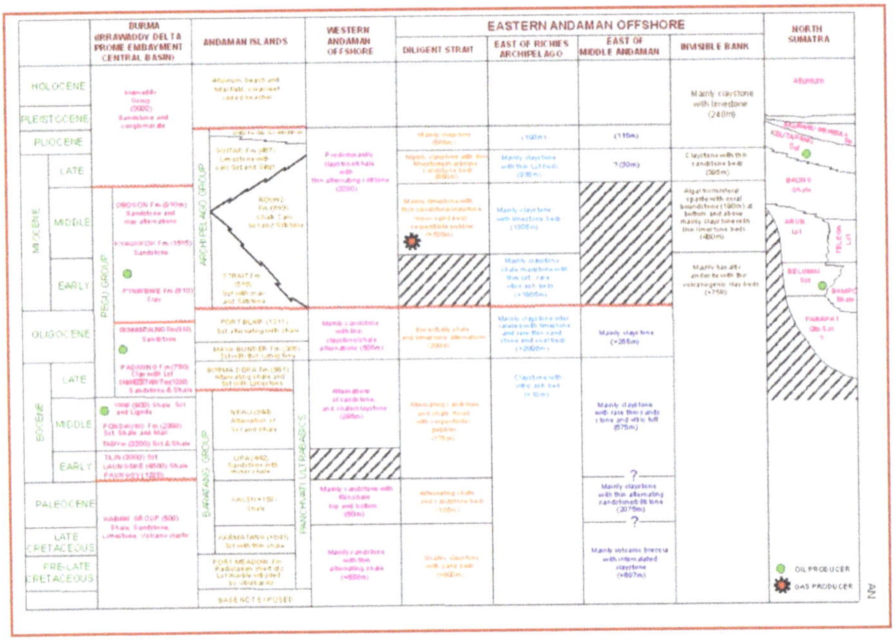

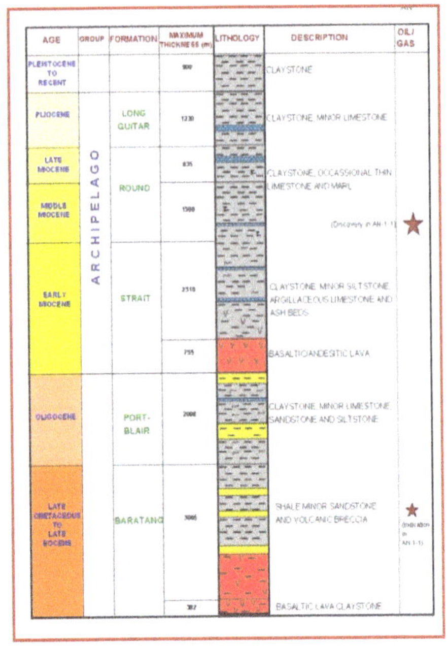

Fig. 6.6.2.a

Fig. 6.6.2.b

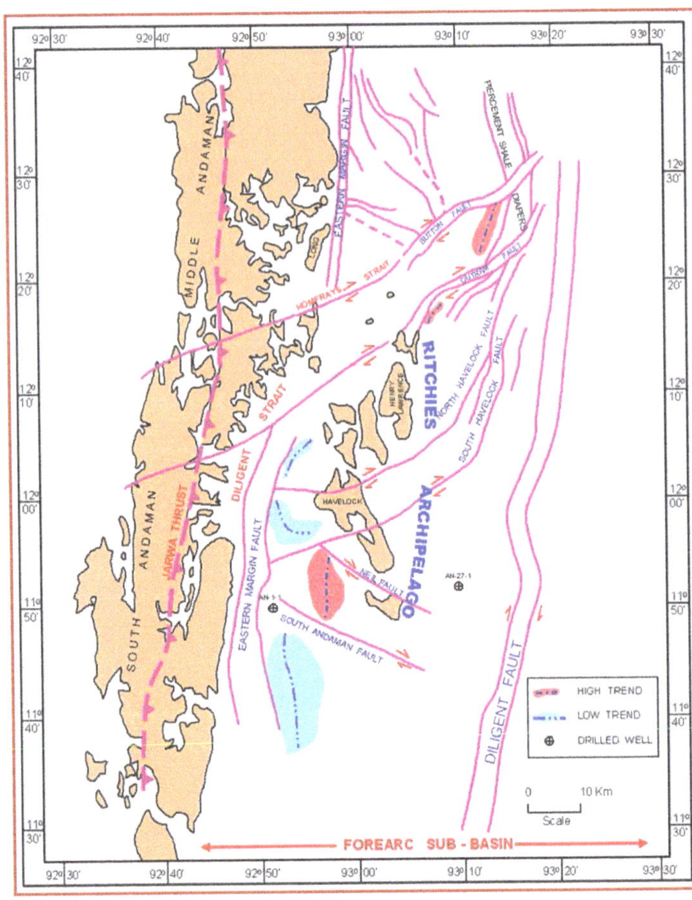

Fig. 6.6.3

Based on speculative seismic surveys, other geological and geophysical data available in Andaman area and East Coast (Bay of Bengal); the deep water exploration blocks were carved out and included in the NELP bidding. The valuable information obtained from the speculative surveys is having direct bearing on the hydrocarbon discoveries made by RIL, ONGC and Cairn Energy in the East coast. The total geophysical data acquisition till date by ONGC, M/s Dresser Olympic, GSI and GECO is of the order of 25679 Sq.Km. This is in addition to speculative surveys carried out by DGH, during 1996 -97. In addition, Gravity-Magnetic data is also available for the entire Andaman- Nicobar basin (both onland and Offshore) Fig. 6.6.5. Sniffer surveys (around 2500 LKM) had been carried out by ONGC in the Eastern Andaman for obtaining geochemical information. ONGC in early 1980s carried out exploratory drilling both on Andaman main Island and Archipelago group of Islands. 12 Exploratory wells have been drilled on different prospects. One exploratory well AN-1-1 drilled on diligent strait is a discovery well for gas (Fig. 6.6.3). Over 3000 mt of sediments ranging in age from Cretaceous to Recent are exposed in this part of the basin and 3-major unconformities have been identified. The main Andaman Nicobar islands expose the rocks of cretaceous, where as Neogene rocks exposed mainly on Ritchies' archipelago group of islands. AN-2-1, drilled on the western offshore part of the main island was drilled to 4088 m which ended in Cretaceous Section.

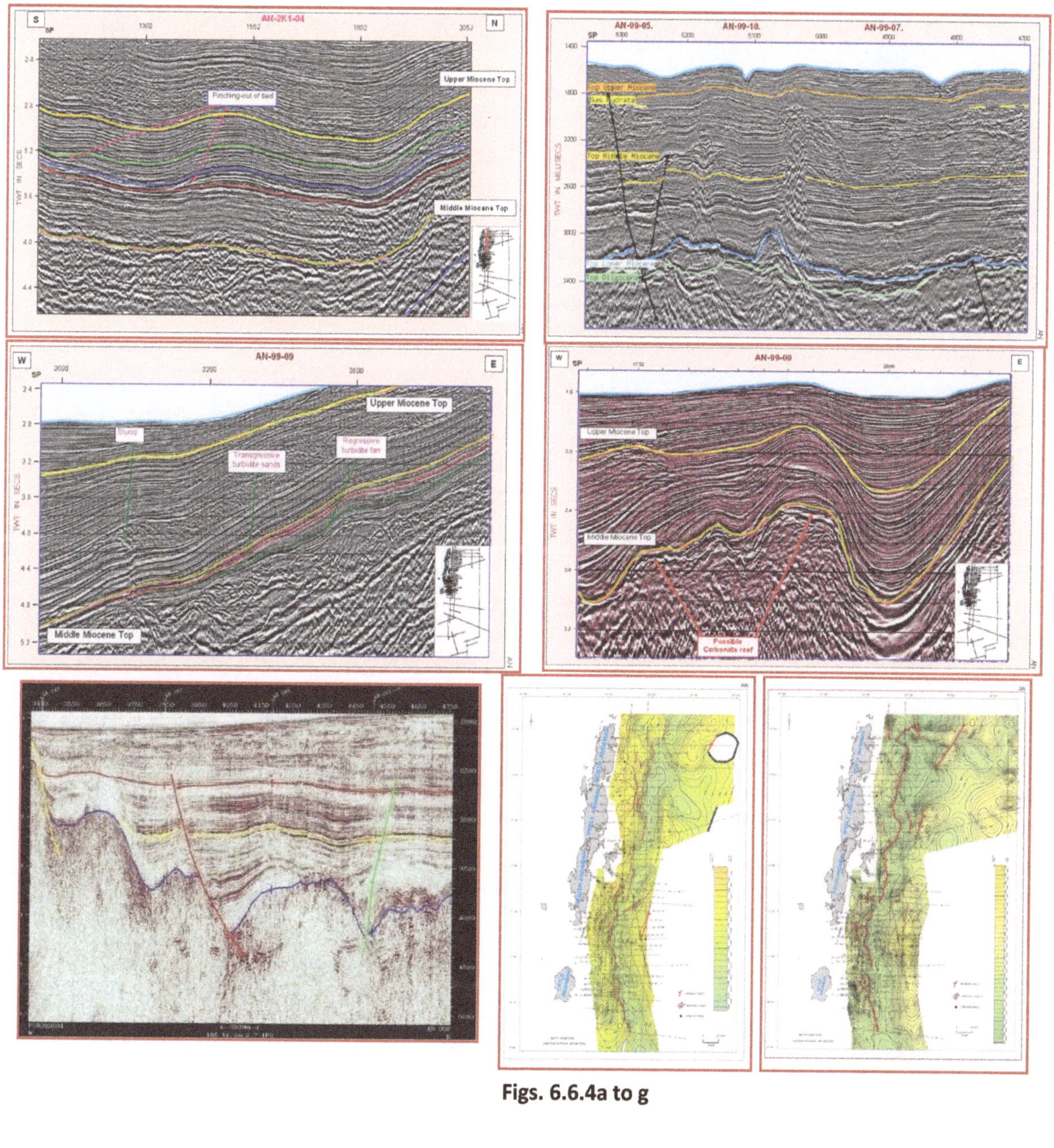

Figs. 6.6.4a to g

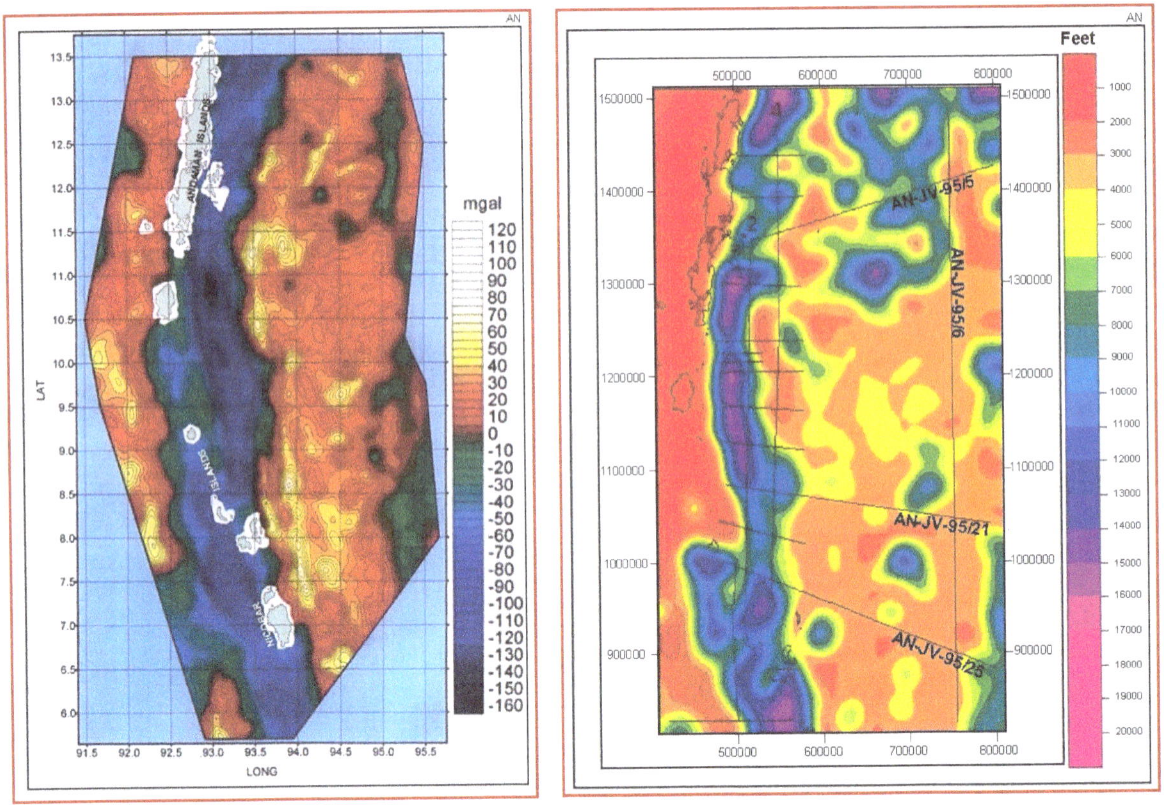

Fig. 6.6.5

Tectonics and Stratigraphy

The Andaman-Nicobar basin is a part of typical island arc system associated with convergent plate boundaries. The island arc system starts from southern part of Indonesia, passes through Andaman-Nicobar group of Islands and extends into Burmese arc. The total length of the Island chain is around 5000 km. The basin came into existence as a result of North ward movement and anti clock wise rotation of the island plate and this under thrusting beneath the Asiatic plates since cretaceous times. Based on the various geotechnical informations, the major tectonic elements recognized in the Andaman basin from west to east are (Fig. 6.6.6).

1. Andaman trench inner slope
2. Outer high/trench slope break
3. Fore arc basin
4. Volcanic Arc
5. Back Arc
6. Mergue terrace.

The above are also referred differently by different authors from East to West as:

1. The Malayan Orogene
2. The Back Deep
3. Inner Volcanic Arc
4. Inter Deep

5. The Outer Non Volcanic Island Arc
6. Fore Deep

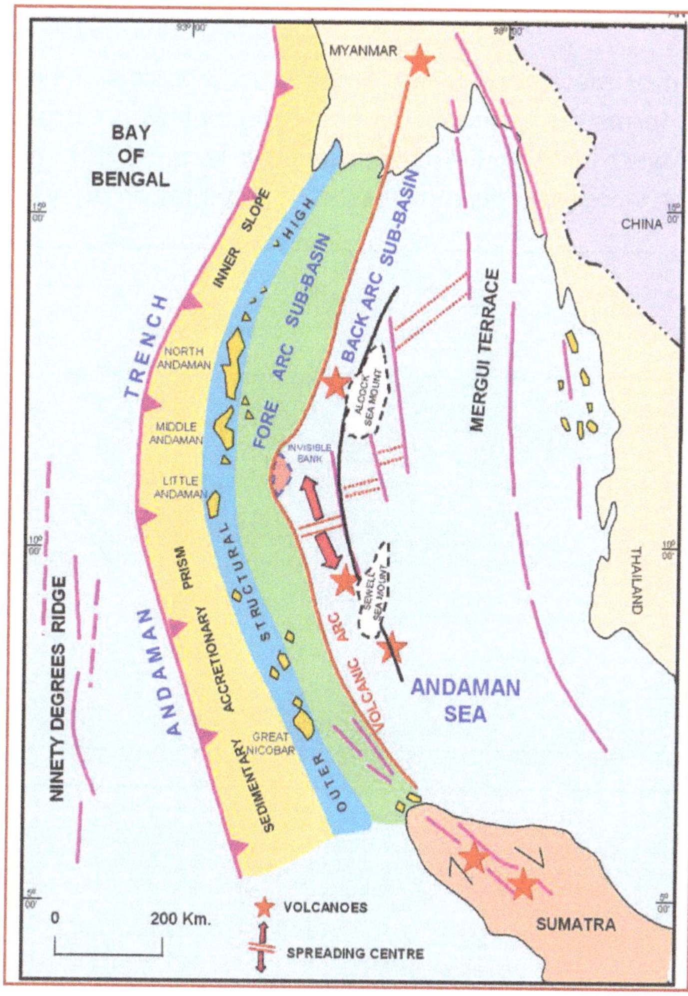

Fig. 6.6.6

The stratigraphy formulated for Andaman-Nicobar group of Islands, on the basis of surface geological mapping and stratigraphic information obtained from shallow well data is shown in Table- 6.6.1.

Andaman–Nicobar Island system forms the outer high with a slope/trench to the west of the island known as Java Trench (Fig. 6.6.7). This trench is mainly filled by Bengal fan sediments. The Eastern Andaman offshore is represented by a long trough known as fore arc basin which was faulted on its western margin, and is almost North–South in direction. This fore arc basin contains numerous N-S trending Neogene folds which are not much disturbed, but preserved their identities as Archipelago group of islands. This fore arc basin is limited to the east by volcanic arc system, where the volcanic activity is represented by Cretaceous ultra basic rocks and Eocene tuffs. Volcanic ash beds have also been recorded from rocks of early & middle Miocene age.The structural pattern in the southern part of the fore arc basin indicated tight folding with high dips and fair amount of deformation. The well known active volcano of the Barren and Narcondam islands represents the part of the volcanic arc. The invisible bank lies to the south east of Andaman Islands and to the east of volcanic arc adjoining it.

To the east of the volcanic arc lies the back arc, which is known to be a region of high heat flow. The new oceanic crust is formed in this region with a spreading center.

The Malayan Orogene

Malayan Orogene composed of metamorphosed Cambrian to Triassic sediments extensively intruded by Mesozoic granites/Volcanic, forms the economic basement for petroleum prospects. An extensive zone of flexures and faulting in the west of Mergui-Archpielageo had resulted in numerous horsts and grabbens with thick prism of sediments, which are the good locales for exploration of hydrocarbon.

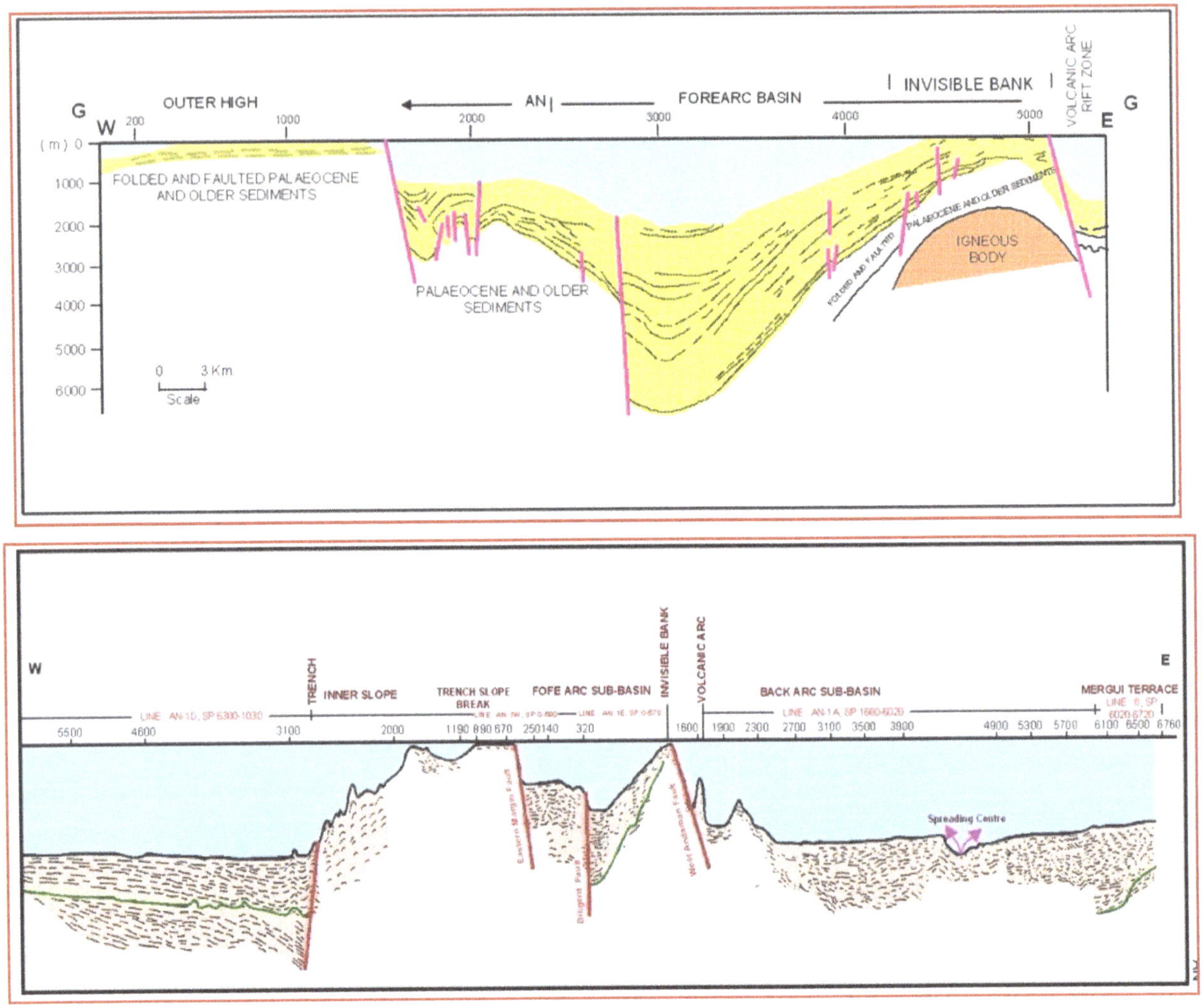

Fig. 6.6.7

The Back Deep

The structural features of back deep are less complex. This geotectonic unit lies between the Malayan orogen in the East and inner volcanic island in the west. The basin was subdivided to a great deal during Late Miocene. Extensional and various periods of extensions can be noticed in this back deep setup. The troughs are expected to contain over 2000 mt of unconsolidated sediments. The period of formation of back deep is thought to be related to the emergence of outer non volcanic system.

In the Irawadi delta of Burma, the back deep is subjected to a large pile of deltaic sequence which is largely confined to the shelf part of Gulf of Martaban.

The north Sumatra Tertiary basin, east of Barison range continues in the back deep. The NW-SE striking Sumatra basin contains numerous sub-basins within the basinal highs. The total sedimentary thickness of this basin is over 3000 mt. Along the barison range, the basin is highly folded and faulted in the offshore part of the back deep where clays and shales are predominant and structures are developed over shale diapirs.

Inner Volcanic Arc

The Narcondam barren island-invisible bank and numerous other volcanic sea mounts capped by Bio clastics sediments is the characteristic features of the tectonic unit. The extension of this if carried out into Burma to Mount Poppa, the oil fields of Burma are located in the troughs to the west of this volcanic unit. Extensive exploration, development and production of hydrocarbons were carried out in the Back arc / inter deep tectonic regime both in Burma and Sumatra region of Indonesia. From this analogy, it is almost certain that the back deep and inter deep areas of Andaman – Nicobar Islands of India play an important role for hydrocarbon exploration.

Inter Deep

This is structurally a negative belt between the inner volcanic arc in the east and Non-Volcanic arc to the west. This negative belt has been reported with sedimentary thickness of the order of 2000mt and several unconformities have been mapped. Close to the volcanic arc, the intensity of folding and faulting is severe. The inter deep and back deep, with their fine clastics and thick sedimentary and high temperature regime should form the "tidal euxinic basin" during late Oligocene to late Miocene, possible, with better maturation of organic matter. Ideal spurse basin conditions of regional nature might have prevailed during most of Oligocene period. The Ritchies Archipelago group of islands to the east of south Andaman are situated with in this geotectonic unit of inter deep. One exploratory well drilled in this tectonic setup, on Henry Lawrence Island proved to be Hydrocarbon bearing. Hence the inter deep and the back deep play an important role for hydrocarbon exploration. More- over the Inter deep and the back deep are in continuation to petroliferous Burmese Tertiary basin. Also the Inter deep extends towards south into Mevlaboh and other basins west of barison range of Indonesia. The back arc continues towards south into the petroliferous basins of north Sumatra.

The Outer Non-Volcanic Arc/Outer Sedimentary Arc

This geotectonic unit includes the main Andaman Nicobar island groups and is correlated with Naga hill ranges of NE India through Arakan Yoma fold belt. The eastern flank is steep and has a narrow shelf. The thick **flysh** of early Tertiary sediments of Baratang formation and part of Port Blair formation rest unconformably on Mesozoic metamorphosed sediments, with associated Ophiolite lavas & intrusions (Port meadow formation). The shearing and under thrusting of Indian plate from the west created this geanticlinal_features with westerly directed, over thrusts. In Andaman-Nicobar Islands, the structural folds in the post Oligocene are aligned nearly N-S. The structural dips in the post oiligocene sequence of Archipelago group of islands are moderate. The incidence of folding appears to have become gradually younger towards Nicobar and Mentavi group of islands. Pre tertiary (Metamorphosed) have been identified and main phases of folding is recognized as late Miocene and early Miocene.

Fore Deep

West of Nicobar and Mentavi islands, these geotectonic units are mapped as Java-Trench to the west of Andaman Islands and Arakan Yoma units. The Pliocene Bengal fan sediments and the clastics derived from Irrawaddy delta (Burma) had filled the topographic lows of Java-Trench in the western shelf of Andaman. Likely sub parallel, another sedimentary arc associated with imbricate structures could be expected with local depressions created during early Miocene, due to the episodic movements (Sastry & Prof. Eremenco etal). This region is expected to be of low temperature regime and the Paleogene sediments are expected to be largely Bathy environment.

1. In the Arkan-Yoma fold belt, Peridotite to Diorite igneous rocks along with the metamorphic schist's, gneisses and phyllites form the basement complex of Pre -upper Cretaceous age. This is interpreted as sheared oceanic crust accompanied by extrusions and intrusions of Ophiolites during upper cretaceous. Sands and shale together with the Ophiolite suite would form the 'Melange unit'.
2. Mud volcanoes and associated gas/oil shows are mapped widely from Baratang formations of Upper Cretaceous to Middle Eocene in the middle Andaman (Baratang Island) and are considered encouraging which may yield hydrocarbons in economic quantities for the entrapment overlying Eocene, Port Blair formations and also Miocene reservoir rocks.
3. During Oligocene, there appears to be a change in the direction of movement of Indian plate towards NNE to NE, and the sediments of the Bengal fan passing on the seduction plate under the Mentawi – Andaman, Arakan-Yoma system.
4. The Andaman basin (Back deep) might have been formed due to crustal extension during Late Tertiary period and during Oligocene-Miocene period. The Andaman basin appears to have been divided into several grabens and horsts in which shallow marine sediments were deposited. The back deep subjected to a great deal of tectonic activity during the Late Miocene, transcurrent faults trending E-W or ENE – WSW, appeared to have displaced some of the submerged banks, from their normal arcuate line. The present non-volcanic island arc system of Andaman Nicobar Island had a varied Neogene history.

Hydrocarbon Prospects

Active pursuit for hydrocarbon exploration in the offshore areas is currently undertaken wherein, the Burmese part of the fore deep i.e., west of Arakan yoma fold belt, Irrawaddy delta area, the deeper waters of Margui terrace of Thailand exist.

Gradually more oil and gas fields are being found associated with mud volcanoes. From the Arakan Yoma block and Gulf of Martaban offshore, several oil/gas/oil discoveries have been reported.

Geotechnically, the oil field basins of Burma have been correlated with the inter deep of Andaman Sea. In the North Sumatra basin, which is in the same geotectonic unit, as back deep of Andaman Sea, the production of oil and gas are both from sandstones and carbonate reservoirs. The principle reservoirs are developed in Lower to Middle Miocene formations of Kutapong and Baong formations. The principle source rocks are shales of Baong formations which are rich in organic matter.

Both inter deep and back deep tectonic units being locales for volcanic eruptions, resulted in generating higher geo thermal gradients, which in turn will aid in early release of hydrocarbons. The Andaman Island arc which forms the part of Arakan-yoma unit of Burmese arc, thrust sheets piled up due to the under thrusting of Indian plate in Java Trench.

The main Andaman Islands (Little, Middle & North Andaman), have a major thrust fault on the west margin of the islands (near Java trench) which hades towards East, over which the Baratang and Port Blair are thrusted. For Nicobar Islands, the Nicobar thrust is on the eastern side of the islands which heads towards west. This tectonic phenomenon can be explained by a major wrench fault in between the Andaman and Nicobar Islands. The Ritchies Archipelago group of islands in 'inter deep' are situated to the east of main Andaman Island, but such type of islands are absent to the east Nicobar Islands.

CHAPTER - 7

DEEP WATER EXPLORATION

Exploration companies are venturing into more and more hostile environments and logistically difficult areas (frontier areas) in search of hydrocarbons. Moving farther away from the near shore, deep water regions of continental shelf and deep sea basins are the main frontiers of exploration today. Any basin situated beyond 400 m water depth is classified as a deep water basin as per the Government of India NELP procedure. Globally, energy demand is gaining phenomenally and at a steady rate while the discovery trends in the onshore and the shallow offshore areas are declining rapidly. Hence, the major E&P companies are focusing their attention for exploration in deeper water areas to meet the energy demands. It is also observed that the reserves / discoveries in deeper waters and the average production from deeper water fields are manyfold higher than that of shallow water areas. Technological advances and innovations, spurting crude prices, improved commercial practices resulting in very good project efficiency in cost reductions etc. will drive the major oil companies to focus their attention for deep water exploration. The major discoveries in deep waters of Campose Basin Brazil, Offshore Angola, North Sea, and Gulf of Mexico and even the recent discoveries in the Krishna-Godavari deep waters encourage the oil companies to move towards Indian deep water basins for exploration. British Petroleum, Exxon Mobil, Shell, Petrobrass of Brazil, ONGC, Reliance, Cairn, Total Fina Elf, Agip-Eni, ChevronTexaco, Stat Oil, Unocal, BG are the top oil companies who are having good reserve holdings in the deep water basins.

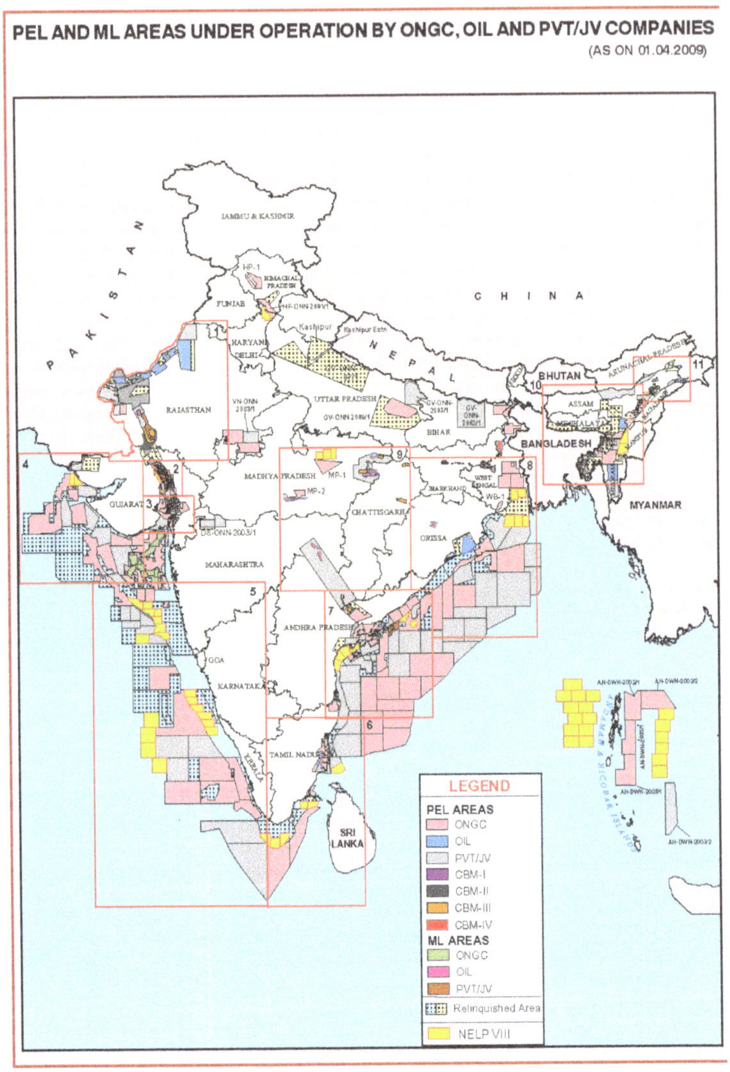

(Courtesy DGH)

Fig. 7.1

The growing demand for hydrocarbons in our country to cater to the needs of fast developing economy has turned the deep waters into focus areas of exploration for enhancing the reserves growth and production. Deep water exploration is gaining momentum not only in India, but also throughout the world, in countries like South East Asia, Africa, Indonesia, South China, Angola, Latin America. The discovery of Giant oil fields offshore Brazil, in the Gulf of Mexico and Norway in water depths beyond 400m has brought out hydrocarbon prospectivity to deep offshore basins.

It is reported that worldwide 3500 exploratary cum delineation wells were drilled in deep waters and around 125 billion barrels of oil equivalent have been discovered, of which are around 47 billion barrels of oil and 428 TCF of gas. It is also observed that in deep waters, the gas and oil discovery will be in the order of 70:30.

Indian Government had offered 90 deep water blocks during NELP-1 to NELP-VII, of which 76 blocks have been awarded covering both West Coast and East Coast. Andaman deep water exploration prospects are identified. The Blocks on offer/awarded are shown in (Fig. 7.1).

NELP	Blocks offered	Blocks awarded	Awardees	Date of Signing/ Effective Date	Present Area (sq.km)	Status
I	--	--	--	--	--	--
II	GS-DWN-2000/1	ONGC (100%-OP)	17-Jul-2001/ 16-Aug-2001	Relinquished	--	
	GS-DWN-2000/2	ONGC (85%-OP), GAIL (15%)	17-Jul-2001/ 16-Aug-2001	Relinquished	--	
	MB-DWN-2000/1	ONGC (85%-OP), IOC (15%),	17-Jul-2001/	Relinquished	--	
	MB-DWN-2000/2	ONGC (50%-OP), GAIL (15%), IOC (15%), OIL (10%), GSPC (10%)	17-Jul-2001	Relinquished	--	
	KK-DWN-2000/1	RIL (100%-OP)	17-Jul-01	Relinquished	--	
	KK-DWN-2000/2	ONGC (85%-OP), GAIL (15%)		Relinquished	--	
	KK-DWN-2000/3	RIL (100%-OP)	17-Jul-01	Relinquished	--	
	KK-DWN-2000/4	ONGC (100%-OP)	17-Jul-01	Relinquished	--	
III	KK-DWN-2001/1	KK-DWN-2001/1	RIL (100%-OP)	04-Feb-2003/ 03-Apr-2003	27315	
	KK-DWN-2001/2	KK-DWN-2001/2	RIL (100%-OP)	04-Feb-2003/ 03-Apr-2003	31515	
	KK-DWN-2001/3	KK-DWN-2001/3	ONGC (100%-OP)	04-Feb-2003/ 12-Mar-2003	21775	
IV	GS-DWN-2002/1	GS-DWN-2002/1	ONGC (100%-OP)	06-Feb-2004/ 17-Mar-2004	RELINQUISHED	
	KK-DWN-2002/1	KK-DWN-2002/1	ONGC (70%-OP), OIL (20%), BPCL (10%)	06-Feb-2004/	20910	

Contd...

NELP	Blocks offered	Blocks awarded	Awardees	Date of Signing/ Effective Date	Present Area (sq.km)	Status
	KK-DWN-2002/2	KK-DWN-2002/2	ONGC (80%-OP), HPCL (20%)	06-Feb-2004/ 17-Mar-2004	22810	
	KK-DWN-2002/3	KK-DWN-2002/3	ONGC (80%-OP), HPCL (20%)	06-Feb-2004/ 17-Mar-2004	20910	
V	KK-DWN-2003/1	KK-DWN-2003/1	RIL (100%-OP)	23-Sep-2005/ 23-Jan-2006	18245	
	KK-DWN-2003/2	KK-DWN-2003/2	RIL (100%-OP)	23-Sep-2005/ 05-Dec-2005	12285	
VI	KK-DWN-2004/1	KK-DWN-2004/1	ONGC (45%-OP), CIL (40%), TATA (15%)	02-Mar-2007/ 09-May-2007	12324	
	KK-DWN-2004/2	--	--	--	--	--
	KK-DWN-2004/3	--	--	--	--	--
VII	MB-DWN-2005/6	--	--	--	--	--
	MB-DWN-2005/7	MB-DWN-2005/7	BHP Billiton Petroleum International Pty.Ltd (26% -OP), GVK Oil and Gas Ltd (74%)	22-Dec-2008/	3324	
	MB-DWN-2005/8	--	--	--	--	--
	MB-DWN-2005/9	MB-DWN-2005/9	BHP Billiton Petroleum International Pty.Ltd (26% -OP), GVK Oil and Gas Ltd (74%)	22-Dec-2008/	3138	
	KK-DWN-2005/1	KK-DWN-2005/1	BHP Billiton Petroleum International Pty.Ltd (26%-OP), GVK Oil and Gas Ltd. (74%)	22-Dec-2008/	14675	
VIII	M B-DW N-2009/1	M B-DW N-2009/1	Cairn Energy India Pty Limited (10%-OP), Cairn India Ltd (90%)	30-Jun-2010/	2961	
	KK-DWN-2009/1	--	--	--	--	--
	KK-DWN-2009/2	--	--	--	--	--
	KK-DWN-2009/3	--	--	--	--	--
	KK-DWN-2009/4	--	--	--	--	--

(Courtecy DDH)

The deep water basinal area constitutes about 40% of the total sedimentary area of our country (i.e., 13, 50,000 sq.kms out of the total sedimentary area of 31, 34,700 sq.kms). About 11 billion tons of hydrocarbon resources have been assessed in the deep water sector of India, of which, the western offshore is estimated at 5.3 billion and East Coast at 5.7 billion oil and OEG. This resource potential of 11 billion tons of O+OEG does not include the Gas Hydrate resource potential. The exploration and exploitation of gas hydrates from the deep waters is gaining attention of the Government of India and also the major oil companies. A number of locations for gas hydrate exploration have already been identified and the blocks are going to be offered for exploration and exploitation by JV participations.

GEOLOGICAL PLAYS : KRISHNA – GODAVARI

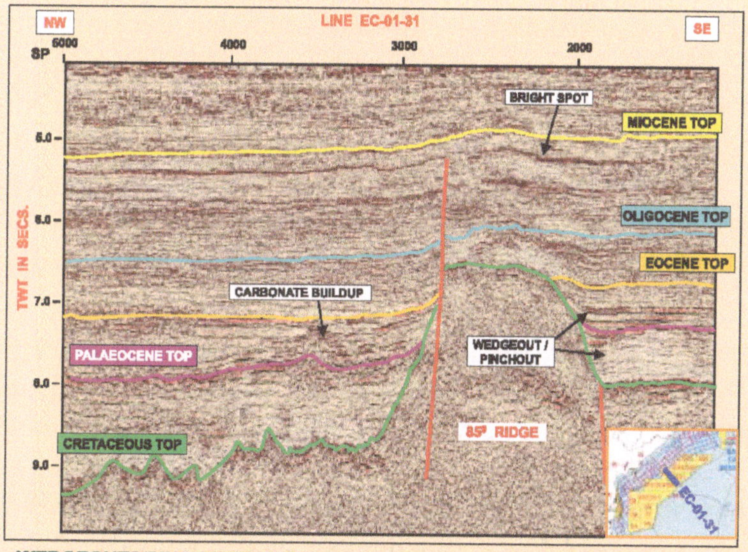

WEDGEOUTS/PINCHOUTS IN PALEOCENE-EOCENE SEQUENCE, CARBONATE BUILDUP IN EOCENE AND BRIGHT SPOT IN MIOCENE

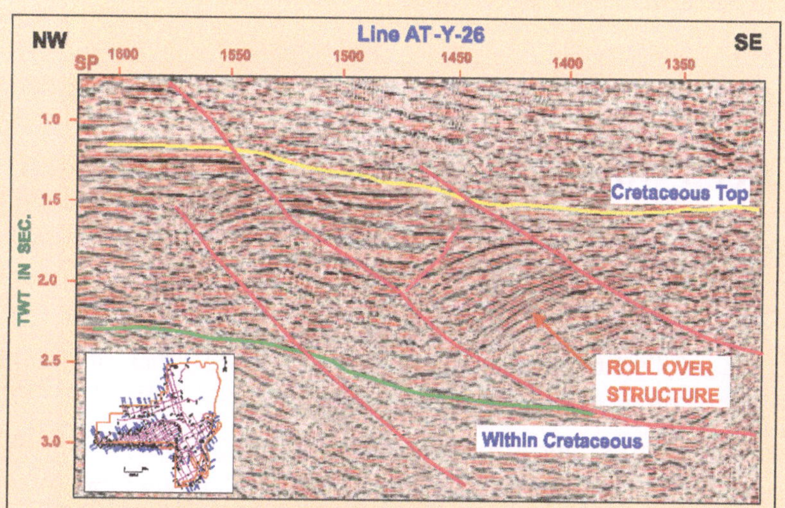

DIP SECTION SHOWING ROLL OVER STRUCTURE WITHIN CRETACEOUS

(Courtesy DGH)

Fig. 7.2

The hydrocarbon plays in the deep waters are much different from the hydrocarbon plays of onshore and shallow waters. Sub-marine canyon flows, turbidite fan deposits, very large size channel deposits play an important role in the hydrocarbon exploration of deep waters, which are essentially stratigraphic, or at the most structural cum stratigraphic rather than purely structural (Fig. 7.2).

ONGC started deep water exploration project way back in 1980 and drilled 2 deep water wells in Krishna-Godavari basin in water depths of 350-400 m (on G1 & G2 structures). Both the wells were successful and produced oil and gas from Mio-pliocene sediments. The structures are situated on growth fault features.

Subsequently, ONGC carried out intensive seismic surveys covering from 200 m to 3000 m water depth in the west coast of India. On the basis of the seismic surveys, a number of prospects have been identified in the deeper waters of the basin. The total area covered is around 6,64,800 sq.kms. On the basis of these seismic surveys 5.3 billion tons of hydrocarbon resources have been estimated for the deep water basins of west coast. Over the years, ONGC acquired 47,487 LKM of seismic data in the deep water areas of west coast and identified 22 prospects for exploration. Another 10,102 LKM of 3D seismic data was acquired over prospect KK-DW-1 and KK-DW-A located in Kerala-Konkan Basin. One location KK-DW-17-1 was drilled in the deep waters of Kerala-Konkan offshore during October 1997 and drilled to a depth of 2991 m. This location was positioned on the flanks of Padua island of Lakshadweep. Even though the well did not give commercial hydrocarbons it had given valuable information for future exploration. The well was drilled with an objective to explore Pre-Paleogene, Paleogene and Neogene clastics and carbonate sequence for its hydrocarbon potential.

Till early 1990's India's deep water exploration activity was limited to half a dozen wells that ONGC drilled. After the NELP-1 licensing and discovery of gas by Reliance in deeper waters, an extensive exploration thrust was given for the deep water exploration. In 2001, Cairn Energy India Limited drilled its first deep water well in Krishna-Godavari block (DWN-98/2) which resulted in the discovery of 7 TCF of gas in place. Immediately thereafter, drilling of 4 wells in the second block KG-DWN-98/3 by Reliance resulted in a very big gas discovery of 13 to 14 TCF, with a number of further leads. These discoveries marked the onset of deep water E&P era for India. ONGC, RIL and Cairn Energy are the three major companies which are active in Indian offshore deep water basins. ONGC is working aggressively on its 60 blocks in the deep waters of Kutch-Saurashtra offshore, Bombay offshore, Kerala-Konkan, Cauvery basin, Krishna-Godavari offshore and Mahanadi offshore basins. The other two private companies, Reliance and Cairn are also putting their maximum efforts to under take deep water exploration activities in their blocks.

Deep Water Exploration and Development

7.1 East Coast Basins

Cairn Energy India Pvt Ltd

Cairn Energy India Pvt. Ltd., drilled 6 wells in deep water offshore block of Krishna-Godavari basin (KG-DWN-98/2) during 2001. The company made 5 major discoveries (Fig. 7.3).

1. (Prospect R) is located in water depths ranging from 750 to 1200 m. Annapurna well on this prospect was drilled during June 2001 in 1023m water depth. The well flowed gas @ 18 MMSCFT/day through 1" choke. The initial reserve estimates are in the range of 1 TCF.
2. Kanakadurga (Prospect P) is drilled in a water depth of 493 m. The interval 2268-2285 m flowed oil @ 3650 bbls of oil / day. The initial estimate of recoverable reserves of oil was in the range of 5.5 to 11.5 MMT. This was drilled in August 2001.

3. Padmavati (Prospect M) was drilled at a water depth of 500 m. Two intervals proved to be hydrocarbon bearing. The interval 2081-2092 m (deeper sand) flowed oil @4500 bopd and gas 2.70 MMCFT). The initial estimates of recoverable reserves were in the range of 5.5 to 11 MMT. DWN-N-1 is another deep well drilled in a water depth of 500 m proved to be hydrocarbon bearing in Mio-Pliocene sands. The initial reserve estimate is 115 MMBOE. Subsequent exploratory efforts carried out by ONGC after acquiring the operatorship from Cairn Energy, lead to major gas discovery of UD-1 in this block.

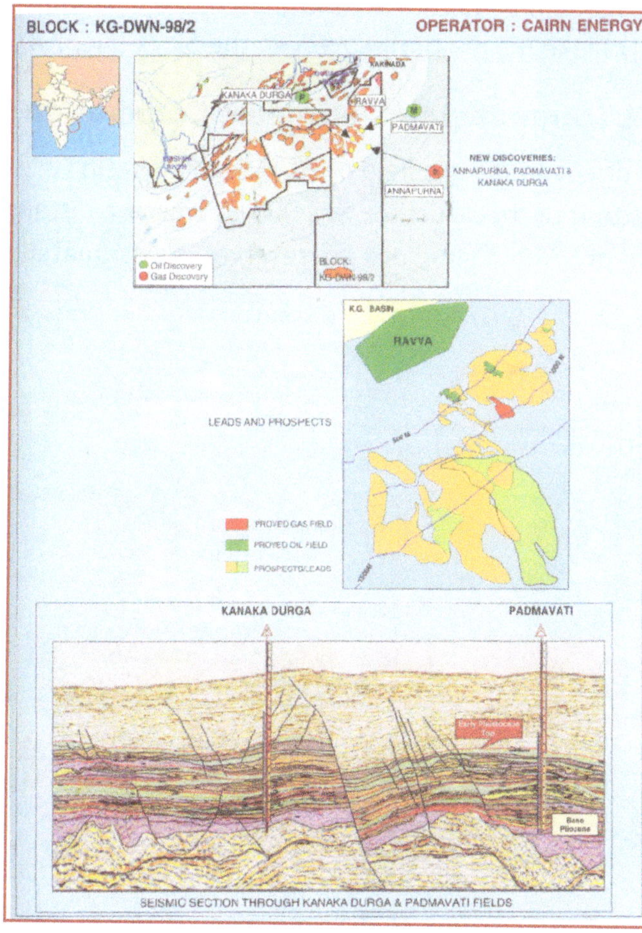

(Courtesy DGH)

Fig. 7.3

Above three discoveries are viable for development and commercial production. Annapurna was estimated to contain 40-60 bcf of gas, Kanakadurga 4280 MMBBLS of oil and Padmavathi 70-80 Billion bbls of oil, as in-place reserves.

ONGC

ONGC, after taking over the operatorship of the block (KG-DWN-98/2), drilled D1 prospect to a depth of 2416 m, in a water depth of 603 m. This prospect is having oil reserve potential of 20-40 MMBOE and gas potential of 300-800 bcf. Mio-Pliocene turbidite reservoir at 2074-2224 m proved to be oil and gas bearing.

Another location A1 in a water depth of 800 m was drilled to a target depth of 2610m which penetrated Pliocene sediments. A zone at 1700 m proved to be gas bearing. It flowed 6.8 million SCMD with a surface pressure of 2600 psi.

Another location U1 was drilled in a water depth of 1200 m to a target depth of 2400 m which flowed 6.8 million SCMD of gas at a surface pressure of 300 psi. Another location W1 was drilled in a water depth of 1280 m to a target depth of 2640 m. Channel sands in lower Pliocene sequence around 2170-2190 m flowed gas around 2.5 million SCMPD through 3/8" choke. Another well D1 was drilled in a water depth of 650 m to a depth of 2556 m during May 2006. Gas was found at 1600 m (8m thick pay) and also at 2350-2390 m. Analysis indicated that the gas is mainly methane (92%) sweet gas. Another well UD1 was drilled by ONGC during 2006 in a water depth of 2840 m at a distance of about 150 Kms from the coast. A well was originally planned to a target depth of 6200 m or 100 m within the basement, with an objective to explore hydrocarbon potential of Eocene and Oligocene clastic reservoirs / carbonate reservoirs and fractured basement. The well reached a depth of 5387 m in which only the first target interval, 5242-5305 m could be penetrated (Oligocene). MDT test indicated around 28 m of gas bearing sand within Oligocene and the Areal extent of the structure at this level is around 140 Sq.Kms. The well potential was primarily estimated at around 21.0 TCF of gas or 60 BCM. As per the reassessment of the seismic data, and VSPs

VSPs conducted indicate that the wells can penetrate, another 4 more interesting zones which could possibly make it a major discovery in the block (Fig. 7.4).

GSPCL (Gujarat state Petroleum Corporation Limited)

GSPCL discovered a gas field of 20 TCF in K-G basin very adjacent to Reliance Industries Ltd. (RIL), block. The Ariel extent of the structure was 600 sq.kms at a distance of 50 kms from the coast. The well was drilled to a depth of 5780 m. In-place gas reserves are estimated around 21 TCF.

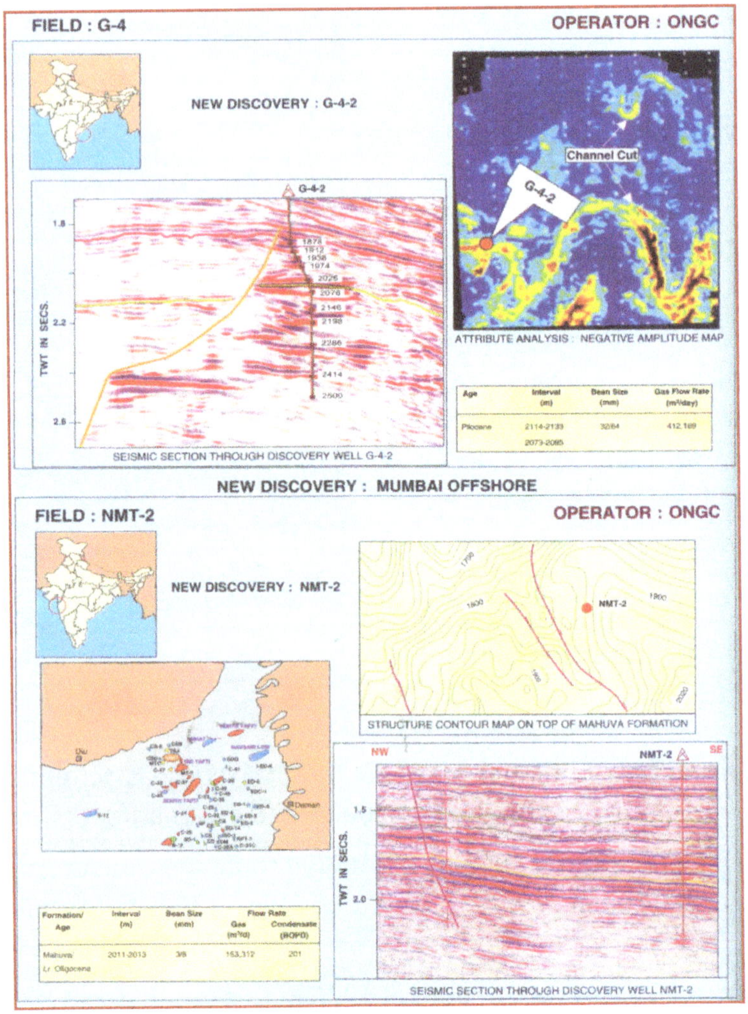

(Courtesy DGH)

Fig. 7.4

RIL (Reliance Industries Limited)

RIL's first successful well A-1 (Dirubhai-1) was drilled to a target depth of 2903 m in a water depth of 640 m, during 2002, in the block KG-DWN-98/3 (D-6 block). This is a major gas discovery of the world, as reported by the company. The well has a gross hydrocarbon pay thickness of about 240-250 m of Upper Miocene age. A zone tested at 2040 m, produced gas around 40 million SCFT/day. It is a sweet gas, mostly methane (99.7%). It drilled 10 exploratory wells, all of them proved to be successful gas finds (Fig 7.5, 7.6, 7.7 & 7.8).

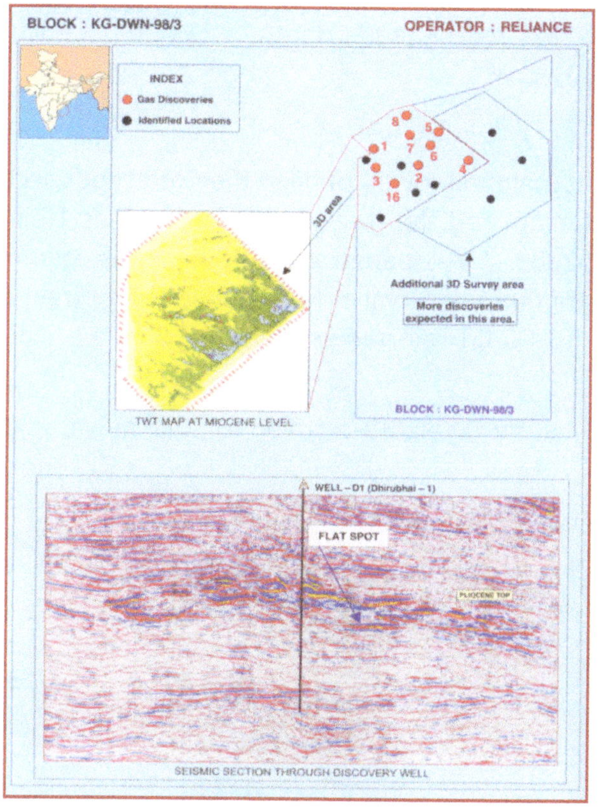

Fig. 7.5

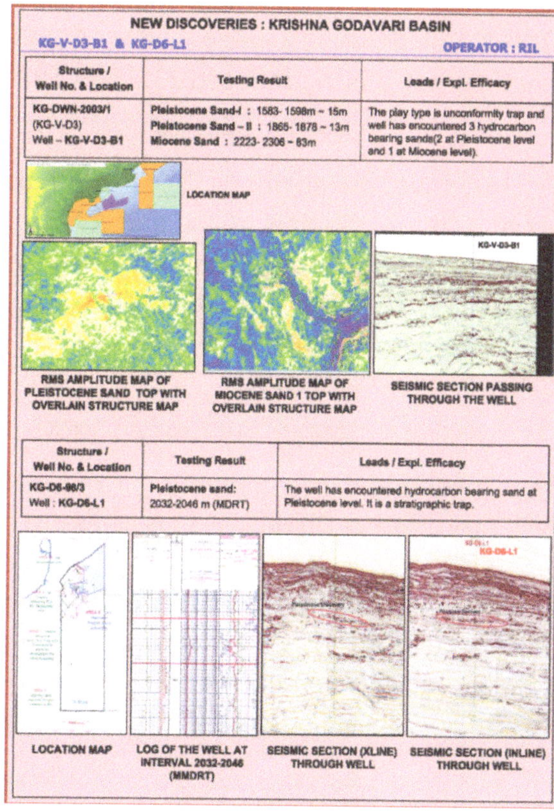

Fig. 7.6

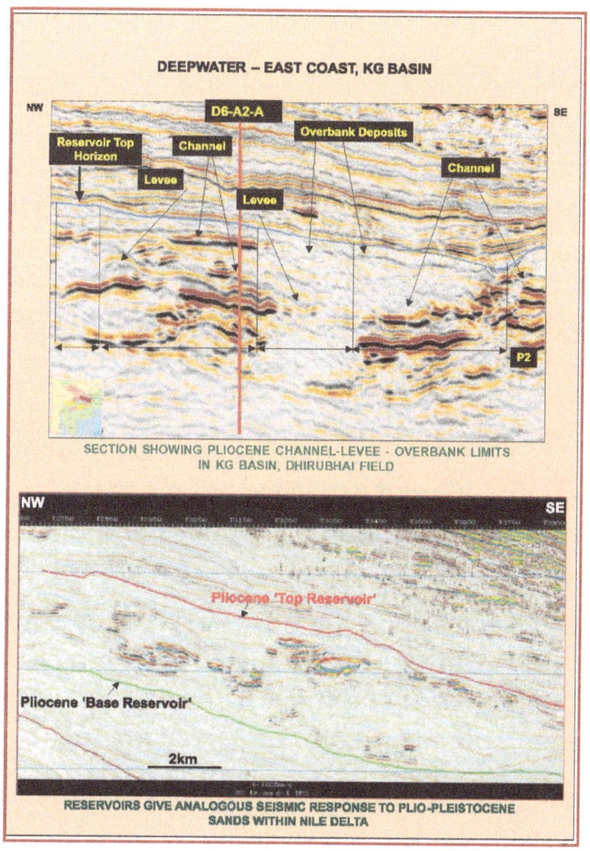

Fig. 7.7

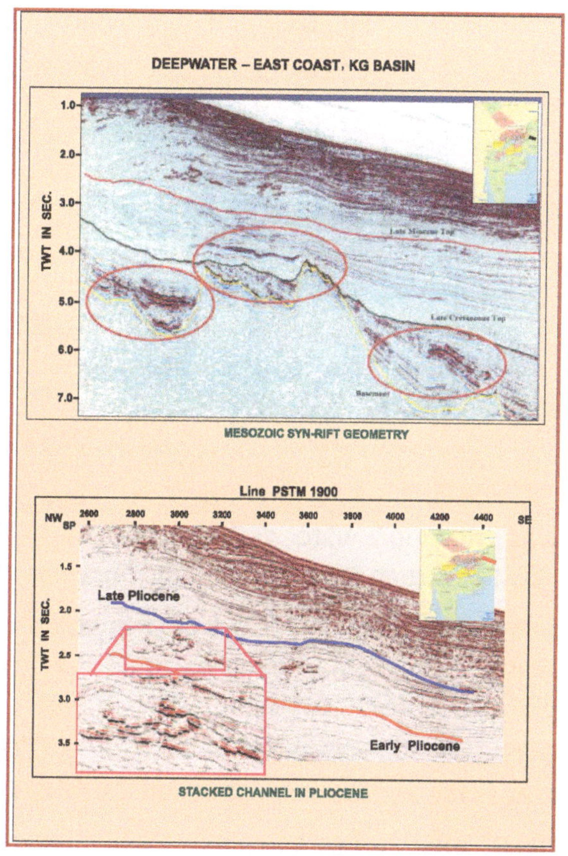

(Courtesy DGH)

Fig. 7.8

D6-M1 was drilled during 2004 in a water depth of 1327 m to a target depth of 2942 m. It encountered 155 m of net gas pay which produced 23.7 MSCFPD at 2750 m.

Subsequently, RIL drilled the wells M-1, G-1A, G-1, D-1, E-1, P-1A, AA-1.

The well MA-1 was drilled during 2006. This is a first major discovery of oil in the Cretaceous section. Oil was struck at 1800 m. Gaffney, Cline and Associates estimated the reserves of the block as 1 billion bbls and the field is put on production for oil during Sept 2008. MA-1 penetrated 170 m of gas column. An oil column of 24 m (42^0 API) is met, in the Cretaceous stratigraphic interval of MA-2, drilled to a target depth of 3581 m. AA-1 well is reported to be a gas discovery. KG-D6-Q1 referred as Dhirubhai 30 is a gas discovery. P2 well is another gas discovery.

MG-1 was drilled to 6000 m but the well proved to be dry. Well MB-1 drilled to 3080 m was abandoned due to technical reasons.

Well ME-1 referred as Dhirubhai-31 is a gas discovery.

Well	Target Depth	Gas pay	Water Depth
A-1 (Dhirubhai-1)	2903 m	140 m	603 m
C-1 (Dhirubhai-2)	1332 m	25 m	
B-1 (Dhirubhai-3)	2340 m	100 m	
B-2	2445 m	67 m	
F-1 (Dhirubhai-4)	Gas well		
D-1 (Dhirubhai-6)	Gas well		
K-1 (Dhirubhai-7)	2531 m	Gas well	1031 m
K-2 –ST-1 (Dhirubhai-8)	2220 m	Gas well	992 m
SH-2 (Dhirubhai-5)	Gas well		

Well R-1 in a water depth of 2010 m drilled to a target depth of 4857 m encountered two gas bearing sands in Mio-Pliocene (3306-3027 m, 4695-4762 m) and the well is referred as Dhirubhai-35 discovery. The discoveries are in Mio-Pliocene submarine channels, lobes and fan complex. The four gas discoveries D-29, 30, 31 and 34 are under appraisal. The well L-1 (Dhirubhai-42) proved to be gas bearing in the Pleistocene submarine channel. The company acquired additional 3D and close grid 2D seismic to finetune the discoveries.

Crude oil production was commenced from this block in Sept 2008 and gas production from April 2009. KG-D6-AP-2 was drilled to a target depth of 4358 m in April 2009 and happened to be a gas discovery. KG-D6-BA2 was drilled in a water depth of 2472 m. In May 2009, KG-D6-MA6H was drilled to a target depth of 6250 m. On MA structure, four oil producing wells were located.

D&M estimated around 11.9 TCF inplace and proven reserves at 5.7 TCF. The reserves have gone up to 12.59 TCF and recoverable gas up to 10.03 TCF. From MA structure the estimated peak production of oil was around 35000 to 40,000 bls/day and the gas production is about 40MSCM/day which is likely to be stepped up further to 80 MSCMPD shortly. The production started in 2008.

- The possible Hydrocarbon plays, off Cauvery-Palar basins are shown in Fig. 7.9, 7.10 & 7.11
- Similarly the Hydrocarbon plays, off Mahanadi and NEC area are shown in Fig. 7.12 & 7.13
- Deep water plays in Andaman and Nichobar Islands are shown in Fig. 7.14 & 15.

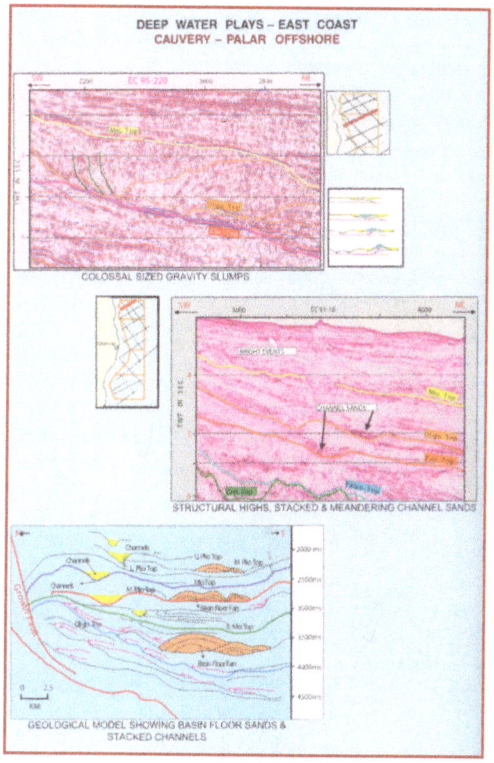

Fig. 7.9

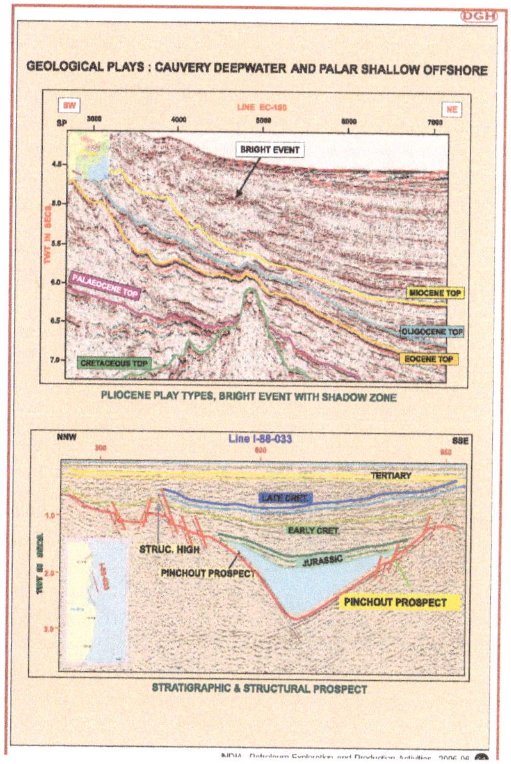

Fig. 7.10

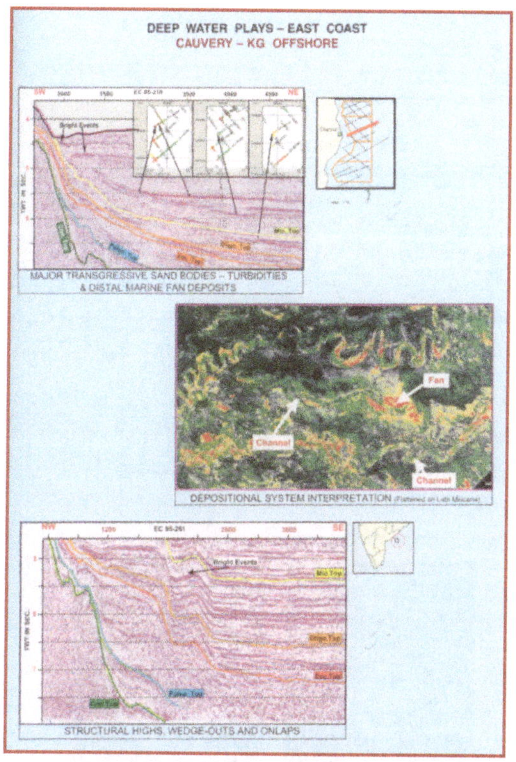

Fig. 7.11

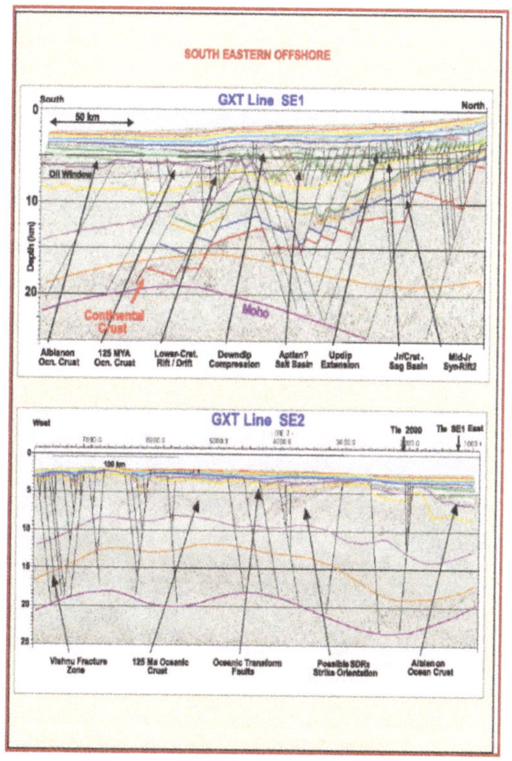

Fig. 7.12

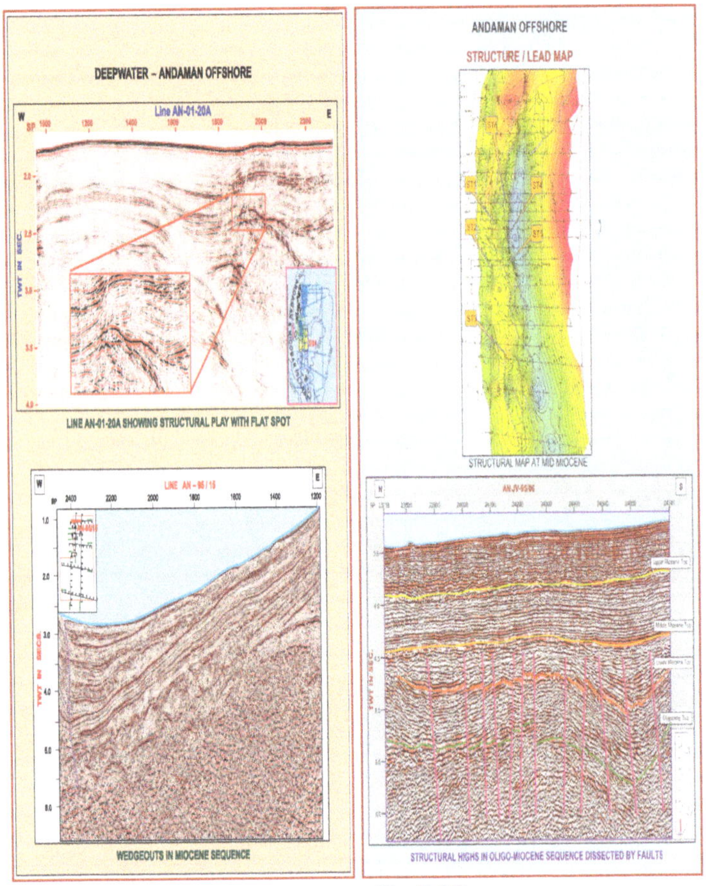

Fig. 7.13

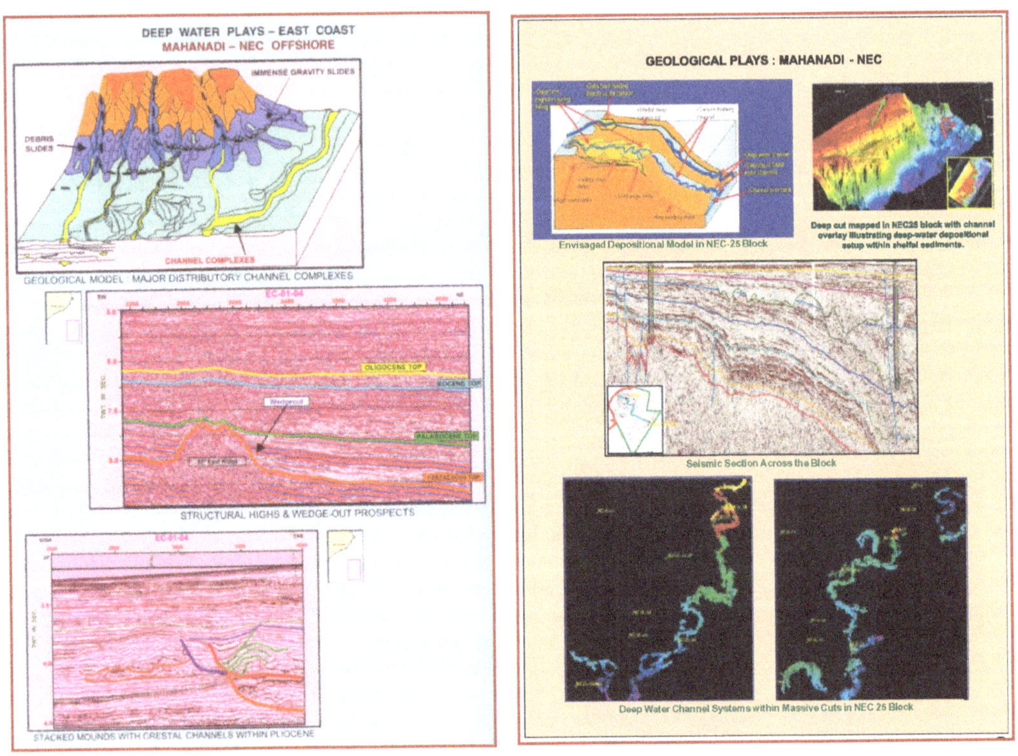

Fig. 7.14

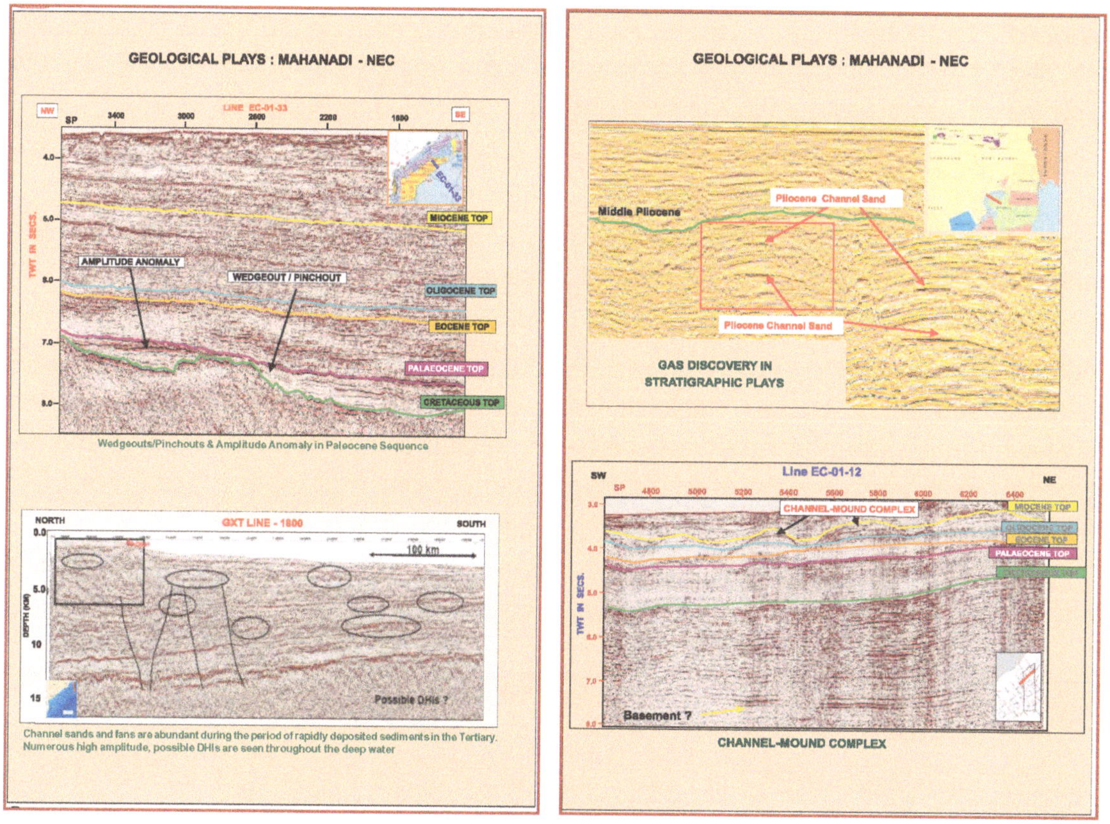

Fig. 7.15

7.2 Western Offshore

ONGC so far drilled 41 deep / ultra deep water wells in both the East and West deep offshore basins. 13 wells were drilled in western deep offshore basin up to a bathymetry of 3008 m (Fig. 7.16) and 28 wells in Eastern deep offshore up to 2850 m bathymetry. In Western offshore of Kerala-Konkan basin ONGC drilled KK-DW.17-1 on a large carbonate bank (Padua). The well was drilled to a target depth of 32000 m over Padua bank which is situated in very deep water regime. The monotonous carbonate section is met with upto 1622 m starting almost from the sea bed. The well penetrated basalt section upto the target. The drilling results indicated presence of highly porous limestone from sea bed up to 1660 m. This stratigraphic section represents from Eocene to Recent. During the drilling of this porous carbonate section, total mud loss was observed. The size of the prospect is 3000 sq.kms. Objective of the well is to explore Pre-Paleogene, Paleogene and Neogene clastics and carbonate sequence for its hydrocarbon potential. Though no commercial hydrocarbons have been obtained, the well gave valuable information for further exploration in the basin. Another well KKDW-A is also planned for drilling with different hydrocarbon play. The prospect represents a turbidite fill, fault closure. The stratigraphic column of KK-DW-17 is shown in Fig. 7.17.

In the western deep offshore hydrocarbons are yet to be proved. Gas analysis on drop core samples from Kutch-Saurashtra area reveals thermogenic gas. Source rock potential of Cretaceous is expected to be fairly good in Kutch and Kerala-Konkan areas. In the Western offshore, the TOC values of Late Cretaceous to Paleocene are < 0.5. The S_2 values also are <2. The organic material type is mainly Type-III and Type-IV.

Indus fan depositon continues to fill lower slopes of western margin from Late Miocene to Recent i.e. almost to the west of Laccadive ridge. Accelerated subsidence of margin led to the creation of deep water basin, by the clastics deposition from shelf and across slope, mainly as turbidite slumps. Growth faulting and related structural pattern is seen in this type of environmental deposition.

In the Eastern deep offshore where Mio-Pliocene section is met with in number of wells. The TOC content varies from 1.1. to almost 1.5% and the SO_2 also ranges from 1.65 to 2.75. The TOC content of Cretaceous samples obtained from various drilled wells of deep waters of west coast indicate 1 to 4% . S_2 (mghc-g) varies from < 2 to almost 4.2. In Early Oligocene, the TOC varies from 1 to 1.5%. Fairly good source potential hasbeen identified in Mio-Pliocene sequence of Mahanadi – Bengal deep waters.

Paleocene / Cretaceous synrift sediments with good source rock potential are the exploration target in eastern offshore deep water basin. Reliance's find of oil in Cretaceous sediments of MA structure strengthens this contention in other parts of the basin. It is also reported that the gas find on East coast show biogenic to mixed horizon. The isotopic studies carried out on produced gas from DST samples in Mio-Pliocene of Mahanadi deep waters also indicate biogenic horizon.

The hydrocarbon plays envisaged of Kerala-Konkan, Mahanadi and Cauvery deep water basins are shown in the Fig. 7.16(a)(b)&(c).

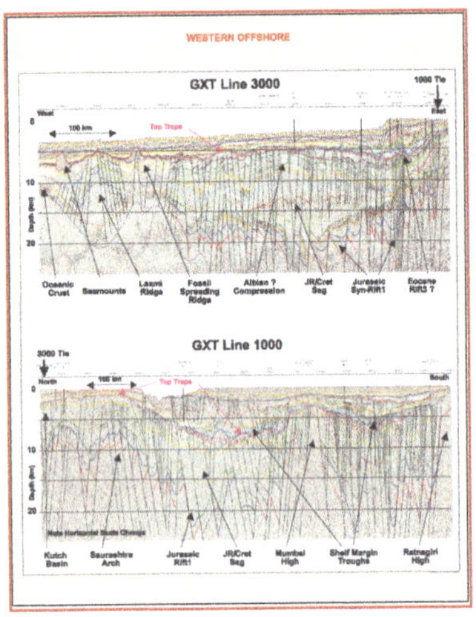

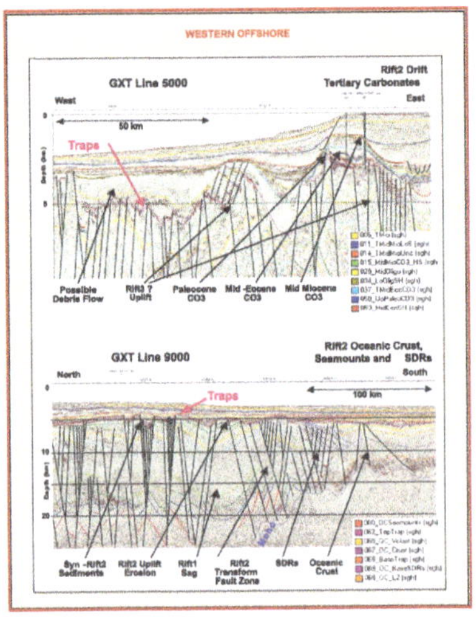

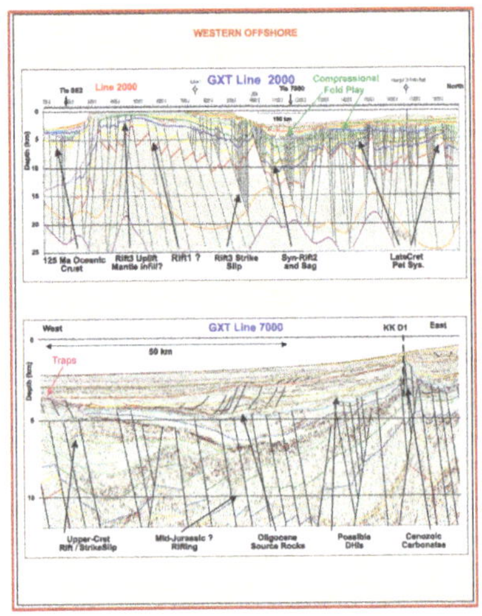

(Courtesy DGH)

Fig. 7.16a **Fig. 7.16b** **Fig. 7.16c**

CHAPTER - 8

GAS HYDRATES

In the deep waters beyond 500 m water depth unconventional Hydrocarbons Resource, "Methane Hydrate" is drawing attention in the exploration quest to meet the country's energy demands.

The prognosticated gas hydrate resources in the country are placed at 1894 TCM. A number of significant prospects have been identified in both Eastern and Western offshores in the water depths ranging from of 500-3000 m. The gas hydrate prospect map is shown in the Fig. 8.1.

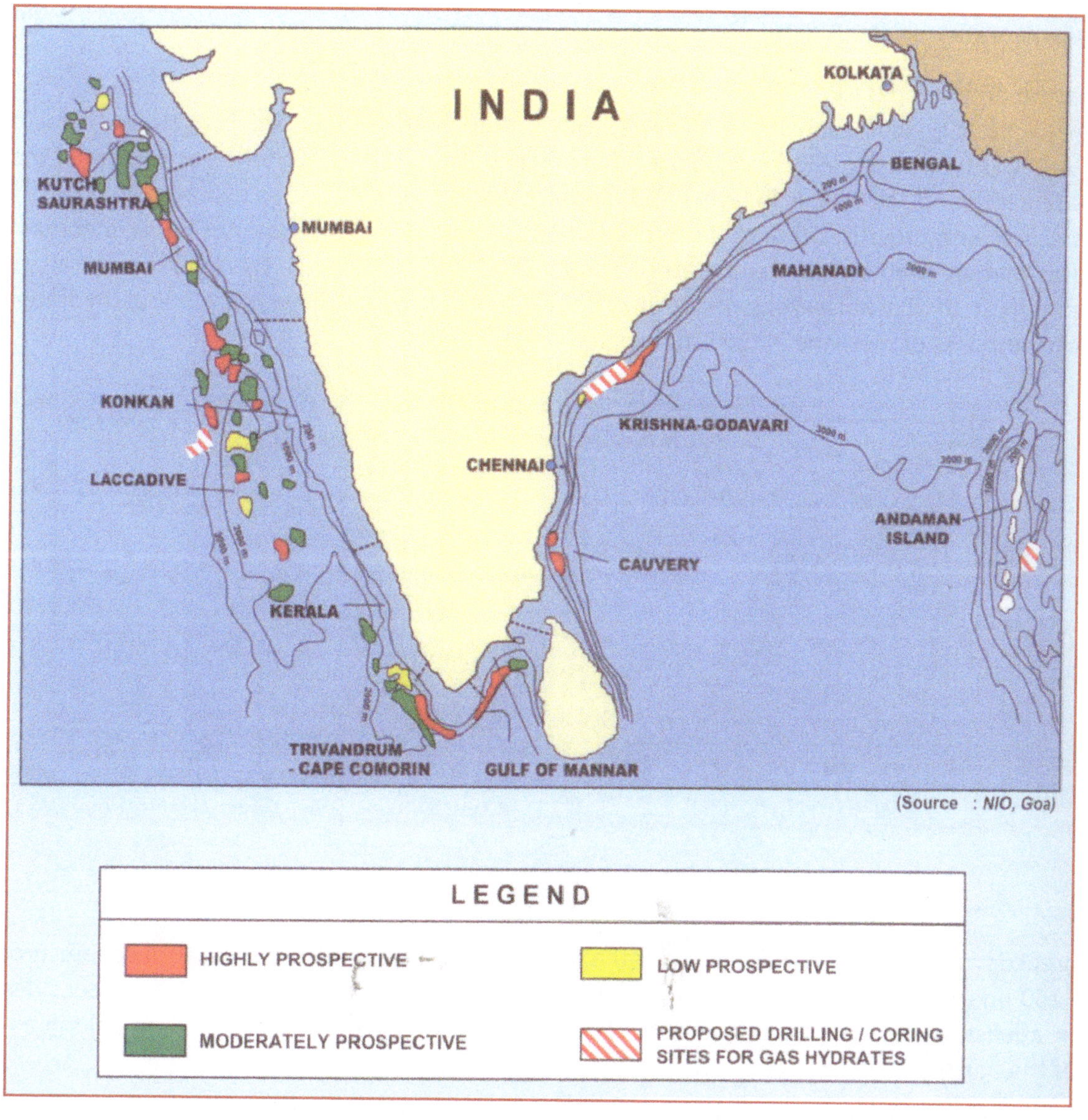

(Cortesy DGH)

Fig. 8.1

Gas hydrate is a crystalline solid and its building block consists of a gas molecule surrounded by a cage of water molecules. It is similar to ice except that, the crystalline structure is stabilized by the guest gas molecules within the cage of water molecules. Many gases have molecular sizes suitable to form hydrate, including such natural occurring gases such as Carbondioxide, Hydrogen Sulphide and several low carbon number hydrocarbons. But most marine gas hydrates that have been analysed are methane hydrates Fig. 8.2 & 8.2a.

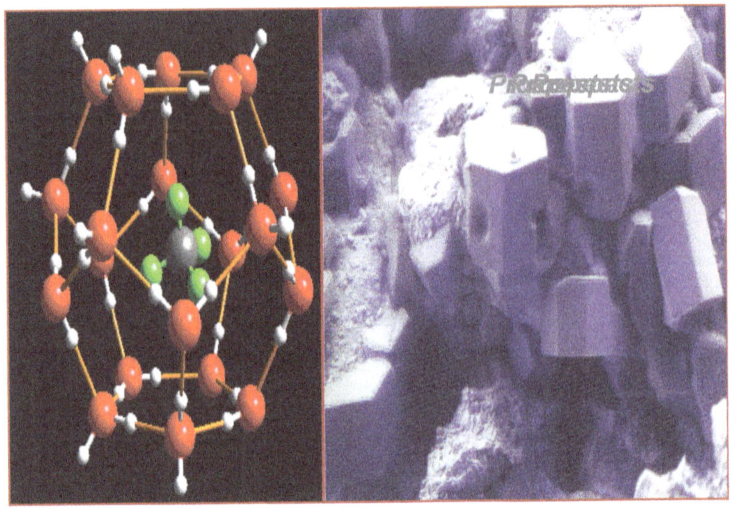

(Courtesy DGH)

Fig. 8.2 & 8.2.a

Pressure and Temperature are the two major factors that control the hydrate, where the methane gas will be stable in the crystal structure. Gas hydrate stability occurrences are shown in Fig. 8.3. Hydrate forms as cement in the pore spaces of sediments, as well as in layers and modules of pure hydrates. Gas hydrates are stable at the temperatures and pressures that occur in ocean floor sediments at water depths >500 m and at those pressures that are stable at the temperatures, about those for ice stability. Gas hydrates are also stable in permafrost of the Polar Regions (both offshore and onshore sediments). Gas hydrates bind immense amounts of methane in the sea floor sediments.

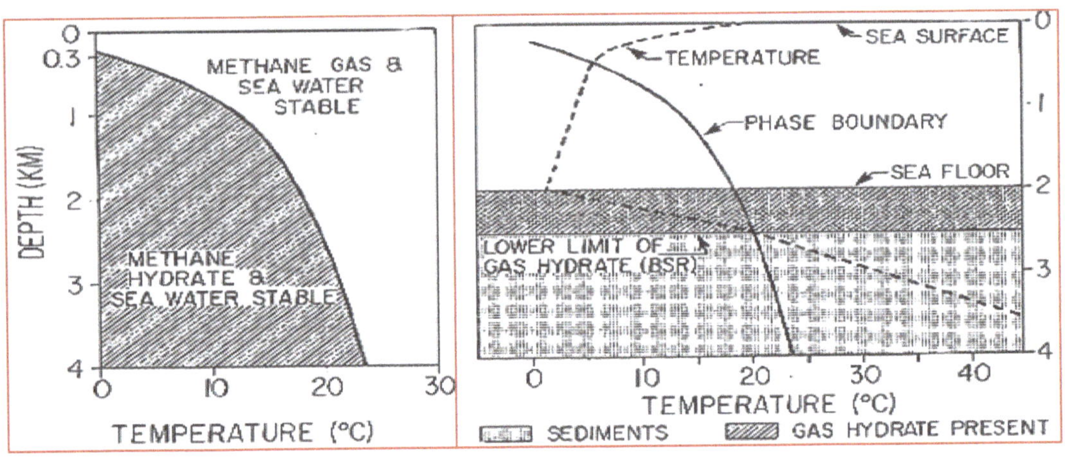

Fig. 8.3

A breakdown of one unit volume of methane hydrate with a pressure of 1 atmosphere produces about 160 unit volumes of gas. The worldwide amounts of methane in gas hydrates are considered to contain atleast 1x 10^{14} gegatons of carbon. This is about twice the amount of carbon held in all fossil fuels of the earth.

Gas hydrate concentration typically occurs at depocenters, probably because the most hydrate gas is Methane of biogenic origin, where there is rapid accumulation of sediments with rich organic detritus (from which bacteria generate methane, protecting from oxidation).

First evidence of naturally occurring gas hydrate deposits was found in Messiyokha field in Russian permafrost region. Subsequently, gas hydrates were also found in shallow marine sediments of Arctic region during ODP legs in Tropical deep water areas, where water depth exceeds 650-750 m and the pressure temperature conditions are favourable for formation of gas hydrates. Gas hydrate accumulation in deep water areas is believed to be due to accumulation of biogenic methane generated in shallow sediments. Areas with higher sedimentation rates are favourable for such accumulation as the organic carbon is preserved due to its rapid burial.

The fact, that 164 m^3 of methane gas is trapped in 1 m^3 of gas hydrate and large areas in deep waters have favorable conditions for formation of gas hydrates makes it a suitable option as an alternative source of energy. In the Indian context, Ministry of Petroleum of Natural gas had rightly launched a National Gas Hydrate Programme in 1997. The Author had an opportunity to work with Western Geophysical Company in London on the speculative surveys carried out in the East coast basins of Bay of Bengal and Andaman-Nichobar basin. During the study, for the first time, the presence of gas hydrates deposits are reported in the deep waters of Bay of Bengal i.e. around deep waters of Andaman-Nichobar islands and K-G deep water basins. During that period itself, International Agencies estimated the hydrate gas reserves in those identified areas are of the order of 13 TCF. From then onwards, an intensified integrated study for identification of gas hydrate prone areas in Indian offshore was taken up.

Seismic data beyond 600 m isobaths was scanned for evidences of gas hydrate based on the attributes of BSR (Bottom Simulating Reflector), like blanking polarity reversals etc. Very clear BSRs cuttings across sedimentary sequences supported by interval feasibility inversions were observed beyond 800-900 m bathymetry of K-G deep offshore. The studies also brought out that the Eastern offshore is more favorable as compared to the Western offshore. Seismic AVO models also suggest that huge amounts of free gas underlie by solid gas hydrate layers. This gas is likely to be over pressured (Fig 8.5).

Directorate General of Hydrocarbons carried out gas hydrate related studies in collaboration with Western Geophysical and LARGE of Russia using Multi Channel Seismic data (MCS) that was acquired for hydrocarbon exploration in the Andaman deep water area during the three different campaigns in 1996-97, 1999 and 2001. The BSRs are seen to cut across the Upper Miocene to Middle Miocene strata and broadly follow the sea bed topography. The processed MCS data show the indications of free gas below the BSRs and the corresponding seismic signatures resembling like anticlinal closures (Fig. 8.4). Seven gas hydrate bearing prospects have been identified below the sea bed in water depths of 850-2000 m. The flat spot identified in few structural prospects beneath the BSRs suggest possible gas water contact.

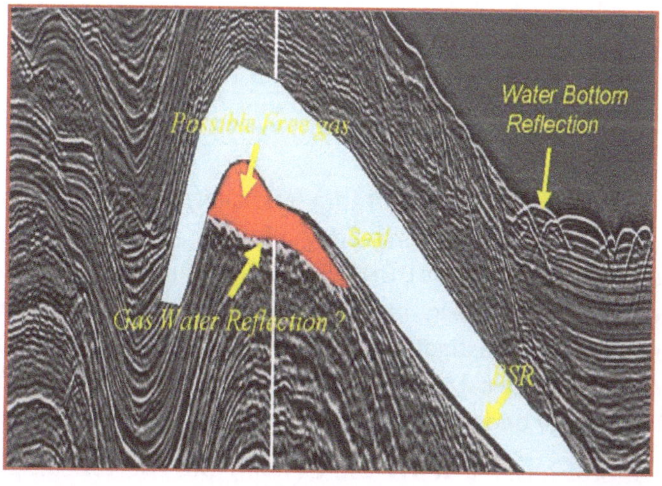

(Courtesy DGH)

Fig. 8.4

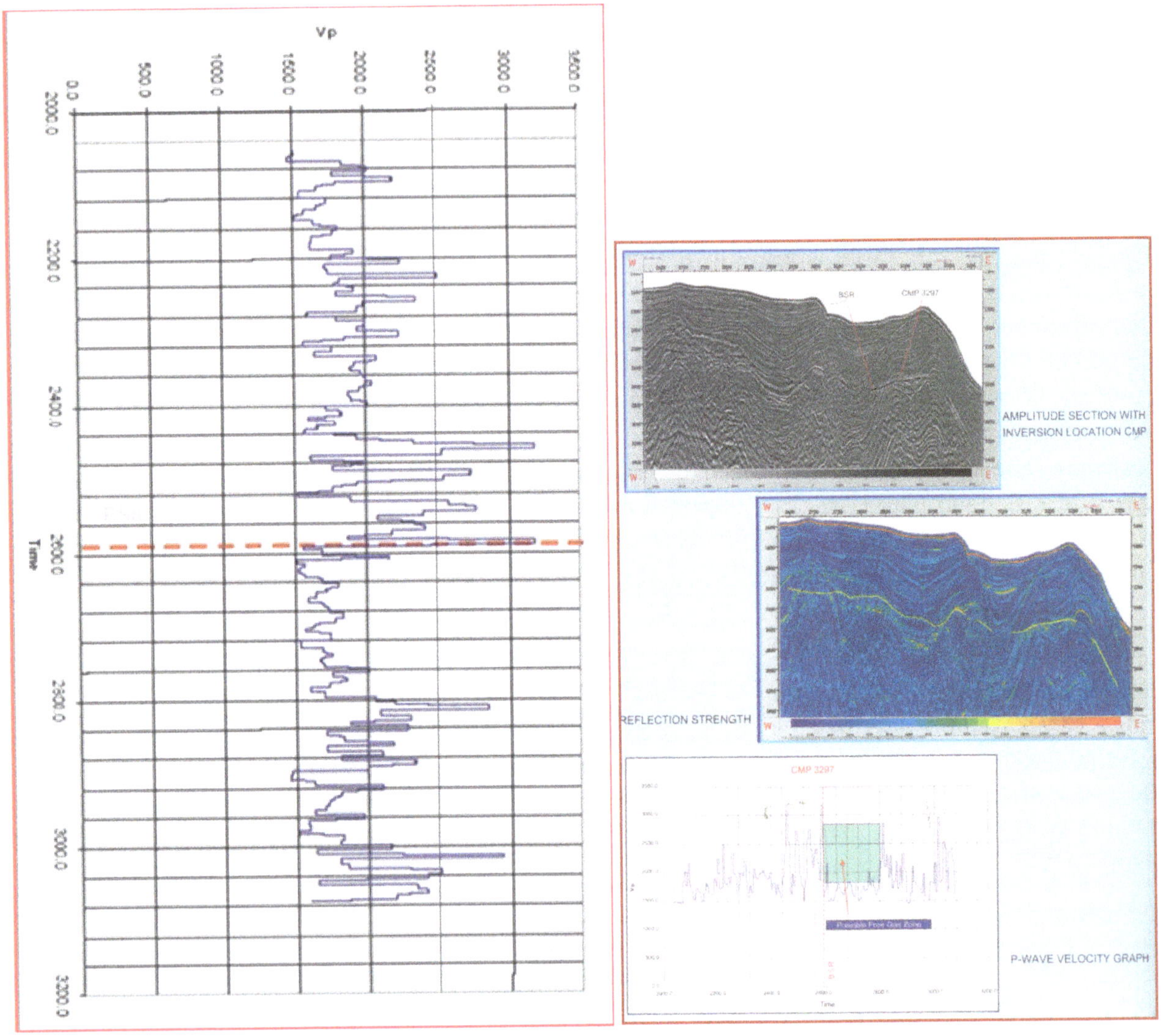

(Courtesy DGH)

Fig. 8.5

Pre Stack Genetic Algorithm Inversion, Full Wave form inversions carried out by Western Geophysical Co. confirmed the seismic anomalies (Fig. 8.6). Based on the results of multi disciplinary studies carried in collaboration with DGH, ONGC, GAIL and NIO, three offshore areas i.e. K-G basin in the East coast, Kerala-Konkan basin in the West coast and Andaman Offshore area were identified for further geocientific investigations. Swath bathymetry, deep TOW digital site scan sonar images, chirp sonar, gravity, geological, geochemical, and micro biological data were generated as a first step to understand the geology and the nature of gas hydrate formations.

In K-G basin, a WNW-ESE trending positive topographic feature (sedimentary) appears to be similar to Blake ridge which needs to be investigated thoroughly. The crest of the ridge rises from the surrounding depths > 1400m to <900 m. The width of the ridge is about 10 Kms (Fig 8.4). In the north western part of the study area, a fault controlled fan like geomorphic feature is also observed. The organic carbon concentration (TOC) varies between 0.35 to > 2.5 wt%. The Kerala-Konkan basin in west coast also appears to be a promising prospect from gas hydrate occurrence point of view. Based on the studies carried out so far on gas hydrate exploration, the probable presence of gas hydrates in the K-G basin and Andaman Offshore are very high, when compared with that of Kerala-Konkan basin.

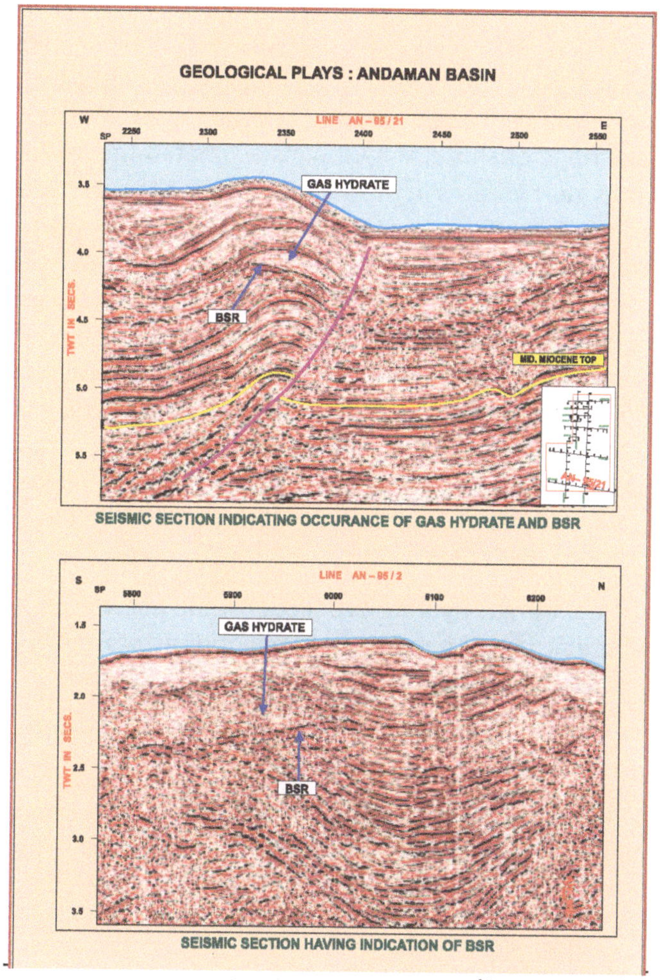

(Courtesy DGH)

Fig. 8.6

An international research program involving the department of USGS has proven that it is technically possible to produce gas from gas hydrates. A symposium was held in Japan where the results of first modern fully integrated production testing of gas hydrates are discussed, by an international gathering of Research Scientists. The international consortium, including the USGS, the Department of Energy, Canada, Japan, India, Gemany and the Energy Industry conducted test drilling at a site know as "Mallik" in the Mackenzie delta of the Canadian Arctic. This location was chosen because it has one of the highest concentrations of known gas hydrates in the world. Even though gas hydrates are known to occur in numerous marine and arctic settings, little was known before the Mallik project, about the technology necessary to produce gas hydrate.

While gas hydrates hold the potential, as an environmental friendly fuel for the next century, the technical challenges of releasing them as a resource are substantial. Additional research is required to understand and develop new techniques to quantify their distribution and production.

Deep pressurization and thermal heating experiments at the Mallik site were extremely successful. The results demonstrated that the gas can be produced from gas hydrates with different concentrations and characteristics, exclusively through pressure stimulations. The data supports the interpretation that the gas hydrates are much more permeable and conducive to flow from pressure stimulation.

In one test, the gas production rates were substantially enhanced by artificially fracturing the reservoir. The technology may take about 10 years from now to develop and tap gas hydrate reserves estimated to be more than double the known reserves of fossil fuel (CNR Rao the founder and Hon. President of Jawaharlal Neharu Centre of Advance Research and Prof. Kuxnetsog, Director Institute of

Oganic Chemistry, Russia). Interest on hydrate E&P has soared in recent years because of growing evidence that more hydrocarbons exist in hydrate deposits than the combined oil, gas and coal reserves worldwide. As per US Energy Information Agency, the recovery of only 1% of hydrates would more than double the domestic gas resource base. A report from Ocean Drilling Program like 164, which investigated the huge blake ridge offshore, the Carolins estimated the US Methane hydrate reserves at 2,00,000 TCF, that really puts gas hydrates in the big league. On the worldwide basis, it is estimated that methane hydrate reserves are 400 MTCF compared with 5000 TCF of known conventional gas reserves. If the technology to extract methane gas hydrates can be made most cost effective, then humanity might need to refrain from using non renewable fossil fuels. However, gas hydrates represent both Scientific and Technological challenges and many remains to be learnt about the Geologic, Engineering and Economic factors controlling the ultimate energy resource potential of gas hydrates. Proposed methods of gas recovery from hydrates generally deal with disassociating gas hydrates insitu by heating the reservoir beyond the temperatures of gas hydrate formation and decreasing the reservoir pressure below hydrate equilibrium. Researchers have spelt down that gas hydrates could eventually be a commercial resource, but yet the technical and economic hurdles for gas hydrate developments have to be overcome.

Inspite of all the plus points for gas hydrate productions, the hazards associated with this have to be kept in mind. The gas hydrate production may lead to major disasters:

1. Excessive methane production is 10 times worse than Carbondioxide as a climate warmer.
2. Uncontrolled drilling for gas hydrates can lead to serious blowouts because of its associated abnormal pressures.
3. The real hazard is that 'bursts of methane released' into the atmosphere, accelerates global warming and the consequences thereof are well known. All this is because, gas hydrates where they are stable; but once dislodged can be highly hazardous due to their great expansion ratio – very volatile nature.

National Gas Hydrate Program

The gas hydrate exploration project was to conduct scientific ocean drilling / logging, analytical activities to assess the geological occurrence and its regional context, and characteristics of gas hydrate deposits along the continental margins of India.

This exploration project was undertaken in collaboration between National Oil Companies, DGH, Ministry of Petroleum and Natural Gas (Government of India) and the U.S. Geological survey (USGS) in association with Overseas Drilling Limited (ODL), Fugro, and also with the association of leading scientists from all over the world.

"The National Gas Hydrate Programme-Exploration -01"started on April 28th 2006 and continued for 104 days. It cored and drilled 39 holes at 21 sites (1 site in Kerala-Konkan, 15 sites in Krishna-Godavari basin, 4 sites in Mahanadi basin and 1 site in Andaman deep offshore). 12 holes were logged with LWD and 13 holes with conventional wire logging. This expedition had given the following information:

1. Gas hydrate bearing marine sediments exist in both passive continental margins and marine accretionary wedges.
2. The BSR deposits are found matching in most places with computed methane stability zones.
3. The gas hydrates were found in combination type reservoirs.
4. The richest gas hydrate concentrations were in K-G basin samples.
5. In Andaman offshore, thick and deep gas hydrate formations were found at 600 m depth below sea floor (gas hydrate bearing volcanic ash).
6. Mahanadi basin has well developed gas hydrate system.

7. The K-G basin gas hydrates are closely associated with large scale structural features, in which the occurance of gas hydrates is controlled by the escaping gas through fractured system.

There is still no proven technology world over to recover directly methane from gas hydrate on commercial scale. Gas hydrates are unconventional hydrocarbon deposits and much of the work being done world over is still in research stage. National gas hydrate project (NGHP) would intensify its efforts to acquire seismic data, drilling, coring and production testing through expedition 02. This expedition is being carried out by the research drill ship Joides Resolution (JR) operated by ODL.

National Gas Hydrate program exploration 01 was designed to study the gas hydrate occurrences of Indian offshore and along the Andaman convergent margin with special emphasis to understand the geologic and geochemical controls on the occurrence of gas hydrates in diverse tectonic settings.

Future Plans for NGHP

The NGHP technical committee is planning to locate an ideal gas hydrate bearing sand in the channel-levee system by drilling several of the identified locations in the areas KG and Mahandi deepwater areas during NGHP Expendition-02. This would be followed up by taking pilot production testing under NGHP-03. The steering committee of NGHP has mandated ONGC to execute the NGHP Expendition-II.

The Geoscientific study has been carried out to identify gas hydrate sweet spots and firm up suitable sites for NGHP-02 expedition and estimate gas hydrate resources in this area. ONGC has identified prospective areas for Gas hydrate bearing sands in two areas in eastern offshore named as GD Block (700 Sq. Km) and HG-DWN-98/2 Block (3500 Sq. Km) in Godavari offshore. Detailed studies were carried out on Seismo-geological data such as post stack processing (structural smoothening) of 3D seismic data, seisimic event/horizon identification and correlation, sub-surface relief map generation; seismic amplitude attribute analysis and identification of seismo-geological features along sea floor and above BSR. Based on the studies at KDMIPE, total six prospective sites were identified.

Total 43 sites were proposed by ONGC to NGHP technical committee and the same were reviewed by the international expert committee. The recommendations for proposed sites were presented to NGHP Steering Committee, chaired by Secretary-P&NG in its meeting held at Mop&NG on 7th October, 2013. The steering committee has approved for the NGHP-02 expedition for coring/drilling/LWD/MWD/WL of 40 wells at 20 sites in Indian Offshore.

In addition, a regional gas hydrate prospect map was generated for KG and Mahanadi basin by integrating the bathymetry derived from 3D seismic data along-with seismic attributes near the sea floor BSR zones with identified prospective sites.

APPENDIX

1. Brain Storming session on focused exploration in Western Offshore : Feb,1998.
2. Comprehensive Exploration Plan: Cauvery Basin.
3. Conceptual Study of Mumbai offshore area-Docket.
4. Course Programme on Modern Deltas, Holocene Godavari Delta (200 bgl), Organized by Alkor Petroo Ltd, Hyderabad:2004.
5. DGH website (http://www/dghindia.org/SedimentaryBasins:aspx:2008-2009).
6. Directorate General of Hydrocarbons (DGH), Government of India. Petroleum Exploration and production activities, India - 2005-2012.
7. Exploration and Production opportunities in India, NELP-VI, Geo Scientific Information, DGH.
8. Geological information on Bangladesh- Docket.
9. Geological information on Krishna Godavari Basin- Docket.
10. Geological information on Mahanadi Offshore Basin-Docket.
11. Geological information on west Bengal offshore Basin- Docket.
12. Information on Bengal Basin – Docket.
13. Integrated Development Plan for a Naturally –Fractured Gas-Bearing Granite Reservoir.
14. Oil & Gas Exploration and Production in India: A Reference Book, Infraline Energy Publication 2009.
15. ONGC Bulletin, Special Issue (30th Year), Volume: 31, Number-2, Dec-1994.
16. Petroliferous Basins of India-II: Petroleum Asia Journal.
17. Petroleum Exploration and Production activities – India: 2002-03.
18. Petroleum Exploration and Production activities – India: 2003-04.
19. Petroleum Exploration and Production activities – India: 2004-05.
20. Petroleum Exploration and Production activities – India: 2005-06.
21. Petroleum Exploration and Production activities – India: 2006-07.
22. Petroleum Exploration and Production activities – India: 2007-08.
23. Petroleum Exploration and Production activities – India: 2008-09.
24. Selected Lectures on Petroleum Exploration: Dec-1969.
25. Status of Exploration in Krishna-Godavari Basin (Onland), Andhra Pradesh by B.T.V Seshavatarm and M.S. Prasad.
26. Symposium of Krishna-Godavari Modern Deltas, Dec- 1996.
27. Tectonic Evolution of Western and Eastern Continental Margins- Docket.
28. Tectonics, Sedimentation & Hydrocarbon prospects of Kerala- Laccadiv Basin- Docket.

REFERENCES

1. Acharya. S.K., 1991: Late Mesozoic- Early Tertiary Basin Evaluation along the Indo-Burmese Range and Andaman Island Arc. Sedimentary Basin of India (Tectonic Context), S.K. Tandon et al (EDs), pp 104-130.
2. A. T. R. Raju, "Geologic evolution of Assam and Cambay Tertiary basins of India".
3. Arthur Holmes (1978), "Principles of Physical Geology".
4. Arthur Holmes, "The Age of the Earth, an Introduction to Geological Ideas" 1927.
5. Arthur Holmes, "A Review of the continental Drift Hypothesis".
6. Arts, K.P., Radhakrishna, M., Murthy, B.V.S., Arora, S.K. and Nambiar, C.G. "Gravity anomalies, structure and tectonics of the Kerala Basin, Southwest Continental margin of India".
7. Authors: W.G. Hazlett, Gemini Solutions, Inc.; P.W. Snow, Mosbacher Energy Co.; C.S. Duncan, Worley International, Inc. Source: SPE Annual Technical Conference and Exhibition, 1-4 October 2000, Dallas, Texas.
8. B. Sahay, Oil & Natural Gas Commission, "Geology and Hydrocarbon Potential of Bombay Offshore Basin".
9. Bastia, R. "Depositional model and Reservoir Architecture of Tertiary Deep Water Sedimentation, Krishna - Godavari offshore Basin, India".
10. Basu, D.N., A. Banerjee, and D.M. Tamhane, 1982, Facies distribution and petroleum geology of Bombay offshore basin, India.
11. Battacharya, G.C., et al 1994. "Evidence for seafloor spreading in the Laxmi Basin, Northeastern Arabian Sea".
12. Beck, R.J. 2004. "Worldwide petroleum industry outlook-2004-2008, Projection to 2013".
13. Beckinsale, R.D., Drury, S.A. & Holt, R.W. (1980) "3,360-Myr Old Gneisses from the South Indian Craton".
14. Bernard, A., M. Munschy, Y. Rotstein, and D. Sauter (2005), "Refined spreading history of the Southwest Indian Ridge for the last 96 Ma, with the aid of satellite gravity data".
15. Bhandari, L.L., Venkatachala, B.S., Kumar, R., Swami, S.N., Garga, P. and Srivastava, D.C. (Eds.), 1983. Petroliferous basins of India. Special Volume.
16. Biswas, B., 1963: Results of Exploration for Petroleum in the western part of Bengal Basin, India: Proc.2nd Symp.Dev.Petrol Res. ECAFE Mins .Res.Dev.Dol.Ser No.18, pt.1, pp. 241-250.
17. Biswas, S.K., 1971: Note on Geology of Kutch: Quart.Jour, Gel.Min.Met.Soc.India. Vol(4); pp.223-225.
18. Biswas, S.K., 1987, "Regional Tectonics Framework, Structure and Evolution of Western Marginal Basins of India".
19. Biswas, S.K.1982. "Rift Basins in the western margin of India and their hydrocarbon prospects".

20. Biswas, S.K., 1988: Stratigraphy and sedimentary evolution of the Mesozoic Basin of Kutch, Western India, Proc. Conference on Sedimentary Basins of India, Kumaon Uni Vol., Nainital, India, Gyanodaya Prakasan, pp. 74-103 (Publ.1991).
21. Biswas, S.K., A review of structure and tectonics of Kutch basin, Western India, with special reference to earth quakes (Department if earth science, Indian Institute of technology Mumbai, India).
22. Chapman, R.E. Petroleum Geology. 1983, "Developments in Petroleum Science".
23. Chetty T R K and Murthy D S N 1998 "Regional tectonic framework of the Eastern Ghat Mobile Belt: A new interpretation; In: Eastern Ghat Mobile Belt".
24. Coffeen, J.A. 1986. Seismic exploration fundamentals and seismic techniques for finding oil.
25. Dill. R.F., 1964, Sedimentation and erosion in Scripps Submarine canyon head: in Miller, R.L., (ed), Papers in Marine Geology: Macmillan Publishing co., New York, P. 23-41.
26. D.S.N.Raju et. Al, "An overview of Litho, Bio and Chrono Sequence Stratigraphy and Sea level changes of Indian Sedimentary Basins".
27. D'Silva, K., D.M. Nathaniel, Bastia R., Verma R., Tenepalli, S., Sreedhar S.V., "Southwestern Continental Margin of India: A Poly-Rift Basin".
28. De Wit, M.J., "Madagascar: Heads it's a continent, Tails it's an Island".
29. Dessai, A.G., and H. Bertrand, 1995, "The 'Panvel Flexure' along the Western Indian continental margin: an extensional fault structure related to Deccan magmatism".
30. Dickkers, A.J. 1985. "Geology in Petroleum Production, Development in Petroleum Science".
31. Douglas Waples, 1984. "Organic geochemistry for exploration geologists".
32. Dr. Pushpendra Kumar, DGM-Head, Gas Hydrates & ACM, IEOT, ONGC, "Gas Hydrate Exploration & Production".
33. E. Dendy Sloan Jr, "Fundamental principles and applications of natural gas hydrates".
34. F.Ricci Lucchi, 1972. Turbidities of the northern Appenines: Introduction of facies analysis (English translation by T.H Nilsen) 1978, International Geology Review, Vol. 20, pp .125-166.
35. F.Ricci Lucchi, 1975. Turbidities facies and facies associations, in Examples of Turbidites facies and facies association from selected formations of the Northern Appenines. Field Trip Guide Book. A-11, International sediment logical Congress IX Nice P.31-36.
36. Fisher, A.C. and Judson, S (Eds.), 1975. "Petroleum and global tectonics".
37. Fuloria, R.C., 1993, "Geology and hydrocarbon prospects of Mahanadi Basin, India".
38. Fuloria, R.C., Pandey, R.N., Bharali, B.R. and Mishra, J.K., 1992, "Stratigraphy, structure and tectonics of Mahanadi Offshore Basin".
39. G.D.Hobson (Ed.). "Developments in Petroleum Geology".
40. Govindan, A., 1984, "Stratigragphy and sedimentation of East-Godavari sub-basin".
41. Guillemot, J. 1991. "Elements of Geology – Oil and gas exploration techniques".
42. Gupta, P.K. and Nandi, P.K. 1994. Well site geological techniques and formation evaluation – "A user manual".
43. H E Le Grand, "Drifting Continents and Shifting Theories, Cambridge University Press, Cambridge", 1988.
44. Hess, H. H. (November 1962). "History of Ocean Basins".

45. Hobson, J.D. and Tirastoo, E.N. 1975. "Introduction to Petroleum Geology".
46. Holt, W. E., et al. (1991), "The active tectonics of the eastern Himalayan syntaxis and surrounding regions".
47. Hughes, Patrick. (April 1, 1994) "The Meteorologist Who Started a Revolution".
48. Hunt, J.M., Petroleum Geochemistry and Geology", 1996.
49. 2007 Interim Results : Hardy Oil and Gas Plc.
50. I. V. S. V. Prasad, A. Banerjee, S. Pahari, and Meenu Varshney, 2004, "Evaluation of petroleum source rocks of Bengal Basin, India".
51. J.R. Curray, D.G. Moore, "Growth of the Bengal deep sea fan and denudation in the Himalaya".
52. J.R. Curray, F.J. Emmel, D.G. Moore, R.W. Raitt, "Structure, tectonics and geological history of the northeastern Indian Ocean".
53. Jagdish Pandey et.al, "Stratigraphy of Indian Petroliferous Basins Jagdish".
54. Jarrard, R. (1986), "Terrane motion by strike-slip faulting of fore-arc systems".
55. Jianchun Dai, Haibin Xu, Fred Snyder, and Nader Dutta, Schlumberger Reservoir Services/Data and Consulting Services, Houston, Texas, U.S, "Detection and estimation of gas hydrates using rock physics and seismic inversion: Examples from the northern deepwater Gulf of Mexico".
56. K.S.Valdiya , "Aspects of tectonics-Focus on South central Asia".
57. Karato, S.-I., and P. Wu (1993), "Rheology of the upper mantle: A synthesis".
58. Katz, M.B., 1978, "Sri Lanka in Gondwanaland and the evolution of the Indian Ocean".
59. Kent C. Condie: Plate tectonics & crustal evaluation.
60. Kind, Rainer (September 2007). "The fastest continent: India's truncated lithospheric roots".
61. Kuene, P.H and C.I. Miglorini, 1950, Turbidity currents as a cause of graded bedding, J.Geol.V.58, pp. 91-127.
62. Kumar, S.P. (1983) "Geology and hydrocarbon prospects of Krishna-Godavari and Cauvery basins".
63. Levorsen, A.I. Geology of Petroleum".
64. Liu, C.-S., J. R. Curray, and J. M. McDonald (1983), New constraints on the tectonic evolution of the eastern Indian Ocean".
65. M.V. Ramana and T. Ramprasad, 2010, "Gas hydrate occurrence in the Krishna–Godavari Basin offshore, East coast of India".
66. Malod J. A., Droz L., Kemal Mustafa B., Patrait P., June 1997. Early spreading and continental to oceanic basement transition beneath the Indus deep-sea fan: northeastern Arabian Sea.
67. McQuarrie, N., J. M. Stock, C. Verdel, and B. P. Wernicke (2003), "Cenozoic evolution of Neotethys and implications for the causes of plate motions".
68. Molnar, P., and P. Tapponnier (1975), "Cenozoic tectonics of Asia: Effects of a continental collision".
69. Morgan, J.P., McIntire, W.G., 1959, "Quaternary geology of the Bengal basin, East Pakistan and India".
70. Mougone, G.J (ed.), 1979. Energy policies of the world. Vo. III.
71. Murthy K.S.R, et.al (1995). Geodynamic aspects of east coast offshore basins.
72. Muthuvairvasamy Ramkumar, Doris Stüben & Zsolt Berner, 2004, "Lithostratigraphy, depositional history and sea level changes of the Cauvery Basin, southern India".

73. Nagy, B and Colombo, V. 1967. Fundamental aspects of petroleum geochemistry".
74. Naini, B.R., Talwani, H., 1982. "Structural framework and evolutionary history of the continental margin of western India".
75. Nair, K.M. and Vijayam, B.E., 1980, "Sedimentology of limestones in Niniyur Formation, Palaeocene, Cauvery basin".
76. Narayana Swamy S. 1975. "Proposal for Charnockites - Khondalite system in the Archaean shield of peninsular India".
77. North, F.K., 1990. "Petroleum Geology".
78. Norton, L.O. and Sclater., (1979), J.G., "A Model for the Evolution of Indian Ocean and Breakup of Gondwanaland".
79. P. J. Cook, Chris Carleton (2000) "Continental Shelf Limits: The Scientific and Legal Interface".
80. Pahari, S; Varshney, M and Mishra, R, (2002), Potential Source Rocks of Kerala-Konkan Offshore".
81. Pangtey, K.S.; Valluri, R.C.; Vedachalam T. and Suryanarayana, L. (1996), "Structural Pattern and Hydrocarbon Prospect Analysis of Kerala–Konkan Basin".
82. Patriat, P., and J. Achache (1984), "India-eurasia collision chronology has implications for crustal shortening and driving mechanism of plates".
83. Pickering, K.T., 1985, Turbudite system, Norway, In A.H. Bouma et al (eds,) submarine fans and related Turbudite systems, Springer Verlag, New York, P.237-244.
84. Prasad K.L., Rangaraju M.K., 1987, "Modern and recent canyon-fan systems in Masulipatnam Bay, Krishna - Godavari Basin".
85. Prof. M.S. Krishnan, 1943, "General Geological Succession in different parts of India and BURMA".
86. Rao, G.N, "Sedimentation, Stratigraphy and Petroleum Potential of Krishna-Godavari basin, East coast of India; AAPG Bulletin, V 85, Bo. 9 (September, 2001).
87. R.D. Hyndman, and S.R. Dallimore, Pacific Division, Terrain Sciences Division, Pacific Geoscience Centre, Geological Survey of Canada, "Natural Gas Hydrate Studies in Canada".
88. Ramam P K and Murthy V N, 1997, "Geology of Andhra Pradesh".
89. Ramanathan, S., 1968. "Stratigraphy of the Cauvery basin with reference to its oil prospects. In: Cretaceous–Tertiary of south India".
90. Reeves Colin and Leven Jim, Jan 2001. "The evolution of the west coast of India from a perspective of global tectonics".
91. Richard, C. Selley, 1998. "Elements of Petroleum Geology".
92. Rowley, D. B. (1996), "Age of initiation of collision between India and Asia: A review of stratigraphic data".
93. S. K. Verma, 2010, "Showcasing Indian Fields – Offshore".
94. Sastri, V.V., Venkatachala, B.S. and Narayanan, V., 1981, "The evolution of the east coast of India".
95. Sastry, M.V.A. & Rao, B.R.J., 1964. "Cretaceous–Tertiary boundary in south India".
96. Scotese, Christopher R. (January 2001). "The collision of India and Asia (90 mya - present)".
97. Seshavataram B.T.V. "Hydrogeology and Hydro Carbon Prospects of Cauvery Basins".
98. Shenai, K.R. & Rao, M. S., 1982, "Offshore exploration of Godavari - Krishna Basin".
99. Singh, N.K. and Lal, N.K., (1993), "Geology and Petroleum Prospects of Konkan-Kerala Basin".
100. Stratigraphy of Indian Petroliferous Basins by Jagadish Pandey and Alok Dave.

101. Subramanyam V et al. (1995): "Structure and Tectonics of the western continental margin of India".
102. Swamy A.S.R & I. B. Singh. Delta Sedimentation-East Coast of India Technology Publication 2005.
103. T. S. Collett (USGS), M. Riedel (IODP), R. Boswell (DOE-NETL), J. R. Cochran (LDEO), P. Kumar (ONGC), A. K. Sethi (DGH India), A. V. Sathe (ONGC) & NGHP Expedition-01 Scientific Party, "Methane Hydrate Newsletter".
104. Tissot, B.P. and Welte, D.H. 1984. Petroleum formation and occurrence, 2nd Edn., Springer-Verlog, Berlin.
105. Turaga, S.P, 2005. An introduction to oil well cements its properties, functions and slurry design techniques for trainees in an oil industry.
106. Turaga, S.P., 2006. "Drilling Fluids, their Composition, Function and Properties".
107. V.V. Sastri, R.N. Sinha, G. Singh, K.V.S. Murthy, "Stratigraphy and tectonics of sedimentary basins on east coast of peninsular India".
108. Vairavan, V., 1993, "Tectonic history and hydrocarbon prospects of Palar and Pennar Basins".
109. Van Hinsbergen, D.; Lippert, P.; Dupont-Nivet, G.; McQuarrie, N.; Doubrivine, P.; Spakman, W.; Torsvik, T. (2012). "Greater India Basin hypothesis and a two-stage Cenozoic collision between India and Asia".
110. Venkatarengan. R. Rao. G.N., Prabhakar. K.N., Singh. D.N. Awasthi A.K., Reddy. P.K., Mishra. P.K., and Roy. S.K., 1993: Lithostratigraphy of Petroliferous Basins, Document VIII, Krishna Godavari Basin, KDMIPE, ONGC Publication, pp1-29.
111. Venkatarengan.R., Prabhakar.K.N., Singh.D.N., Awasthi A.K., Reddy.P.K., Mishra.P.K., Roy.S.K., and Palkshi.K., 1983: Lithostratigraphy of Petroliferous Basins, Document VII, Cauvery Basin, KDMIPE, ONGC publication, pp1-31.
112. Waggen, W.1872: Jurrassic fauna of Cutch Pal.Ind.(G.S.I) Ser.9,Vol.1.
113. Wandrey, C.J., 2004, "Bombay geologic province Eocene to Miocene composite total petroleum system, India".
114. Wegener, Alfred (1966). "The origin of continents and oceans".
115. Welte, D.H. Harsfield, B. and Baker, D.R. 1997. (Eds.). "Petroleum and Basin Evolution".
116. Wilson, J.L., 1975, "Carbonate Facies in Geologic History".
117. Wynne, A.B.,1872, "Memoir on the Geology of Kutch, to Accompany" the map compiled by A.B. Wynne and F.Fedden during the season 1867-68,1868-69, mem.Geol.Sur.Ind. Vol.19(2),PP.269.
118. Yadagiri, K. and Govindan, A., 2000, "Cretaceous carbonate platforms in Cauvery basin: Sedimentology, depositional setting and subsurface signatures".
119. Zutshi, P.L., Sood A., Mohapatra P., Ramani K.K.V., Dwivedi A.K., and Srivastava H.C., 1993," Bombay offshore basin-lithostratigraphy of Indian petroliferous basins".
120. Zutshi, P.L., and M.S. Panwar, 1997, "Geology of Petroliferous Basins of India": Oil and Natural Gas Corporation, Ltd., KDM Institute of Petroleum Exploration.
121. Zutshi, P.L., Sood, A., Mahapatra, P. Ramani, K.K.V Divedi, A.K. and Srivastava H.C. 1993: Lithostratigraphy of Indian Petroliferous Basins, Document-V, Bombay Offshore Basin, KDMIPE, ONGC, Publication, PP 1-383.

INDEX

1/2" choke, 74
3d seismic surveys, 112
85 °E ridge, 5, 11, 136
90 °E ridge, 115

A

A fault, 24
Accumulations, 38
Active volcano, 169
Africa, 14
Albian, 93
Alibag formation, 36
Allepy platform, 92
Alps mountains, 11
Ambe field, 55
Amplitude, 24
Andaman nicobar basin, 10, 165
Andaman trench, 168
Andaman, 165
Androth, 85
Antarctica, 14
Anticlinal closure, 23
Anticlinal reversals, 38
Anticline, 24
API gr. At 60°f, 69
API gravity, 38, 60
Aquitanian, 80
Arabia, 14
Arabian abyssal plains, 16
Arabian sea, 21, 83
Arakan yoma fold belt, 149
Arakan-yoma, 172, 173
Aravalli trend, 16
Aravalli, 13
Aravalli-delhi, 13
Arch, 28
Archaean era, 13
Archaean fault lineaments, 14
Archean trend, 113
Archipelago, 165
Arctic region, 189
Areal extent, 23
Ariyalur formations, 93
Ariyalur-pondicherry graben, 96
Arkan yoma fold belt, 172
Artificially fracturing, 191
Asaphaltene, 60
Asiatic plates, 168
Aspheltenes, 69
Asymmetric anticline, 60
Australia, 11
Axis plunges, 22

B

Back arc, 171
Back deep, 168, 170
Badrak shallow basement block, 138
Bangladesh, 154
Banni graben, 19
Banskantha, 21
Bantumilli grabens, 119
Baong formations, 172
Bapatla horst, 114, 119
Barail formation, 154
Baratang formation, 171
Baratang island, 172
Barisal-chandpur high, 148, 149, 162
Barison range, 171
Barren and narcondam islands, 169
Basal clastics sandstone reservoir, 38
Basalt, 150
Base of crude, 69
Basement arches, 16
Basin area, 30
Basin evolution, 1, 138
Basin margin fault zone, 146
Basin type, 30
Basin, 4
Basinal area (sq. Km), 4
Basins of india, 2
Bassein field, 60
Bassein limestone, 63
Bathymetry, 82
Bay of bengal ocean, 14
Below the sea bed in water, 189
Bengal basin, 10, 149
Bengal fan sediments, 169
Bengal fan, 100, 115, 162
Bijawars, 13
Biogenic origin, 188
Biogenic, 186
Bombay arch, 16
Bombay high field, 56
Bombay high platform, 9, 33
Bombay high-DCS, 37
Bombay offshore basin, 32
Bombay offshore, 9

Bombay platform, 28, 37
Bouguer gravity, 95
Brahmaputra, 15, 161
BSR (bottom simulating reflector), 189
Burdigallian, 80
Burial history, 41
Burmese arc, 173

C

Calcium chloride, 99
Calcutta mymensingh high, 148
Cambay basin, 9
Cambay graben, 21
Cambay gulf, 9
Cambrian, 13
Campose basin brazil, 174
Canadian arctic, 191
Cap rocks, 38, 143
Carbonate build-ups, 86
Carbonate platform, 32, 44
Carbonate sequence, 23
Carbonate, 185
Carbondioxide, 188
Category I basins, 2
Category II basins, 2
Category III basins, 2
Category IV basins, 2
Category, 4
Caucasus range, 11
Cauvery basin, 10
Cauvery sub-basin, 10
Cenomonion, 14
Central depression, 39
Central graben, 35
Chandpur high, 149
Channel cuts, 162
Channels, 68
Chirp sonar, 190
Chittagong, 154
Clastic basin, 32
Clastic wedge, 22
Closure, 24
Coal bands, 42
Coastal bars, 36
Coastal plains, 92, 117
Co-genetic, 14
Collided, 11
Commercial production, 46
Concion (late cretaceous), 14
Condensate formation, 41
Condensate, 20
Configuration, 14
Conglomerates, 13
Consortium, 63
Contiguous, 9
Continental crust, 136
Continental environment, 150
Continental shelf, 9, 17, 174
Convergent margins, 1
Convergent plate, 17
Cooling, 13
Cori-comorin depression, 17
Cori-comorin ridge, 17
Coromandal coast, 14
Corresponding, 9
Cost effective, 46
CP viscosity, 60
Craton, 11
Cratonic horst, 20
Cretaceous, 11, 21
Crustal extension, 172
Crustal shortening, 15
Crystalline solid, 188
Crystalline, 13
Cuddapah, 13
Culminations, 64
Cuttack-chand – bali depression, 138
Cycles of deposition, 118

D

Dahanu & surat depression, 82
Dahanu, 53
Damodar, 150
Deccan lava flows, 13
Deccan trap, 15, 21
Deccan volcanism, 14
Deep basin, 146
Deep canyon, 148
Deep sea basins, 174
Deep water basins, 17
Deep water exploration, 174
Deeper waters, 15
Deform, 11
Deformation, 13
Delhi – aravali trend, 21
Delineation / development plans, 23
Delineation wells, 74
Delta front, 100
Delta plains, 117
Delta, 100
Deltaic environment, 150
Deltaic facies, 14
Deltaic progradational sediments, 162
Density, 69
Denudational surfaces, 22
Depocenters, 14, 38, 188
Depositional environments, 149
Depositional system tracks, 93
Detritus, 188
Devagarh depression, 43
Development of the field, 62
Development plan, 62
Dhamodar trend, 162
Dharwar, 16
Dharwarian group, 13
Diligent strait, 167
Distributary channels, 36
Divergent passive continental margins, 1
Diverse tectonic settings, 193
Domal structure, 23
Dominant role, 87
Doubly plunging anticlinal structure, 48
Doubly plunging anticline, 68
Drape, 38

Drapover structures on shale diapirs, 143
Drift phases, 99
Drift stage, 92, 114
Drift, 13
Drifted, 11
DSDP, 79
Durian, 14

E

Early miocene, 15
Early mio-pliocene, 110
Early paleozoic, 13
Early rift stage, 14
Early tertiary, 13
East coast basins, 10
East coast of india, 110
East coast, 15
East godavari sub-basin, 114, 119
Eastern ghat trend, 14, 113
Eastern homocline, 33
Eastern passive continental margin, 110
Embayment, 21
Entrapment, 99, 172
Episodic movements, 172
EPS – tahara, 70
Ethane, 161
Eurasian plate, 11, 161
Eustatic rise in sea level, 37
Evolution, 11
Exploration efforts, 20
Exploration targets, 86
Exploration, 1
Exploratory drilling, 20
Exploratory wells, 26, 74
Exposure, 13, 110

F

Factor, 60
Fall and rise of sea levels, 100
Fan complex, 182
Fault closures, 38
Faults, 14, 72
Few drain holes, 59
Field parameters, 22
Flanks, 70
Flasti point, 69
Flat spot, 189
Flood plains, 117
Fluvial systems, 15
Fluvial, 15
Fluvio- deltaic, 20
Flysh, 171
Fore arc basin, 168, 169
Fore deep, 169
Foreland block', 14
Formation volume, 60
Fracture zone, 84, 86, 87
Frontier areas, 174

G

Gaffney, cline and associates, 182
Ganges / brahmaputra river, 151
Ganges, 161
Ganges-padma, 161
Gas analysis, 53
Gas composition, 49
Gas hydrate exploration, 177
Gas hydrates, 1, 177
Gauri fields, 55
Generation of hydrocarbons, 39
Generation, 23
Genetic algorithm inversion, full wave form inversions, 190
Geochemical studies, 38
Geochemistry, 96
Geo-dynamic aspects, 118
Geological evolution, 82
Geological history, 150
Geological reserves, 17, 50
Geological surveys, 103
Geo-morphology, 117
Geo-tectonic activity, 161
Geothermal gradients, 100
Giant, large and medium sized fields, 47
Glaciations, 13
Glacio-fluvial deposits, 13
Gneisses, 13
GOC & OWC, 60
Godavari grabens, 14
Godavari river, 114
Godavari sand spit, 117
Gondwana grabben, 106
Gondwana sediments, 106
Gondwana, 11
Gowthami, 114
Graben blocks, 78
Graben, 16
Grabens, 14
Granites, 13
Gravity magnetic surveys, 20, 103
Gravity, 26
Great nicobars, 165
Growth faulting, 186
Growth faults, 143
Gudivada, 119
Gulf of cambay, 21, 82
Gulf of mannar block, 10
Gulf of mannar, 92, 96
Gulf of mexico, 174

H

Habitat of oil and gas, 38, 151
Hazards, 192
Heera-bassein-panna, 33
Heera-panna-bassein-neelam, 9
Heterogeneous, 68
Higher hydrocarbons, 161
Himalayan mountains, 11
Hinge zone, 18, 22, 146
Hinge, 84
Horizontal wells, 59
Horst and graben tectonics, 96
Horst, 16

Horsts & grabens, 21
Hot spot, 14
Humic, 41
Humid, 13
Hydrate equilibrium, 192
Hydro geological conditions, 99
Hydro geology, 96
Hydrocarbon accumulations, 111
Hydrocarbon discoveries, 103
Hydrocarbon generation, 38
Hydrocarbon habitat, 99
Hydrocarbon occurrence, 121
Hydrocarbon plays, 23
Hydrocarbon prospect, 87
Hydrocarbon reserves, 26, 46
Hydrocarbon resource potential, 26
Hydrocarbon, 46
Hydrocarbons, 1, 68, 189
Hydrogen sulphide, 188

I

Indian offshore basins, 1
Indian plate, 14, 172
Indian platform, 13
Indian stratigraphy, 5
Indo- australian plate, 11
Indus fan, 186
Infiltrated, 99
Infiltration processes, 99
Innovations, 174
Innovative, 46
Integrated development, 52
Inter deep, 168
Inter-granular porosity, 68
Intergranular, 38
International statigraphic nomenclature, 8
Intertrappean, 82
Intra-cratonic basin, 15
Intrusions, 13
Invisible bank, 169
Irawaddi, 162
Island arc, 165
Islands, 78, 165
Isobath, 9, 110
Iso-maturation contours, 97
Isotopic studies, 186

J

Jakhau, 24
Java-trench, 172
JOIDES, 78
Joint oceanographic institution for deep earth sampling, 78
Jurassic period, 11
Jurassic, 13
Juxtaposition, 14
Juxtra position, 161

K

Kakinada bay, 117
Kanakadurga, 130
Karaikal ridge, 96
Karaikal, 95
Karakoram, 13
Karwar-kasargod area, 79
Kasargod basin, 78
Kasargod block, 9
Kathiawar uplift, 22
Kerala basin, 17,78
Kerala offshore block, 9
Kerala-konkan-lakshadweep basin, 78
Kerala-konkan, 9
Kerala-laccadiv depression, 82
Kerogens, 43
Khari series, 22
Kinematic visc, 69
Kolkata mymensingh high, 149
Kommugudem, 119
Konark uplift, 138
Konkan basin, 16, 17, 78, 84
Konkan block, 9
Kori-comorin depression, 16, 78
Kori-comorin ridge, 16, 78
Krishna graben, 114
Krishna river, 118
Krishna-godavari basin, 10
Krishna-godavari, 110
Kumbakonam shiyali ridge, 96
Kutapong, 172
Kutch – saurashtra, 17
Kutch basin, 17
Kutch onshore, 9
Kutch-saurashtra, 9
Kutubdia, 161

L

Laccadive continental, 14
Laccadive depression, 17
Lagoon, 86, 117
Lakshadweep basin, 78
Lakshadweep block, 9, 178
Lakshadweep, 85
Lakshdeep islands, 86
Lapotin, 41
Large grabens, 13
Late albian, 15
Late carboniferous, 14
Late cretaceous, 13
Late drift stage, 114
Late jurassic, 14
Late paleozoic, 13
Lateritic clay, 42
Latitude, 78
Laurasia, 11
Laxmi basin, 14
Laxmi laccadive depression, 78
Laxmi laccadive ridge, 17, 78
Laxmi, 55
Laxmi-laccadive depression, 16
Laxmi-laccadive ridge, 16
Lenticular sands, 39
Lignite, 42
Limestone / dolomite, 38
Limestone reservoirs, 38

Limestones, 151
Litho facies, 93
Litho stratigraphy, 80, 139
Lobes, 182
Lower cretaceous, 13
Lower indus basin, 18

M

Maastrichtian, 93
Mackenzie delta, 191
Magnetic anomalies, 95
Magnetic surveys, 113
Magnetic, 26
Mahanadi basin, 192
Mahanadi including nec block, 10
Mahanadi offshore basin, 136
Mahanadi, 14
Main land beach and dunes, 117
Malayan orogene, 168, 170
Man grove swamps, 117
Mandapam delft, 96
Mandapeta troughs, 119
Mannargudi devakottai ridge, 96
Marginal fields, 47
Marginal sag basins, 14
Marine fan deposits, 100
Marine intercalations, 14
Marine regression, 162
Marine transgression, 15
Martaban, 165
Mascarene basin, 14
Masked, 22
Maturation, 171
Maturation, 41
Maturity curves, 38
Maturity, 83
Mayyand formation, 80
Medagaskar, 11
Mentavi islands, 172
Mergue terrace, 168
Mesozoic sequence, 99
Mesozoic, 13, 18
Messiyokha, 189
Methane hydrate, 187
Methane, 161
Mevlaboh, 171
Micro fractures, 68
Middle-jurassic, 13
Mineralization of waters, 99
Miocene age, 55
Miocene, 18
Modules, 188
Molds, 68
Monotonous, 68
Morphological features, 118
Morphology, 118
Mozamique, 13
Mud loss, 84
Mud volcanoes, 172
Mukta and B-126 fields, 61
Multi channel seismic data, 189

N

Nagapattinam graben, 100
Nagar-parker uplift, 22
Naidupeta ridge, 106
Naramada graben, 21
Narmada and tapti, 87
Narmada-cambay-kutch, 20
Narmada-son rift (satpura and son), 14
Narmada-son tectonic lineament, 14
National gas hydrate programme in 1997, 189
Natural gas, 1, 161
Natural levees, 117
Neelam field, 63
NELP, 26
Neocomian, 14
Neogene clastics, 185
Neogene tectonic play, 22
Neogene tectonic stage, 114
Neogene tectonic trend, 113
Neogene tectonics, 87
Nicobar group, 165
Nimapara – balikuda up lift, 138
North gujarat, 21
North sea, 174

O

Ocean-continent boundary, 119
Offshore angola, 174
Offshore, 4, 103
Oil and gas discoveries/strikes, 46
Oil and gas pools, 61
Oil and gas shows, 23
Oil window, 41
Oligocene, 15
Onland, 4, 103
On-laps, 100
Onshore, 9
Ordovician - late jurassic, 14
Ordovician, 14
Organic carbon concentration, 83
Organic carbon content, 154
Organic matter, 171
Organic matter, 41
Orogenic, 11, 20
Oscillations in the basin, 162
Outer high, 168
Outer non volcanic island arc, 169
Overlying, 24

P

Padmavathi, 130
Padua island, 178
Palar basin, 10, 106
Paleo - current, 14
Paleo sand ridges, 117
Paleo shelf lies, 87
Paleo slope, 15
Paleocene carbonates, 28
Paleogene 172, 185
Paleogene carbonates 18
Paleo-highs, 35, 39
Paleosink, 41

Palk bay-ramnad, 96
Pangaea, 11
Pangea, 13
Panna field, 66
Panna formation, 35
Paradap depression, 138
Paraffinic, 38
Passive continental margin, 87
Passive divergent margin, 106
Passive margin, 15, 21
Pay zone, 23
Pebbly sandstones, 93
Peninsula, 13
Peninsular block, 14
Peninsular india, 14
Peri-cratonic basins, 15
Pericratonic passive margin, 92
Permafrost, 188
Permeability, 59
Permian marine, 13
Permocarboniferous, 13
Petroleum plays, 43
Petroleum potential, 121
Petroleum prospects, 23
Petroleum system elements, 100
Petroleum system, 38, 99
Pinchouts, 142
Plate movement, 11
Plate tectonics, 1, 11
Platforms, 64
Pliocene/pleistocene, 80
Plunging, 16
Polar regions, 188
Polarity reversals, 189
Polycyclic basin, 161
Porbandar, 17
Porosity, 59, 64
Port blair formation, 171
Post-rift thermal, 15
Pour point at 60 °C, 69
Pour point, 60
Pranhitha godavari, 113
Precambrian, 22
Pre-paleogene, 185
Present is the key to the past, 118
Pressure, 188
Primary, 38
Production sharing contract, 31
Production testing, 49
Prognosticated gas, 187
Prognosticated resources, 17, 48, 110, 136
Progradation of delta, 151
Prograded, 83, 110
Propane, 161
Prospects, 48, 178
Proto brahmaputra deltas, 161
Proven petroliferous basin, 110
Puri depression, 138
Purnea grabens, 150

Q

Quaternary transgression, 22
Quilon limestones, 80
Quilon, 80

R

Radhanpur- barmar, 22
Rajmahal group, 150
Rajmahal volcanics, 150
Ramnad palk bay block, 10
Range, 13
Rann of kutch, 17
Rapid subsidence, 162
Ratnagiri block, 33
Ratnagiri, 16
Ravva field, 126
Reactivation, 16
Recent flood plains, 117
Recoverable reserves, 50
Red bed system, 122
Regional profiles, 86
Regression, 22
Regressive phase, 22
Reliable reflections, 24
Reservoir rock, 38, 100
Resins, 69
Resources, 5
Rift grabens, 20
Rift stage, 114
Rift, 11, 13, 99
Rifted grabens, 14
Rifted volcanic continental margin, 14
Right lateral transform movement, 14
River channels and estuary, 117
River ganges, 15
Rollover structures, 143

S

Sandstone reservoirs, 38
Sandstones, 13
Sangu field, 154
Satellite gravity, 75
Satellite, 26
Satpura trend, 16, 28
Satpura weak zone, 14
Satyavedu depression, 106
Saurashtra arch, 16
Saurashtra offshore block, 28
Saurashtra onshore, 9
Scan sonar images, 190
Sea-floor spreading, 15
Secondary porosities, 38
Sediemtnary systems, 115
Sedimentary basins, 1, 9
Sedimentary environments, 117
Sediments and provinance, 82
Sediments, 20
Seismic data, 14
Seismic surveys, 20, 28
Seismo stratigraphic, 93
Semi arid climatic, 13
Sequence stratigraphy, 119

Shale, 28
Shales of baong, 172
Shales, 38, 151
Shelf break, 17
Shelf margin area, 37
Shelf margin block, 33
Shelf margin, 86
Shelf width, 78
Shelf zone, 146
Shelf, 16, 84, 186
Shield, 14
Slope, 100, 186
Sniffer survey, 143
Sodium bicarbonate, 99
Solidification, 13
Solution vugs, 68
Solution, 68
Somalia, 13
Source rock potential, 41
Source rock, 23, 38, 100
Southern africa, 11
Sp.GR at 60°F, 69
Speculative surveys, 113,167
Spreading, 13
Sriperumbadur, 106
Stratigraphic play, 100
Stratigraphic prospects, 70
Stratigraphic section, 21
Stratigraphic, 54
Stratigraphy, 5, 18, 66, 138
Structural cum stratigraphic plays, 39
Stratigraphic cum structural prospect, 54
Structural features, 14
Structural styles, 22
Structural traps, 110
Structure contour, 24
Sub aqueous ridges, 115
Sub-marine canyon fans, 110, 143
Submarine channels, 182
Submarine complex, 84
Subsidence history, 42, 100
Subsidence, 15
Subsurface stratigraphy, 93
Sumatra, 171
Super continent, 11
Super deep wells, 28
Surat depression, 35, 37
Sutured, 11
Swatch of no ground, 148, 149
Syn rift stage, 114
Syn rift, 38, 92
Synrift clastics, 44
Syn-rift stage, 35

T

Tanjore-nagapattinam depression, 96
Tanuku horst, 114
Tapti daman offshore block (surat depression), 48
Tapti-daman basin, 16
Tapti-daman, 9, 33
Targets for exploration, 28
Technological advances, 174
Technological scheme, 62
Tectonic history, 33
Tectonic style, 22
Tectonics, 1, 118
Teleseis, 24
Tellicherry, 17
Temperature, 59, 188
Terrestial sources, 41
Terriginous influx, 83
Tethyan orogeny, 11, 161
Tethys ocean, 11
Thermal maturity, 42
Thrust sheets, 173
Tibetan plateau, 11
Tidal channels, 117
Tidal deltas, 36
Tidal euxinic basin, 171
Tidal flats, 117
Tillites, 13
Tilted fault blocks, 39
Tiruchinapally, 93
TOC ranges, 38
Topographic lows, 172
Total organic carbon, 42
Tranquebar depression, 96, 100
Tranquebar graben, 100
Transgression, 13, 150
Transgressive phase, 22
Transverse, 16
Traps and entrapments, 38
Trend, 21
Triassic time, 14
Tri-junction, 21
Triple junctions, 14
Turbidite fan complex, 84
Turbidite fan systems, 143
Turbidite fans, 110
Turbidite slumps, 186
Turbidites, 100
Turonian, 93
Type, 42

U

Unconfirmed surfaces, 22
Unconformities, 93
Unconformity, 86
Unconventional hydrocarbon, 193
Unconventional, 187
Under thrusting, 173
Upland plains, 117
Uplift of himalayas, 161
Upper curst, 13
Upper jurassic, 13
Utattur, 93

V

Vainateya, 114
Vasishta, 114
Vedaranyam-sri lanka high, 96
Vengrula arch, 78, 84

Vengurla, 17
Vertical closure, 23
Vertical displacements, 48
Vijayadurg graben, 35
Vindhyan basins, 13
Vinjkan shale, 22
Vitrinite reflectance, 43, 97
Vitrinite reflections, 151
Volcanic arc, 168, 169
Volcanic episode, 15
Volcanic eruptions, 13
Volcanic ridges, 115
Volcanic trap, 18

W

Warakali beds, 80
Water depth, 24, 51
Water injection, 58, 59
Wax, 69
Wedge outs, 86, 100, 110, 142
Well data, 30
West coast offshore basins, 9
West godavari sub-basin, 114, 115
Western continental margin, 13
Western offshore, 16, 167
Wrench fault, 173

9 789352 300839

Printed at Repro India Ltd.